T0236518

Lecture Notes in Computer Science   8382

Commenced Publication in 1973
Founding and Former Series Editors:
Gerhard Goos, Juris Hartmanis, and Jan van Leeuwen

More information about this series at http://www.springer.com/series/7409

Andreas Nürnberger · Sebastian Stober
Birger Larsen · Marcin Detyniecki (Eds.)

Adaptive Multimedia Retrieval

Semantics, Context, and Adaptation

10th International Workshop, AMR 2012
Copenhagen, Denmark, October 24–25, 2012
Revised Selected Papers

 Springer

Editors

Andreas Nürnberger
Sebastian Stober
Otto-von-Guericke-Universität Magdeburg
Magdeburg
Germany

Marcin Detyniecki
Université Pierre et Marie Curie
Paris
France

Birger Larsen
Royal School of Library and Information
  Science
Copenhagen
Denmark

ISSN 0302-9743   ISSN 1611-3349 (electronic)
ISBN 978-3-319-12092-8 ISBN 978-3-319-12093-5 (eBook)
DOI 10.1007/978-3-319-12093-5

Library of Congress Control Number: 2014953271

LNCS Sublibrary: SL3 – Information Systems and Applications, incl. Internet/Web, and HCI

Springer Cham Heidelberg New York Dordrecht London

Printed on acid-free paper

Springer is part of Springer Science+Business Media (www.springer.com)

Preface

This book comprises invited surveys and a selection of revised contributions that were initially submitted to the 10th International Workshop on Adaptive Multimedia Retrieval (AMR 2012). This time, the workshop was organized at the Royal School of Library and Information Science in Copenhagen, Denmark, during October 24–25, 2012.

Systems for searching and organizing multimedia information have matured during the last few years. However, retrieving specific media objects is still a challenging task, especially if the query can only be vaguely defined or refers to different types of media. The main difficulties in multimedia search are still, on the one hand, the users' incapacity in specifying their interests in the form of a well-defined query due to insufficient support from the interface, and on the other hand, the problem of extracting relevant (semantic) features from the multimedia objects itself. Ideally, user-specific interests should also be considered when ranking or automatically organizing result sets. To improve today's retrieval tools and thus the overall satisfaction of a user, it is necessary to develop advanced techniques able to support the user in the interactive retrieval process. The general goal of the AMR workshops is to promote the exchange of ideas between the different research communities involved in this topic.

This book includes contributions ranging from theoretical work to practical implementations and its evaluation. Moreover, it is a combination of state-of-the-art surveys and focused contributions. The first part of the book includes overview chapters covering three important aspects of adaptive multimedia retrieval challenges: The first chapter focuses on discovery based on language, the second on evaluation in the particular case of music retrieval, and the third on concepts of recommender systems. All chapters provide a fundamental introduction to the general concepts and the state of the art in these fields.

In the submitted contributions to the workshop three main topics emerged. The first topic is the recurring issue of semantics when dealing with multimedia data. Several papers address the problem of how to annotate images or audio. Furthermore, a lot of work is still tackling issues of feature extraction, data and source identification, and classification, which are important tasks to iteratively close the gap between low-level features and high-level semantics.

The second important topic that emerged is the exploitation of the context for improved retrieval. A good example of this are approaches exploiting location information. Other important aspects explored in these works are the type of media (music, image, natural language text) and the source of data (e.g., the crowd). The context also plays an important role in the contributions presenting recent work on facets of human computer interfaces.

Finally, the crucial problem of (dynamic) adaptation is addressed by the majority of papers. The quantity and diversity of these papers reflect the importance of the topic and its transversality. Adaptation can be understood based on two aspects: What is being modified and what it is based on? The multiplicity of crossing of these two objects of interest, treated in this book, make us believe that it is still a great ground for ongoing cross-fertilization of the research fields involved.

We would like to thank all members of the Program Committee for supporting us in the reviewing process, the workshop participants for their willingness to revise and extend their papers for this book, the sponsors for their financial help, and Alfred Hofmann from Springer Verlag for his support in the publishing process.

May 2014

Andreas Nürnberger
Sebastian Stober
Birger Larsen
Marcin Detyniecki

Organization

Program Chairs

Andreas Nürnberger   Otto-von-Guericke-University Magdeburg,
Germany
Sebastian Stober   Otto-von-Guericke-University Magdeburg,
Germany
Birger Larsen   Aalborg University Copenhagen, Denmark
Marcin Detyniecki   CNRS, Laboratoire d'Informatique de Paris 6,
France

Program Committee

Jakob Abeßer   Fraunhofer IDMT, Germany
Jenny Benois-Pineau   University of Bordeaux, LaBRI, France
Stefano Berretti   Università degli Studi di Firenze, Italy
Susanne Boll   University of Oldenburg, Germany
Jesús Chamorro   Universidad de Granada, Spain
Juan Cigarrán   Universidad Nacional de Educación a Distancia,
Spain
Ana M. García Serrano   Universidad Nacional de Educación a Distancia,
Spain
Bogdan Gabrys   Bournemouth University, UK
Daniel Gärtner   Fraunhofer IDMT, Germany
Fabien Gouyon   INESC Porto, Portugal
Christian Hentschel   Hasso Plattner Institute, Germany
Xian-Sheng Hua   Microsoft Research, China
Alejandro Jaimes   Telefónica R&D, Spain
Philippe Joly   Université Paul Sabatier, France
Gareth Jones   Dublin City University, Ireland
Joemon Jose   University of Glasgow, UK
Peter Knees   Johannes Kepler University Linz, Austria
Stefanos Kollias   National Technical University of Athens, Greece
Anna Kruspe   Fraunhofer IDMT, Germany
Stéphane Marchand-Maillet   University of Geneva, Switzerland
Trevor Martin   University of Bristol, UK
José María Martínez Sánchez   Universidad Autónoma de Madrid, Spain
Bernard Merialdo   Institut Eurécom, France
Gabriella Pasi   Università degli Studi di Milano-Bicocca, Italy
Valery Petrushin   Accenture Technology Labs, USA

Daniel Racoceanu	IPAL, UMI CNRS 2955, Singapore
Stefan Rüger	The Open University, UK
Simone Santini	Universidad Autonoma de Madrid, Spain
Raimondo Schettini	Università degli Studi di Milano-Bicocca, Italy
Ingo Schmitt	University of Cottbus, Germany
Nicu Sebe	University of Amsterdam, Netherlands
Alan F. Smeaton	Dublin City University, Ireland
Arjen De Vries	CWI, Netherlands

Supporting Institutions

Royal School of Library and Information Science, Copenhagen, Denmark
Otto-von-Guericke-University Magdeburg, Germany
Laboratoire d'Informatique de Paris 6 (LIP6), France
University Pierre and Marie Curie (UPMC), France
Centre national de la recherche scientifique (CNRS), France

Contents

State-of-the-Art Contributions

Defining and Applying a Language for Discovery

Tony Russell-Rose[1(✉)], Joe Lamantia[2], and Stephann Makri[3]

[1] UXLabs Ltd., London, UK
tgr@uxlabs.co.uk
[2] Oracle, 101 Main St., Cambridge, USA
jlamantia@oracle.com
[3] UCL Interaction Centre, University College London,
Gower St., London WC1E 6BT, UK
s.makri@ucl.ac.uk

Abstract. In order to design better search experiences, we need to understand the complexities of human information-seeking behaviour. In this paper, we propose a model of information behaviour based on the needs of users across a range of search and discovery scenarios. The model consists of a set of *modes* that that users employ to satisfy their information goals.

We discuss how these modes relate to existing models of human information seeking behaviour, and identify areas where they differ. We then examine how they can be applied in the design of interactive systems, and present examples where individual modes have been implemented in interesting or novel ways. Finally, we consider the ways in which modes combine to form distinct chains or patterns of behaviour, and explore the use of such patterns both as an analytical tool for understanding information behaviour and as a generative tool for designing search and discovery experiences.

1 Introduction

Classic IR (information retrieval) is predicated on the notion of users searching for information in order to satisfy a particular 'information need'. However, much of what we recognize as search behaviour is often not informational per se. For example, Broder [2] has shown that the need underlying a given web search could in fact be navigational (e.g. to find a particular site) or transactional (e.g. through online shopping, social media, etc.). Similarly, Rose and Levinson [12] have identified the consumption of online resources as a further common category of search behaviour.

In this paper, we examine the behaviour of individuals across a range of search scenarios. These are based on an analysis of user needs derived from a series of customer engagements involving the development of customised search applications.

The model consists of a set of 'search modes' that users employ to satisfy their information search and discovery goals. It extends the IR concept of information-seeking to embrace a broader notion of discovery-oriented problem solving, addressing a wider range of information interaction and information use behaviours. The overall structure reflects Marchionini's framework [8], consisting of three 'lookup' modes

© Springer International Publishing Switzerland 2014
A. Nürnberger et al. (Eds.): AMR 2012, LNCS 8382, pp. 3–28, 2014.
DOI: 10.1007/978-3-319-12093-5_1

(*locate, verify, monitor*), three 'learn' modes (*compare, comprehend, evaluate*) and three 'investigate' modes (*explore, analyze, synthesize*).

The paper is structured as follows. In Sect. 2 we discuss the modes in detail and their relationship to existing models of information seeking behaviour. Section 3 describes the data acquisition and the analysis process by which the modes were derived. In Sect. 4 we investigate the degree to which the model scales to accommodate diverse search contexts (e.g. from consumer-oriented websites to enterprise applications) and discuss some of the ways in which user needs vary by domain. In addition, we explore the ways in which modes combine to form distinct chains or patterns, and reflect on the value this offers as a framework for expressing complex patterns of information seeking behaviour.

In Sect. 5 we examine the practical implications of the model, discussing how it can be applied in the design of interactive applications, at both the level of individual modes and as composite structures. Finally, in Sect. 6 we reflect on the general utility of such models and frameworks, and explore briefly the qualities that might facilitate their increased adoption by the wider user experience design community.

2 Models of Information Seeking

The framework proposed in this study is influenced by a number of previous models. For example, Bates [1] identifies a set of 29 search 'tactics' which she organised into four broad categories, including *monitoring* ("to keep a search on track"). Likewise, O'Day and Jeffries [11] examined the use of information search results by clients of professional information intermediaries and identified three categories of behaviour, including *monitoring a known topic or set of variables over time* and *exploring a topic in an undirected fashion*. They also observed that a given search scenario would often evolve into a series of interconnected searches, delimited by triggers and stop conditions that signalled transitions between modes within an overall scenario.

Cool and Belkin [3] proposed a classification of interaction with information which included *evaluate* and *comprehend*. They also proposed *create* and *modify,* which together reflect aspects of our *synthesize* mode.

Ellis and his colleagues [4–6] developed a model consisting of a number of broad information seeking behaviours, including *monitoring* and *verifying* ("checking the information and sources found for accuracy and errors"). In addition, his *browsing* mode ("semi-directed searching in an area of potential interest") aligns with our definition of *explore*. He also noted that it is possible to display more than one behaviour at any given time. In revisiting Ellis's findings among social scientists, Meho and Tibbo [10] identified *analysing* (although they did not elaborate on it in detail). More recently, Makri et al. [8] proposed *searching* ("formulating a query in order to locate information"), which reflects to our own definition of *locate*.

In addition to the research-oriented models outlined above, we should also consider practitioner-oriented frameworks. Spencer [14] suggests four modes of information seeking, including *known-item* (a subset of our *locate* mode) and *exploratory* (which mirrors our definition of explore). Lamantia [7] also identifies four modes, including *monitoring*.

In this paper, we use the characteristics of the models above as a lens to interpret the behaviours expressed in a new source of empirical data. We also examine the combinatorial nature of the modes, extending Ellis's [5] concept of mode co-occurrence to identify and define common patterns and sequences of information seeking behaviour.

3 Studying Search Behaviour

3.1 Data Acquisition

The primary source of data in this study is a set of 381 information needs captured during client engagements involving the development of a number of custom search applications. These information needs take the form of 'micro-scenarios', i.e. a brief narrative that illustrates the end user's goal and the primary task or action they take to achieve it, for example:

- *Find best offers before the others do so I can have a high margin.*
- *Get help and guidance on how to sell my car safely so that I can achieve a good price.*
- *Understand what is selling by area/region so I can source the correct stock.*
- *Understand a portfolio's exposures to assess investment mix.*
- *Understand the performance of a part in the field so that I can determine if I should replace it.*

The scenarios were collected as part of a series of requirements workshops involving stakeholders and customer-facing staff from various client organisations. A proportion of these engagements focused on consumer-oriented site search applications (resulting in 277 scenarios) and the remainder on enterprise search applications (104 scenarios).

The scenarios were generated by participants in breakout sessions and subsequently moderated by the workshop facilitator in a group session to maximise consistency and minimise redundancy or ambiguity. They were also prioritised by the group to identify those that represented the highest value both to the end user and to the client organisation.

This data possesses a number of unique properties. In previous studies of information seeking behaviour (e.g. [5, 10]), the primary source of data has traditionally been interview transcripts that provide an indirect, verbal account of end user information behaviours. By contrast, the current data source represents a self-reported account of information needs, generated directly by end users (although a proportion were captured via proxy, e.g. through customer facing staff speaking on behalf of the end users). This change of perspective means that instead of using information behaviours to infer information needs and design insights, we can adopt the converse approach and use the stated needs to infer information behaviours and the interactions required to support them.

Moreover, the scope and focus of these scenarios represents a further point of differentiation. In previous studies, (e.g. [8]), measures have been taken to address the limitations of using interview data by combining it with direct observation of information seeking behaviour in naturalistic settings. However, the behaviours that this approach reveals are still bounded by the functionality currently offered by existing systems and working practices, and as such do not reflect the full range of aspirational or unmet user needs encompassed by the data in this study.

Finally, the data is unique in that is constitutes a genuine practitioner-oriented deliverable, generated expressly for the purpose of designing and delivering commercial search applications. As such, it reflects a degree of realism and authenticity that interview data or other research-based interventions might struggle to replicate.

3.2   Data Analysis

These scenarios were manually analyzed to identify themes or modes that appeared consistently throughout the set, using a number of iterations of a 'propose-classify-refine' cycle based on that of Rose and Levinson [14]. Inevitably, this process was somewhat subjective, echoing the observations made by Bates [1] in her work on search tactics:

> *"While our goal over the long term may be a parsimonious few, highly effective tactics, our goal in the short term should be to uncover as many as we can, as being of potential assistance. Then we can test the tactics and select the good ones. If we go for closure too soon, i.e., seek that parsimonious few prematurely, then we may miss some valuable tactics."*

In this respect, the process was partially deductive, in applying the insights from existing models to classify the data in a top-down manner. But it was also partially inductive, applying a bottom-up, grounded analysis to identify new types of behaviour not present in the original models or to suggest revised definitions of existing behaviours.

A number of the scenarios focused on needs that did not involve any explicit information seeking or use behaviour, e.g. *"Achieve a good price for my current car"*. These were excluded from the analysis. A further number were incomplete or ambiguous, or were essentially feature requests (e.g. "Have flexible navigation within the page"), and were also excluded.

The process resulted in the identification of nine primary search modes, which are defined below along with an example scenario (from the domain of consumer-oriented search):

1. **Locate:** *To find a specific (possibly known) item*, e.g. "Find my reading list items quickly". This mode encapsulates the stereotypical 'findability' task that is so commonly associated with site search. It is consistent with (but a superset of) Spencer's [14] *known item* search mode. This was the most frequent mode in the site search scenarios (120 instances, which contrasts with just 2 for enterprise search).

2. **Verify:** *To confirm that an item meets some specific, objective criterion*, e.g. "See the correct price for singles and deals". Often found in combination with locating, this mode is concerned with validating the accuracy of some data item, comparable to that proposed by Ellis et al. [5] (39 site search instances, 4 for enterprise search).

3. **Monitor:** *Maintain awareness of the status of an item for purposes of management or control,* e.g. "Alert me to new resources in my area". This activity focuses on the state of asynchronous responsiveness and is consistent with that of Bates [1], O'Day and Jeffries [11], Ellis [4], and Lamantia [7] (13 site search instances, 17 for enterprise search).

4. **Compare:** *To identify similarities & differences within a set of items,* e.g. "Compare cars that are my possible candidates in detail". This mode has not featured prominently in most of the previous models (with the possible exception of Marchionini's), but accounted for a significant proportion of enterprise search behaviour [13]. Although a common feature on many ecommerce sites, it occurred relatively infrequently in the site search data (2 site search instances, 16 for enterprise search).

5. **Comprehend:** *To generate independent insight by interpreting patterns within a data set,* e.g. "Understand what my competitors are selling". This activity focuses on the creation of knowledge or understanding and is consistent with that of Cool and Belkin [3] and Marchionini [9] (50 site search instances, 12 for enterprise search).

6. **Evaluate:** *To use judgement to determine the value of an item with respect to a specific goal,* e.g. "I want to know whether my agency is delivering best value". This mode is similar in spirit to *verify,* in that it is concerned with validation of the data. However, while *verify* focuses on simple, objective fact checking, our conception of *evaluate* involves more subjective, knowledge-based judgement, similar to that proposed by Cool and Belkin [3] (61 site search instances, 78 for enterprise search).

7. **Explore:** *To investigate an item or data set for the purpose of knowledge discovery,* e.g. "Find useful stuff on my subject topic". In some ways the boundaries of this mode are less prescribed than the others, but what the instances share is the characteristic of open ended, opportunistic search and browsing in the spirit of O'Day and Jeffries [11] *exploring a topic in an undirected fashion* and Spencer's [14] *exploratory* (110 site search instances, 16 for enterprise search).

8. **Analyze:** *To examine an item or data set to identify patterns & relationships,* e.g. "Analyze the market so I know where my strengths and weaknesses are". This mode features less prominently in previous models, appearing as a sub-component of the processing stage in Meho and Tibbo's [10] model, and overlapping somewhat with Cool and Belkin's [3] organize. This definition is also consistent with that of Makri et al. [8], who identified analysing as an important aspect of lawyers' interactive information behaviour and defined it as "examining in detail the elements or structure of the content found during information-seeking." (p. 630). This was the most common element of the enterprise search scenarios (58 site search instances, 84 for enterprise search).

9. **Synthesize:** *To create a novel or composite artefact from diverse inputs,* e.g. "I need to create a reading list on celebrity sponsorship". This mode also appears as a sub-component of the *processing* stage in Meho and Tibbo's [10] model, and involves elements of Cool and Belkin's [3] *create* and *use.* Of all the modes, this one is the most commonly associated with information *use* in its broadest sense (as opposed to information *seeking*). It was relatively rare within site search (5 site search instances, 15 for enterprise search).

Although the modes were generated from an independent data source and analysis process, we have retrospectively explored the degree to which they align with existing frameworks, e.g. Marchionini's [8]. In this context, *locate*, *verify*, and *monitor* could be described as lower-level 'lookup' modes, *compare*, *comprehend*, and *evaluate* as 'learn' modes and *explore*, *analyze*, and *synthesize* as higher-level 'investigate' modes.

4   Mode Sequences and Patterns

The modes defined above provide an insight into the needs of users of site search and enterprise search applications and a framework for understanding human information seeking behaviour. But their real value lies not so much in their occurrence as individual instances but in the patterns of co-occurrence they reveal. In most scenarios, modes combine to form distinct chains and patterns, echoing the transitions observed by O'Day and Jeffries [11] and the combinatorial behaviour alluded to by Ellis [5], who suggested that information behaviours can often be nested or displayed in parallel.

Typically these patterns consist of chains of length two or three, often with one particular mode playing a dominant role. Site search, for example, was characterized by the following patterns:

1. **Insight-driven search:** (Explore-Analyze- Comprehend): This patterns represents an exploratory search for insight or knowledge to resolve an explicit information need, e.g. *"Assess the proper market value for my car"*.
2. **Opportunistic search:** (Explore-Locate-Evaluate): In contrast to the explicit focus of Insight-driven search, this sequence represents a less directed exploration in the prospect of serendipitous discovery e.g. *"Find useful stuff on my subject topic"*.
3. **Qualified search:** (Locate-Verify): This pattern represents a variant of the stereotypical findability task in which some element of immediate verification is required, e.g. *"Find trucks that I am eligible to drive"*.

By contrast, enterprise search was characterized by a larger number of more diverse sequences, such as:

4. **Comparative search:** (Analyze-Compare- Evaluate) e.g. *"Replace a problematic part with an equivalent or better part without compromising quality and cost"*.
5. **Exploratory search:** (Explore-Analyze-Evaluate) e.g. *"Identify opportunities to optimize use of tooling capacity for my commodity/parts"*.
6. **Strategic Insight:** (Analyze-Comprehend-Evaluate) e.g. *"Understand a lead's underlying positions so that I can assess the quality of the investment opportunity"*.
7. **Strategic Oversight:** (Monitor-Analyze-Evaluate) e.g. *"Monitor & assess commodity status against strategy/plan/target"*.
8. **Comparison-driven Synthesis:** (Analyze-Compare-Synthesize) e.g. *"Analyze and understand consumer-customer-market trends to inform brand strategy & communications plan"*.

A further insight into these patterns can be obtained by presenting them in diagrammatic form. Figure 1 illustrates sequences 1–3 above plus other commonly found site search patterns as a network (with sequence numbers shown on the arrows).

It shows how certain modes tend to function as "terminal" nodes, i.e. entry points or exit points for a given scenario. For example, *Explore* typically functions as an opening, while *Comprehend* and *Evaluate* function in closing a scenario. *Analyze* typically appears as a bridge between an opening and closing mode. The shading indicates the mode 'level' alluded to earlier: light tones indicate 'lookup' modes, mid tones are the 'learn' modes, and dark tones are the 'investigate' modes.

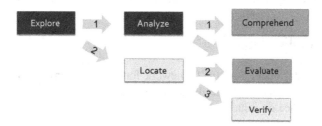

Fig. 1. Mode network for site search

Figure 2 illustrates sequences 4–8 above plus other commonly found patterns in the enterprise search data.

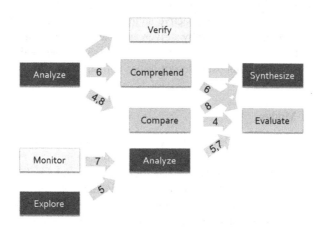

Fig. 2. Mode network for enterprise search

The patterns described above allow us to reflect on some of the differences between the needs of site search users and those of enterprise search. Site search, for example, is characterized by an emphasis on simpler "lookup" behaviours such as *Locate* and *Verify* (120 and 39 instances respectively); modes which were relatively rare in enterprise search (2 and 4 instances respectively). By contrast, enterprise search is characterized by higher-level "learn" and "investigate" behaviours such as *Analyze* and *Evaluate* (84 and 78 instances respectively, compared to 58 and 61 for site search).

Interestingly, in neither case was the stereotype of 'search equals findability' borne out: even in site search (where *Locate* was the most common mode), known-item search was accountable for no more than a quarter of all instances.

But perhaps the biggest difference is in the composition of the chains: enterprise search is characterised by a wide variety of heterogeneous chains, while site searched focuses on a small number of common trigrams and bigrams. Moreover, the enterprise search chains often displayed a fractal nature, in which certain chains were embedded within or triggered by others, to create larger, more complex sequences of behaviour.

5   Design Implications

Although the model offers a useful framework for understanding human information seeking behaviour, its real value lies in its use as a practical design resource. As such, it can provide guidance on issues such as:

- the features and functionality that should be available at specific points within a system;
- the interaction design of individual functions or components;
- the design cues used to guide users toward specific areas of task interface.

Moreover, the model also has significant implications for the broader aspects of user experience design, such as the alignment between the overall structure or concept model of a system and its users' mental models, and the task workflows for various users and contexts. This broader perspective addresses architectural questions such as the nature of the workspaces required by a given application, or the paths that users will take when navigating within a system's structure. In this way, the modes also act as a generative tool for larger, composite design issues and structures.

5.1   Individual Modes

On their own, each of the modes describes a type of behaviour that may need to be supported by a given information system's design. For example, an online retail site should support *locating* and *comparing* specific products, and ideally also *comprehending* differences and *evaluating* tradeoffs between them. Likewise, an enterprise application for electronic component selection should support *monitoring* and *verifying* the suitability of particular parts, and ideally also *analyzing* and *comprehending* any relevant patterns and trends in their lifecycle. By understanding the anticipated search modes for a given system, we can optimize the design to support specific user behaviours. In the following section we consider individual instances of search modes and explore some of their design implications.

Locate. This mode encapsulates the stereotypical 'findability' task that is so commonly associated with site search. But support for this mode can go far beyond simple keyword entry. For example, by allowing the user to choose from a list of candidates, auto-complete transforms the query formulation problem from one of recall into one of recognition (Fig. 3).

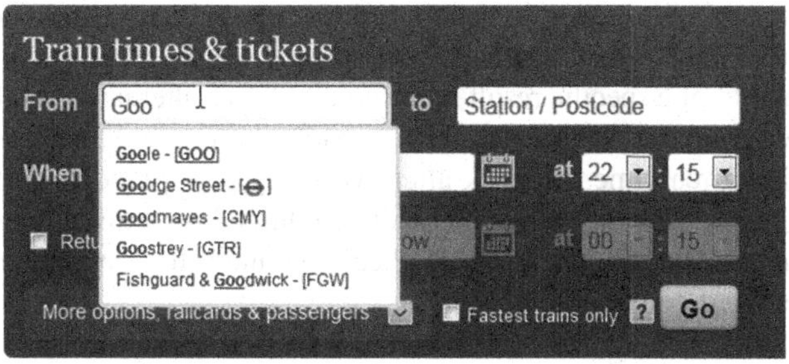

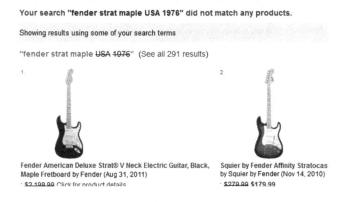

Fig. 3. Auto-complete supports locating

Likewise, Amazon's partial match strategy deals with potentially failed queries by identifying the keyword permutations that are likely to produce useful results. Moreover, by rendering the non-matching keywords in strikethrough text, it facilitates a more informed approach to query reformulation (Fig. 4).

Your search "fender strat maple USA 1976" did not match any products.

Showing results using some of your search terms

"fender strat maple ~~USA 1976~~" (See all 291 results)

Fender American Deluxe Strat® V Neck Electric Guitar, Black, Maple Fretboard by Fender (Aug 31, 2011)
· $2,199.99 Click for product details

Squier by Fender Affinity Stratocas
by Squier by Fender (Nov 14, 2010)
· $279.99 $179.99

Fig. 4. Partial matches support locating

Verify. In this mode, the user is inspecting a particular item and wishing to confirm that it meets some specific criterion. Google's image results page provides a good example of this (see Fig. 5).

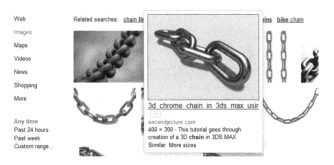

Fig. 5. Search result previews support verification

On mouseover, the image is zoomed in to show a magnified version along with key metadata, such as filename, image size, caption, and source. This allows the user to verify the suitability of a specific result in the context of its alternatives. Likewise, there may be cases where the user needs to verify a particular query rather than a particular result. In providing real-time feedback after every key press, Google Instant supports verification by previewing the results that will be returned for a given query (Fig. 6). If the results seem unexpected, the user can check the query for errors or try alternative spellings or keyword combinations.

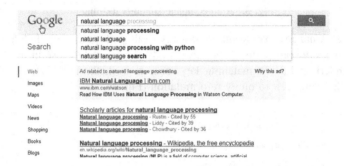

Fig. 6. Instant results supports verification of queries

Compare. The Compare mode is fundamental to online retail, where users need to identify the best option from the choices available. A common technique is to provide a custom view in which details of each item are shown in separate columns, enabling rapid comparison of product attributes. Best Buy, for example, supports comparison by organising the attributes into logical groups and automatically highlighting the differences (Fig. 7).

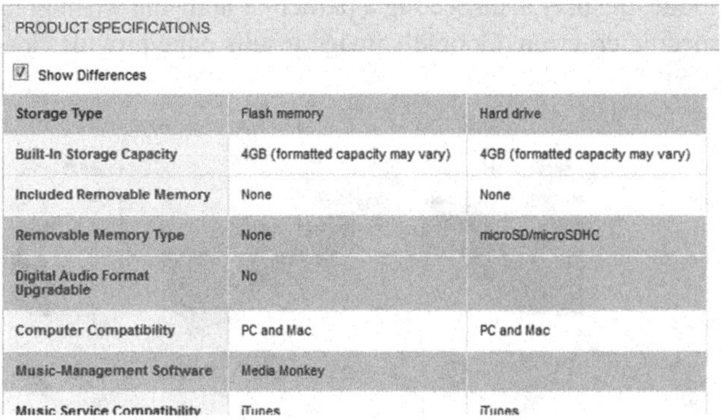

PRODUCT SPECIFICATIONS		
☑ Show Differences		
Storage Type	Flash memory	Hard drive
Built-In Storage Capacity	4GB (formatted capacity may vary)	4GB (formatted capacity may vary)
Included Removable Memory	None	None
Removable Memory Type	None	microSD/microSDHC
Digital Audio Format Upgradable	No	
Computer Compatibility	PC and Mac	PC and Mac
Music-Management Software	Media Monkey	
Music Service Compatibility	iTunes	iTunes

Fig. 7. Separate views support product comparison

But comparison is not restricted to qualitative attributes. In financial services, for example, it is vital to compare stock performance and other financial instruments with industry benchmarks. Google Finance supports the comparison of securities through a common charting component (Fig. 8).

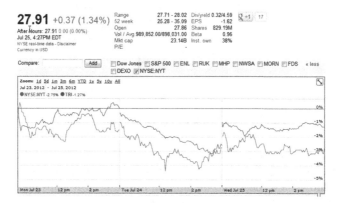

Fig. 8. Common charts allow comparison of quantitative data

Explore. A key principle in exploring is differentiating between *where you are going* and *where you have already been*. In fact, this distinction is so important that it has been woven into the fabric of the web itself; with unexplored hyperlinks rendered in blue by default, and visited hyperlinks shown in magenta. Amazon takes this principle a step further, through components such as a 'Recent Searches' panel showing the previous queries issued in the current session, and a 'Recent History' panel showing the items recently viewed (Fig. 9).

Fig. 9. Recent history supports exploration

Another simple technique for encouraging exploration is through the use of "see also" panels. Online retailers commonly use these to promote related products such as accessories and other items to complement an intended purchase. An example of this can be seen at Food Network, in which featured videos and products are shown alongside the primary search results (Fig. 10).

Fig. 10. 'See Also' panels support exploration

A further technique for supporting exploration is through the use of auto-suggest. While auto-complete helps users get an idea out of their heads and into the search box, auto-suggest throws new ideas into the mix. In this respect, it helps users explore by formulating more useful queries than they might otherwise have thought of on their own. Home Depot, for example, provides a particularly extensive auto-suggest function consisting of product categories, buying guides, project guides and more, encouraging the discovery of new product ideas and content (Fig. 11).

Fig. 11. Auto-suggest supports exploratory search

Analyze. In modes such as exploring, the user's primary concern is in understanding the *overall* information space and identifying areas to analyze in further detail. Analysis, in this sense, goes hand in hand with exploring, as together they present complementary modes that allow search to progress beyond the traditional confines of information retrieval or 'findability'.

A simple example of this could be found at Google patents (Fig. 12). The alternate views (Cover View and List View) allow the user to switch between rapid exploration (scanning titles, browsing thumbnails, looking for information scent) and a more detailed analysis of each record and its metadata.

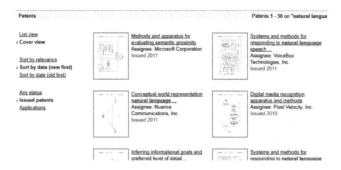

Fig. 12. Alternate views support mode switching between exploration and analysis

In the above example the analysis focuses on *qualitative* information derived from predominantly textual sources. Other applications focus on *quantitative* data in the form of aggregate patterns across collections of records. NewsSift, for example, provided a set of data visualizations which allowed the user to *analyze* results for a given news topic at the aggregate level, gaining an insight that could not be obtained from examining individual records alone (Fig. 13).

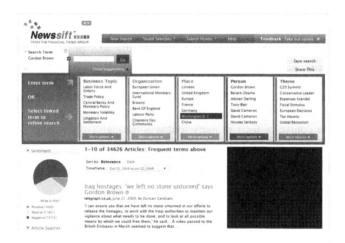

Fig. 13. Visualizations support analysis of quantitative information

5.2   Composite Patterns

The examples above represent instances of *individual* modes, showing various ways they can be supported by one or more aspects of a system's design. However, a key feature of the model is its emphasis on the *combinatorial* nature of modes and the patterns of co-occurrence this reveals [12]. In this respect, its true value is in helping designers to address more holistic, larger scale concerns such as the appropriate structure, concept model, and organizing principles of a system, as well as the functional and informational content of its major components and connections between them.

Design at this level relies on translating composite modes and chains that represent sense-making activities – often articulated as *user journeys* through a task and information space – into interaction components that represent meaningful combinations of information and discovery capabilities [13]. These components serve as 'building blocks' that designers can assemble into larger composite structures to create a user experience that supports the anticipated user journeys and aligns with their users' mental models [14].

The popular micro-blogging service twitter.com provides a number of examples of the correspondence between composite modes and interaction components assembled at various levels to provide a coherent user experience architecture.

Twitter.com: Header Bar. The header bar at the top of most pages of twitter.com combines several informational and functional elements together in a single component that supports a number of modes and mode chains (Fig. 14). It includes four dynamic status indicators that address key aspects of twitter's concept model and the users' mental models:

- the presence of new tweets by people the user follows
- interactions with other twitter users such as following them or mentioning them in a tweet
- activity related to the user's profile, such as their latest tweets and shared media
- people, topics, or items of interest suggested by the systems recommender functions.

These status indicator icons update automatically and provide links to specific pages in the twitter.com application architecture that provide further detail on each area of focus. The header bar thus enables Monitoring of a user's activity within the full scope of the twitter.com network; i.e. its content, members, their activities, etc. The header bar also enables Monitoring activity within almost all the workspaces that users encounter in the course of their primary journeys through twitter.com.

Fig. 14. Twitter.com header bar

The Strategic Oversight chain (Monitor – Analyze - Evaluate) is a fundamental sequence for twitter users, repeated frequently with different aspects of the user's profile. The header bar supports the first step of this chain, in which users Monitor the network for content and activity of interest to them, and then transition to Analysis and Evaluation of that activity by navigating to destination pages for further detail.

The header bar also includes a search box featuring auto-complete and auto-suggest functionality, which provides support for the Qualified Search mode chain (Locate - Verify). The search box also enables users to initiate many other mode chains by supporting the Explore mode. These include Exploratory Search (Explore – Analyze - Evaluate), Insight-driven Search (Explore – Analyze - Comprehend), and Opportunity-driven Search (Explore - Locate - Evaluate). All these mode chains overlap by sharing a common starting point. This is one of the most readily recognizable kinds of composition, and often corresponds to a single instance of a particular interaction component.

The header bar includes support for posting or Synthesizing new tweets, reflecting the fact that the creation of new content is probably the second most important individual mode (after Monitoring). A menu of links to administrative pages and functions for managing one's twitter account completes the content of the header bar.

Twitter.com: Individual Tweets. The individual tweets and activity updates that make up the stream at the heart of the primary workspace are the most important interaction components of the twitter experience, and their design shows a direct correspondence to many composite modes and chains (Fig. 15). Individual items provide the content of a tweet along with the author's public name, their twitter username, profile image, and the time elapsed since the tweet's creation. Together, these details allow users to Compare and Comprehend the content and significance of tweets in their own stream. As users read more tweets and begin to recognize authors and topics, they can Compare, Analyze, and Evaluate them. The indicators of origin and activity allow users to Compare and Comprehend the topics and interests of other twitter users.

Fig. 15. Individual tweet

Options to invoke a number of functions that correspond to other discovery modes are embedded within the individual items in the stream. For example, if an update was retweeted, it is marked as such with the original author indicated and their profile page linked. It also shows the number of times the tweet has been retweeted and favorited, with links that open modal previews of the list of users who did so. This supports Monitoring, Exploration and Comprehension of the significance and attention an individual tweet has received, while the links support Location, Verification and Monitoring of the other users who retweeted or favorited it.

Public profile names and usernames are linked to pages which summarize the activities and relationships of the author of a tweet, enabling users to Locate and Verify authors, then transition to Monitoring, Exploring and Comprehending their activities, interests, and how they are connected to the rest of the twitter network.

Hashtags are presented with distinct visual treatment. When users click on one, it initiates a search using the hashtag, allowing users to Locate, Explore, Comprehend, and Analyze the topic referred to, any conversations in which the tag is mentioned, and the users who employ the tag.

Fig. 16. Expanded tweet

Longer tweets are truncated, offering an 'Expand' link which opens a panel displaying the number of retweets and favourites and the images of the users who did so, along with the date and time of authoring and a link to a 'details' page for a permanent URL that other users and external services can reference (Fig. 16). This sort of truncation enables users to more easily Explore the full set of tweets in a stream and Locate individual items of interest. Conversely, the 'Expand' panel allows the user to more easily Explore and Comprehend individual items.

Tweets that contain links to other tweets offer a 'View tweet' link, which opens a panel displaying the full contents of the original tweet, the date and time of posting, the number of retweets and favorites and a preview list of the users who did so. The 'View tweet' link thus supports the Locate, Explore, and Comprehend modes for individual updates.

Tweets that contain links to digital assets such as photos, videos, songs, presentations, and documents, offer users the ability to preview these assets directly within an expanded display panel, providing support for the Locate, Explore, and Comprehend modes. These previews link to the source of the assets, enabling users to Locate them. Users can also 'flag' media for review by twitter (e.g. due to violation of policies about sensitive or illegal imagery) – which is a very specific form of Evaluation.

Tweets that contain links to items such as articles published by newspapers, magazines, and journals, or recognized destinations such as Foursquare and Google + pages, offer a 'Summary' link (Fig. 17). This link opens a panel that presents the first paragraph of the article or destination URL, an image from the original publisher, and a list of users who have retweeted or favorited it, thus supporting Location, Exploration and Verification of the linked item.

A text input field seeded with the author's username allows users to reply to specific tweets directly from an individual update. Users can also 'retweet' items directly from the list. Both functions are forms of Synthesis, and encourage users to create further content and relationships within the network.

Users can mark tweets as 'favorites' to indicate the importance or value of these tweets to others; a clear example of the Evaluation mode. Favorites also allow users to build a collection of tweets curated for retrieval and interpretation, enabling the Locate, Compare, Comprehend, and Analyze modes for tweets as individual items or as groups.

Fig. 17. Tweet displaying a photo

A 'More' link opens a menu offering 'Email Tweet' and 'Embed Tweet' options, allowing users to initiate tasks that take tweets outside the twitter environment. These two functions support information *usage* modes, rather than *search* and *discovery* modes, so their distinct treatment – invoked via a different interaction than the other functions – is consistent with the great emphasis the twitter experience places on discovery and sense making activities.

If the tweet is part of a conversation, a 'View this conversation' link allows readers to open a panel that presents related tweets and user activity as a single thread, accompanied by a reply field. This provides support for the Locate, Explore, Comprehend, Analyze, Evaluate and Synthesize modes (Fig. 18).

The informational and functional content presented by individual items in their various forms enables a number of mode chains. These include Strategic Oversight, in which users maintain awareness of conversations, topics, other users, and activities; Strategic Insight, wherein users focus on and derive insight into conversations, topics, and other users; and Comparative Synthesis, in which users realize new insights and create new content through direct engagement with conversations, topics, and other users.

In a manner similar to the search box, this interaction component serves as an initiation point for a number of mode chains, including Exploratory Search, Insight-driven

Fig. 18. Tweet showing a conversation

Search, and Opportunity-driven Search. Individual tweets thus combine support for many important modes and mode chains into a single interaction component. As a consequence, they need to be relatively rich and 'dense', compacting much functionality into a single interaction component, but this reflects their crucial role in the user journeys that characterize the twitter experience.

Twitter.com: Primary Workspaces and Pages. In the previous section we reviewed the correspondence between groups of modes and the interaction components of a user experience. In this section, we review the ways in which modes and chains impact the composition and presentation of the next level of UX structure within the system: work spaces.

The primary workspaces of twitter.com all emphasize interaction with a stream of individual updates, but the focus and content vary depending on the context. On the Home page, for example, the central stream consists of tweets from people the user follows, while on the 'Me' page the stream consists of the tweets created by the user (Fig. 19). However, the layout of these pages remains consistent: the workspace is dominated by a single central stream of individual updates. The primary interaction mode for this stream is Monitoring, evident from the count of new items added to the network since the last page refresh.

The placement of the header bar at the top of all of the primary workspaces is a design decision that reflects the primacy of Monitoring as a mode of engagement with

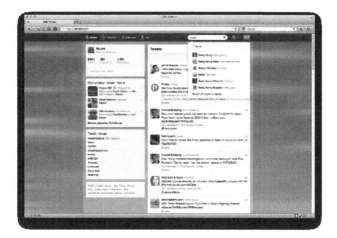

Fig. 19. Twitter.com home workspace

the twitter service; supporting its role as a persistent 'background' mode of discovery independent of the user's current point in a task or journey, and its role as a common entry point to the other mode chains and user journeys.

The consistent placement of the 'Compose new Tweet' control in upper right corner of the workspace reflects known interaction design principles (corners are the second most easily engaged areas of a screen, after the centre) and the understanding that Synthesis is the second most important single mode for the twitter service.

The content of the individual updates attracts and retains users' attention very effectively: the majority of the actions a user may want to take in regard to a tweet (or any of the related constructs in twitter's concept model such as conversations, hash tags, profiles, linked media, etc.) are directly available from the interaction component. In some cases, these actions are presented via modal or lightbox preview, wherein the user's focus is 'forced' onto a single element – thus maintaining the primacy of the stream. In others, links lead to destination pages that switch the user's focus to a different subject – another user's profile, for example – but in most of these cases the structure of the workspace remains consistent: a two column body surmounted by the ubiquitous header bar. There is little need to look elsewhere in the workspace, unless the user needs to check the status of one of the broader aspects of their account, at which point the header bar provides appropriate functionality as discussed above.

The absence of a page footer – scrolling is 'infinite' on the primary pages of twitter. com – reflects the conscious decision to convey updates as an endless, dynamic stream. This encourages users to continue scrolling, increasing Exploration activity, and enhancing users' Comprehension of additional updates – which benefits twitter's business by increasing the attention users direct toward the service.

Although the two-tier, stream-centred structure of twitter's primary workspaces remains consistent, there are variations in the composition of the left column (Fig. 20). On the Home page, for example, the left column offers four separate components. The first is a summary of the user's profile, including a profile image, a link to their profile

Fig. 20. Twitter home page: Left column

page, counts of their tweets, followers, and the people they follow, and a 'compose new tweet' box. This is another example of a component supporting a composite of modes.

The core purpose is to enable users to Monitor the most important aspects of their own account via the counts. The links provide direct Locate functionality for followers, tweets, and accounts the user follows; and also serve as a point of departure for the same mode chains that can be initiated from the header bar. The 'compose new tweet' function encourages users to create updates, underlining the importance of Synthesis as the source of new content within the twitter network.

Twitter.com: User Experience Architecture. The twitter.com experience is intended to support a set of user journeys consisting largely of search and discovery tasks which correspond with specific monitoring and search-related mode chains. Further, we can see that patterns of recurrence, intersection, overlap, and sequencing in the aggregate set of search and discovery modes are substantially reflected in twitter's user experience architecture.

From a structural design perspective, the core [16] of the twitter.com user experience architecture is a set of four interaction consoles, each of which focuses on monitoring a distinct stream of updates around the most important facets of the twitter.com concept model: the content and activities of people in the user's personal network (Home); interactions with other users (Interactions); the user's profile (@Me); and a digest of content from all users in the twitter.com network (Discover) (Fig. 21).

The core monitoring consoles are supported by screens that assist and encourage users to expand their personal networks through location and exploration tools; these include 'Find friends', 'Who to follow' 'Browse categories', and the search results page.

Fig. 21. Twitter.com discover workspace

Specific landing pages provide monitoring and curation tools for the different types of relationships users can establish in the social graph: follow and un-follow, followers and following, public and private accounts, list memberships, etc. A small set of screens provides functionality for administering the user's account, such as 'Settings'.

Underlying this user experience architecture is a concept model consisting primarily of a small set of social objects – tweets, conversations, profiles, shared digital assets, and lists thereof – linked together by search and discovery verbs. A relatively simple information architecture establishes the set of categories used to identify these objects by topic, similarity, and content (Fig. 22).

In its holistic and granular aspects, the twitter user experience architecture aligns well with users' mental models for building a profile and participating in an ongoing stream of conversations. However, what emerges quite quickly from analysis of the twitter concept model and user experience architecture is the role of search and discovery modes in both atomic and composite forms at every level of twitter's design. Rather than merely subsuming modes as part of some larger activity, many of the most common actions users can take with twitter's core interaction objects correspond directly to modes themselves.

The individual tweet component is a prime example: the summaries of author profiles and their recent activity are a composite of the Locate, Explore and Comprehend modes (Fig. 23). Evidently, the presentation, labelling, and interaction design may reflect adaptations specific to the language and mental model of the twitter environment,

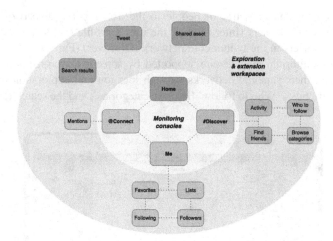

Fig. 22. Twitter.com user experience architecture

but the activities are clearly recognizable. The 'Show conversation' function discussed above also reflects direct support to Locate, Explore and Comprehend a conversation object as a single interaction.

Because the twitter.com experience is so strongly centred on sense-making, search and discovery modes often directly constitute the activity paths connecting one object to another within the user experience architecture. In this sense, the modes and chains could be said to act as a 'skeleton' for twitter.com, and are directly visible to an unprecedented degree in the interaction design built on that skeleton.

Fig. 23. Twitter profile summary

6   Discussion

The model described in this paper encompasses a range of information seeking behaviours, from elementary lookup tasks through to more complex problem-solving activities. However, the model could also be framed as part of a broader set of information behaviours, extending from 'acquisition' oriented tasks at one end of the spectrum to 'usage' oriented activities at the other (Fig. 24). In this context, modes can span more than one phase. For example, Explore entails a degree of *interaction* coupled with the anticipation of further discovery, i.e. *acquisition*. Likewise, Evaluate implies a degree of *interaction* in the pursuit of some higher goal or purpose to which the output will be put, i.e. *usage*.

It would appear that with the possible exception of *synthesize*, there are no exclusively usage-oriented behaviours in the model. This may suggest that the model is in some senses incomplete, or may simply reflect the context in which the data was acquired and the IR-centric processes by which it was analysed.

Reducing the 'scope' of the model such that modes serve only as descriptors of distilled sense-making activity independent of context (such as the user's overall goal and the nature of the information assets involved) may help clarify the relationship between acquisition, interaction and usage phases. In this perspective, there appears to be a form of 'parallelism' in effect; with users simultaneously undertaking activities focused on an overall goal, such as Evaluating the quality of a financial instrument, while also performing activities focused on narrower information-centred objectives such as Locating and Verifying the utility of the information assets necessary for them to complete the Evaluation. These 'parallel' sets of activities – one focused on information assets in service to a larger goal, and the other focused on the goal itself – can be usefully described in terms of modes, and what is more important, seem intertwined in the minds of users as they articulate their discovery needs.

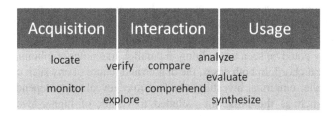

Fig. 24. From information acquisition to information use

A key feature of the current model is its emphasis on the combinatorial nature of search modes, and the value this offers as a framework for expressing complex patterns of behaviour. Evidently, such an approach is not unique: Makri (2008), for example, has also previously explored the concept of mode chains to describe information seeking behaviours observed in naturalistic settings. However, his approach was based on the analysis of complex tasks observed in real time, and as such was less effective in revealing consistent patterns of atomic behaviour such as those found in the current study.

Conversely, this virtue can also be a shortcoming: the fact that simple repeating patterns can be extracted from the data may be as much an artefact of the medium as it is of the information needs it contains. These scenarios were expressly designed to be a concise, self-contained deliverable in their own right, and applied as a simple but effective tool in the planning and prioritisation of software development activities. This places a limit on the length and sophistication of the information needs they encapsulate, and a natural boundary on the scope and extent of the patterns they represent. Their format also allows a researcher to apply perhaps an unrealistic degree of top-down judgement and iteration in aligning the relative granularity of the information needs to existing modes; a benefit that is less readily available to those whose approach involves real-time, observational data.

A further caveat is that in order to progress from understanding an information need to identifying the information behaviours required to satisfy those needs, it is necessary to *speculate* on the behaviours that a user *might* perform when undertaking a task to satisfy the need. It may transpire that users actually perform different behaviours which achieve the same end, or perform the expected behaviour but through a combination of other nested behaviours, or may simply satisfy the need in a way that had not been envisaged at all.

Evidently, the process of inferring information behaviour from self-reported needs can never be wholly deterministic, regardless of the consistency measures discussed in Sect. 3.1. In this respect, further steps should be taken to operationalize the process and develop some independent measure of stability or objectivity in its usage, so that its value and insights can extend reliably to the wider research community.

The compositional behaviour of the modes suggests further open questions and avenues for research. One of these is the nature of compositionality itself: one the one hand it could be thought of as a pseudo-linguistic grammar, with bigrams and trigrams of modes that combine in turn to form larger sequences, analogous to coherent "sentences". In this context, the modes act as verbs, while the associated objects (users, information assets, processes etc.) become the nouns. The occurrence of distinct 'opening' and 'closing' modes in the scenarios would seem to further support this view. However, in some scenarios the transitions between the modes are far less apparent, and instead they could be seen as applying in parallel, like notes combining in harmony to form a musical chord. In both cases, the degree and nature of any such compositional rules needs further empirical investigation. This may reveal other dependencies yet to be observed, such as the possibility alluded to earlier of higher-level behaviours requiring the completion of certain lower level modes before they themselves can terminate.

The process of mapping from modes to design interventions also reveals further observations on the utility of information models in general. Despite their evident value as analytical frameworks and their popularity among researchers (Bates' Berrypicking model has been cited over 1,000 times, for example), few have gained significant traction within the design community, and fewer still are adopted as part of the mainstream working practices of system design practitioners.

In part, this may be simply a reflection of imperfect channels of communication between the research and design communities. However, it may also reflect a growing conceptual gap between research insights on the one hand and corresponding design interventions on the other. It is likely that the most valuable theoretical models will need to strike a balance between flexibility (the ability to address a variety of domains and problems), generative power (the ability to express complex patterns of behaviour) and an appropriate level of abstraction (such that design insights are readily available; or may be inferred with minimal speculation).

7 Conclusions

In this paper, we have examined the needs and behaviours of individuals across a wide range of search and discovery scenarios. We have proposed a model of information seeking behaviour which has at its core a set of modes that people regularly employ to satisfy their information needs. In so doing, we explored a novel, goal-driven approach to eliciting user needs, and identified some key differences in user behaviour between site search and enterprise search.

In addition, we have demonstrated the value of the model as a framework for expressing complex patterns of search behaviour, extending the IR concept of information-seeking to embrace a broader range of information interaction and use behaviours. We propose that our approach can be adopted by other researchers who want to adopt a 'needs first' perspective to understanding information behaviour.

By illustrating ways in which individual modes are supported in existing search applications, we have made a practical contribution that helps bridge the gap between *investigating* search behaviour and *designing* applications to support such behaviour. In particular, we have demonstrated how modes can serve as an effective design tool across varied levels of system design: Concept model, UX architecture, interaction design, and visual design.

References

1. Bates, M.J.: Information search tactics. J. Am. Soc. Inf. Sci. **30**, 205–214 (1979)
2. Cool, C., Belkin, N.: A classification of interactions with information. In: H. Bruce (ed.) Emerging Frameworks and Methods: CoLIS4: Proceedings of the 4th International Conference on Conceptions of Library and Information Science, Seattle, WA, USA, pp. 1–15, 21–25 July 2002
3. Ellis, D.: A behavioural approach to information retrieval system design. J. Documentation **45**(3), 171–212 (1989)
4. Ellis, D., Cox, D., Hall, K.: A comparison of the information-seeking patterns of researchers in the physical and social sciences. J. Documentation **49**(4), 356–369 (1993)
5. Ellis, D., Haugan, M.: Modelling the information-seeking patterns of engineers and research scientists in an industrial environment. J. Documentation **53**(4), 384–403 (1997)
6. Hobbs, J.: An introduction to user journeys. Boxes and arrows (2005). www.boxesandarrows. com/an-introduction-to-user-journeys/
7. Kalbach, J.: Designing screens using cores and paths. Boxes and arrows (2012). www. boxesandarrows.com/designing-screens-using-cores-and-paths/
8. Lamantia, J.: 10 Information retrieval patterns JoeLamantia.com (2006). www.joelamantia. com/information-architecture/10-information-retrieval-patterns

9. Lamantia, J.: Creating successful portals with a design framework. Int. J. Web Portals (IJWP) **1**(4), 63–75 (2009). doi:10.4018/jwp.2009071305

10. Makri, S., Blandford, A., Cox, A.L.: Investigating the information-seeking behaviour of academic lawyers: From ellis's model to design. Inf. Process. Manage. **44**(2), 613–634 (2008)

11. Marchionini, G.: Exploratory search: From finding to understanding. Commun. ACM **49**(4), 41–46 (2006)

12. Meho, L., Tibbo, H.: Modeling the information-seeking behavior of social scientists: Ellis's study revisited. J. Am. Soc. Inform. Sci. Technol. **54**(6), 570–587 (2003)

13. O'Day, V., Jeffries, R.: Orienteering in an information landscape: How information seekers get from here to there. INTERCHI **1993**, 438–445 (1993)

14. Rose, D., Levinson, D.: Understanding user goals in web search. In: Proceedings of the 13th International Conference on World Wide Web, New York, NY, USA (2004)

15. Russell-Rose, T., Lamantia, J., Burrell, M.: A taxonomy of enterprise search and discovery. In: Proceedings of HCIR 2011, California, USA (2011)

16. Russell-Rose, T., Makri, S.: A model of consumer search behavior. In: Proceedings of EuroHCIR 2012, Nijmegen, The Netherlands (2012)

17. Spencer, D.: Four modes of seeking information and how to design for them. boxes and arrows (2006). www.boxesandarrows.com/view/four_modes_of_seeking_information_and_how_to_design_for_them

A Survey of Evaluation in Music Genre Recognition

Bob L. Sturm[(⊠)]

Audio Analysis Lab, AD:MT, Aalborg University Copenhagen,
A.C. Meyers Vænge 15, 2450 Copenhagen SV, Denmark
bst@create.aau.dk

Abstract. Much work is focused upon music genre recognition (MGR) from audio recordings, symbolic data, and other modalities. While reviews have been written of some of this work before, no survey has been made of the approaches to evaluating approaches to MGR. This paper compiles a bibliography of work in MGR, and analyzes three aspects of evaluation: experimental designs, datasets, and figures of merit.

1 Introduction

Despite much work [1–467], music genre recognition (MGR) remains a compelling problem to solve by a machine. In addition to many background chapters of master's theses [39, 79, 113, 132, 153, 154, 188, 193, 239, 361, 367, 371, 418] and doctoral dissertations [9, 141, 146, 280, 284, 290, 320, 341, 342, 381, 427, 447] at least five reviews are devoted specifically to MGR [23, 85, 123, 241, 373], and 19 other reviews discuss related aspects [24, 25, 51, 71, 84, 100, 101, 152, 181, 198, 224, 233, 270, 282, 315, 398, 423, 441, 442]. Many of these reviews compile the variety of feature extraction methods and classification algorithms that have been applied to MGR, and some compare system performance using specific figures of merit (FoM) on particular benchmark datasets. There have also been no fewer than 10 campaigns to formally evaluate and compare state-of-the-art algorithms for MGR [170, 171, 293–299, 316]. However, the variety of approaches used for evaluating performance in MGR has yet to be surveyed. *How does one measure the capacity of a system — living or not — to recognize and discriminate between abstract characteristics of the human phenomenon of music?*

There currently exists at least eleven works [77, 78, 116, 117, 246, 320, 404, 409, 410, 433, 449] that address the difficult but clearly relevant question of how to evaluate the performance of MGR systems, not to mention how to properly create a dataset from which a machine is to learn an abstract and high-level concept such as genre [468, 469]. A few works critically address evaluation in MGR. For instance, [77, 78, 409, 410] argue for more realistic approaches than having a system apply a single label to music, and comparing against a "ground truth" — which itself can be quite wrong [404, 408]. Furthermore, [77, 78, 246, 449] argue for measuring performance in ways that take into account the natural ambiguity arising from genre.

© Springer International Publishing Switzerland 2014
A. Nürnberger et al. (Eds.): AMR 2012, LNCS 8382, pp. 29–66, 2014.
DOI: 10.1007/978-3-319-12093-5_2

In this paper, we take a different direction to answer the question we pose above. We review a significant portion of published research touching upon aspects of evaluation in MGR. We consider all work that is based upon recorded music, and/or symbolic representations of the music, e.g., MIDI, and/or other modalities, e.g., lyrics, album covers, user tags, movie scenes, etc. We do not, however, consider work addressing the more general problem of "tagging," e.g., [470]. While we consider both "genre" and "style," and make no attempt to differentiate them, we do not include "mood" or "emotion," e.g., [471]. We are herein interested only in the ways systems for MGR are evaluated, be they algorithms, humans [79,169, 201,258,261,262,278,290,366,367,370,381,383,460], pigeons [347], sparrows [439, 440], koi [58], primates [278] or rats [317]. To facilitate this survey, we created a spreadsheet summarizing every relevant paper we found in terms of its experimental design, details of the datasets it uses, and the figures of merit it reports. This resource provides a simple means to delimit sets of references sharing particular aspects of evaluation. The bibliography file we assembled for this work is available here: http://imi.aau.dk/~bst/research/MGRbibliography.bib; and the spreadsheet we created identifying the characteristics of the references is available here: http://imi.aau.dk/~bst/research/MGRspreadsheet.xlsx.

Figure 1 shows how the number of the works we reference is distributed since the 1995 work of Matityaho and Furst [271] — before which we have only found the 1984 work of Porter and Neuringer [347]. Many papers allude to the 2002 article of Tzanetakis and Cook [426] as the beginning of research in automatic MGR. We find their manuscript (received Nov. 2001 and growing from [425]) is preceded by seventeen works [44,53,83,89,91,132,148,204,270,271,346,348,350, 401,443,472,473], and is contemporary with nine works [22,79,176,193,202,218, 351,385,448]. The dataset created by Tzanetakis and Cook for [426], however, is the first "benchmark" MGR dataset to have been made publicly available, and as a result continues to be the most used public dataset for MGR.

In our analysis, we do not include [474–479] as they are written in Turkish, and [472] as it is written in German, and we can read neither. We could not

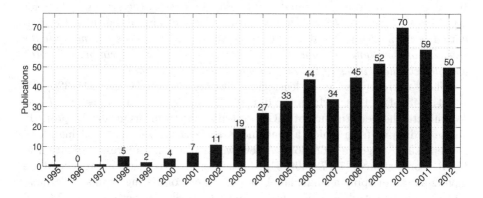

Fig. 1. Annual numbers of publications in this survey.

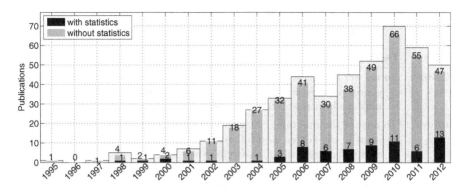

Fig. 2. Annual numbers of publications in this survey having an experimental component (top number), and which use any form of statistical testing for making comparisons (bottom number).

obtain [473, 480, 481], and so do not include them in the analysis. Finally, we neither analyze nor cite seven published works because of plagiarism.

2 Evaluation Approaches in Music Genre Recognition

We now catalogue approaches to evaluation in MGR along three dimensions — experimental design, datasets, and figures of merit (FoM) — and present summary statistics of each. Experimental design is the method employed to answer a specific hypothesis, e.g., in the case of MGR, "System A recognizes 'Blues'." The dataset is simply the collection of data used in the experiment. A FoM quantifies the performance of a system in the experiment, e.g., accuracy. Figure 2 shows how the number of works having an experimental component is distributed over the years. Compared to Fig. 1, the remaining works are reviews, or primarily concerning evaluation.

2.1 Experimental Design

Table 1 describes the ten different experimental designs we find, along with their appearance in the literature. We see that the most common experimental design to test MGR systems is *Classify*. More than 91 % (397)[1] of the referenced work having an experimental component (435) uses such a design [1–9, 11–21, 26–43, 45, 47–50, 52–56, 58–60, 62–65, 68–70, 72–76, 79–83, 86–90, 92–99, 102–106, 108–122, 124–135, 137–148, 150, 151, 153–165, 167–172, 174–180, 182–197, 199–202, 204–217, 219–223, 225–232, 234, 235, 238–240, 242–261, 263–269, 271–281, 283–287, 289–301, 303, 305–307, 309–313, 317, 319–333, 335–349, 352–359, 361–364, 366–372, 375–383, 385–393, 395–397, 399, 401–403, 405–407, 409–412, 414, 416, 418–422, 424–433, 435–440, 443–445, 447, 448, 450–465, 467]. For instance, Matityaho and Furst [271] test

---

[1] Numbers in parentheses are the number of works in the references.

Table 1. Ten experimental designs of MGR, and the percentage of references having an experimental component (435) in which they appear

Design	Description	%
Classify	To answer the question, "How well does the system identify the genres used by music?" The system applies genre labels to music, which researcher then compares to a "ground truth"	91.3
Features	To answer the question, "At what is the system looking to identify the genres used by music?" The system ranks and/or selects features, which researcher then inspects	32.6
Generalize	To answer the question, "How well does the system identify genre in varied datasets?" *Classify* with two or more datasets having different genres, and/or various amounts of training data	15.9
Robust	To answer the question, "To what extent is the system invariant to aspects inconsequential for identifying genre?" The system classifies music that researcher modifies or transforms in ways that do not harm its genre identification by a human	6.9
Eyeball	To answer the question, "How well do the parameters make sense with respect to identifying genre?" The system derives parameters from music; researcher visually compares	6.7
Cluster	To answer the question, "How well does the system group together music using the same genres?" The system creates clusters of dataset, which researcher then inspect	6.7
Scale	To answer the question, "How well does the system identify music genre with varying numbers of genres?" *Classify* with varying numbers of genres	6.7
Retrieve	To answer the question, "How well does the system identify music using the same genres used by the query?" The system retrieves music similar to query, which researcher then inspects	4.4
Rules	To answer the question, "What are the decisions the system is making to identify genres?" The researcher inspects rules used by a system to identify genres	3.7
Compose	To answer the question, "What are the internal genre models of the system?" The system creates music in specific genres, which the researcher then inspects	0.7

a neural network trained to discriminate between classical and pop music. They extract features from the audio, input them to the neural network, and compare the output labels against those they assigned to the excerpts of their dataset. Almost all of the experimental work that applies *Classify* employ a single-label approach, but at least ten employ a multi-labeling approach [31, 255, 258, 280, 367–369, 373, 415, 437]. For instance, McKay [280] looks at how genres at both root and leaf nodes are applied by his hierarchical approach to classification.

One extension of the *Classify* experimental design is *Generalize* [1, 2, 9, 13, 15, 16, 19, 30, 31, 38, 45, 55, 58, 62, 66, 75, 82, 94, 97, 128, 137, 146, 153, 161–164, 199, 200, 209, 214, 218, 222, 223, 225, 226, 231, 238, 240, 249, 262, 268, 269, 285, 287, 289, 290, 302, 303, 307, 312, 313, 319, 320, 327, 335, 341, 347, 349, 361, 364, 374, 378, 426, 427, 435, 439, 440, 461]. For instance, Porter and Neuringer [347] test whether pigeons trained to discriminate between music by J. S. Bach and Stravinsky are able to discriminate between music by composers contemporary with J. S. Bach (Buxtehude and Scarlatti) and Stravinsky (Carter and Piston). Another extension of *Classify* is *Scale* [9, 13, 14, 19, 45, 48, 49, 53, 62, 68, 83, 94, 132, 144, 199, 225, 226, 238, 261, 275, 280, 303, 320, 336, 337, 339–341, 390]. For instance, Chai and Vercoe [53] test their system on various class pairs from their dataset of three folk music genres, as well as on all three classes together.

The second most-used experimental design is *Features* [1, 7, 9, 16, 17, 26, 27, 33–35, 37, 43, 48, 49, 53, 68, 69, 72, 93, 95, 102, 103, 105, 109, 115, 122, 126, 127, 139, 141, 143, 144, 146, 153, 157–160, 179, 182–184, 187–189, 192, 196, 197, 199, 200, 211–213, 219–221, 226–230, 232, 236–238, 240, 242, 244, 245, 247, 249–252, 272–277, 280, 281, 283–287, 289–291, 300–303, 307, 309, 320, 327, 333, 337, 340, 341, 345, 361, 362, 364, 370, 371, 376, 377, 379, 385, 387, 390, 393, 396, 397, 401, 403, 406, 417, 420, 421, 425–427, 430, 432, 434, 436, 438, 443, 447, 451, 454–456, 460, 462, 464, 465, 467]. We do not include in this experimental design work that performs feature selection without an interpretation of the results. For instance, Tzanetakis et al. [425] use *Classify* in comparing rhythmic features (statistics of an autocorrelation of wavelet decomposition) and timbral features (spectral centroid, rolloff, etc.). On the other hand, Yoon et al. [459] explore two different feature selection approaches using *Classify*, but do not discuss or list the selected features. Akin to *Features* is a fifth design, *Rules*, which appears in at least sixteen works [3, 5, 13, 14, 26, 42, 43, 70, 94, 98, 139, 303, 308, 340, 341, 434]. For instance, Bickerstaffe and Makalic [43] look at a decision stump that discriminates "rock" and "classical" music. As another example, Abeßer et al. [5] provide the details of a decision tree algorithmically built for discriminating between 13 genres.

Another experimental design is *Cluster* [22, 33, 66, 67, 72, 107, 126, 136, 189, 196, 218, 236, 237, 242, 253, 261, 301, 302, 304, 318, 320, 334, 350, 351, 365, 415, 417, 430, 438]. For instance, Rauber and Frühwirth [350] employ the self-organizing map method with features extracted from 230 music excerpts, and analyze the contents of the resulting groupings. We find that both *Classify* and *Cluster* are used in about 2.6 % (12) of the experimental work [33, 72, 126, 189, 196, 242, 253, 261, 301, 320, 430, 438]. A seventh experimental design is *Retrieve*, which appears in at least 19 works [10, 46, 57, 61, 86, 118, 119, 121, 203, 222, 232, 262, 320, 348, 384, 388, 446, 447, 466]. For instance, Kuo and Shan [203] incorporate style recognition into their music retrieval system.

An eighth experimental design is *Eyeball*, which appears in at least 29 works [26, 29, 44, 83, 91, 105, 146, 149, 155, 166, 173, 189, 218, 242, 259, 261, 288, 300, 302, 304, 310, 314, 320, 358, 360, 402, 403, 413, 463]. For instance, Dannenberg et al. [83] visually inspect class separability for a few pairs of features to explore the reason for a discrepancy in performance in identifying style between an expert approach and machine learning approach. Bigerelle and Iost [44] visually compare means of fractal dimensions computed from several musical excerpts of various genres.

A ninth experimental design is *Robust* [3,10,21,27,38,48,49,52,55,58,75,79, 131,142,200,235,247,267,268,290,313,320,333,347,387,388,401,409,428,439]. For instance, Porter and Neuringer [347] test whether pigeons that have been taught to discriminate between music by J. S. Bach and Hindemith demonstrate their ability regardless of excerpt content and loudness. Soltau et al. [401] investigate the variability of their system using *Classify* by using features computed from excerpts of several durations. Burred and Lerch [48] consider the effect of noise and filtering in feature extraction using *Classify*.

The final experimental design we consider is *Compose*, which appears in only three works [80,82,409]. For instance, Cruz-Alcáza and Vidal-Ruiz [80,82] invert their music style identification system to compose music in the styles it has learned, which the authors then qualitatively evaluate. While Cruz-Alcáza and Vidal-Ruiz do not directly use this as a means to assess the extent to which their system has learned a style, [409] shows by a formal listening test that excerpts composed to be genre-representative by two high-accuracy MGR systems embody little in common with what is commonly held to be characteristic of those genres.

The bias that results from training and testing MGR systems using music data from the same artist and/or excerpted from the same album are well-documented, e.g., [117–119,319]. Among the 435 works that include experimental work, we find that only 8.3 % (36) explicitly mention the use of an artist or album filter [30,57,74–76,117–119,153,174,187,194,209,222,225,239,254,262,266,319, 320,349,353,355,367–369,376,378,381,383,384,401,418,447,461], or attempt to apply one to datasets without known artists [382]. The earliest article applying an artist filter is from 1998 by Soltau et al. [401].

We find that at least twelve works use human evaluation in the analysis of the experiment [22,46,80,82,83,260,320,347,409,410,434,447]. For instance, Dannenberg et al. [83] discuss the performance of their system in a live-performance context. Cruz-Alcáza and Vidal-Ruiz [80] rate the quality of the melodies composed by their style recognition system. And Pampalk [320] uses a formal listening test to show genre labels are strongly correlated with perceptual similarity.

Figure 1 shows the number of experimental works employing formal statistics over each year. Only 16.5 % (72) of the experimental work we survey contains formal statistical testing [9,15,25,27,37,44,46,58,68,75,79,114–117,121,122,124, 131,132,145,146,169,174,201,221,252,258,272,273,275,277,278,283–285,289– 291,295–299,304,308–310,314,317,320,333,337,341,349,357,377,384,395,397, 409–412,422,434,439,440,444,447,457,466]. For instance, Flexer [116] provides excellent argumentation for the need for statistical testing in music information research, and provides a demonstration of its use in comparing the performance between two MGR systems. We find half of the work using living subjects (11 of 22) employ formal statistical tests [58,79,131,169,201,258,278,290,317,439, 440]. For instance, Chase [58] uses a one-tailed paired t-test of percentages of non-responses of koi fish to test the null hypothesis that the koi are unable to discriminate between music that uses Blues or Classical genres.

Nearly half (213) of the experimental work we survey employs only one experimental design from Table 1. For instance, in several formal MGR challenges [170, 171, 293–299], performance is evaluated only by *Classify*. We find about 32 % (142) of the work we survey employ two experimental designs. For instance, Golub [132] uses *Classify* to test his MGR system for a three-genre problem, and then uses *Scale* to observe how its behavior changes when he augments the dataset with four other genres. More than 18 % (80) employ more than two experimental designs. For instance, the only two experimental designs not used by Pampalk [320] are *Rules* and *Compose*.

2.2 Datasets

We find that of the works we survey having experimental components (435) over 58 % (253) use private data [1–5,7–11,13–16,18,19,22,26,28–31,34,40,43–49,53, 56–58,62–70,72,73,79–83,87–89,91–99,104,105,109–111,118–120,125–128,130–138,142–146,148–151,154,156–160,163,164,166,169,172,173,175,176,178–180, 184–188,190,191,193,196,197,199–205,207,209–211,217–221,225,226,228,229,231, 232,242,243,245–253,255,257–261,266,271–275,277,281,287–292,300–305,308, 312,313,317–320,327–331,334–342,344,346–348,350,358,360,361,363–365,372, 374,385–387,389,390,401,413,416,418,425–435,437–440,443–448,452,453,455,458, 459,462–465,467]. Of those works that use private data, we find over 75 % (191) exclusively use private data. Some work provides a detailed description of the composition of the data such that one can recreate it. For instance, Tsatsishvili [418] lists the 210 names of the albums, artists, and songs in his dataset. Schedl et al. [374] provide a URL for obtaining the list of the artists in their dataset, but the resource no longer exists. Mace et al. [258] also provide a list, but since they only list the song and artist name uncertainty arises, e.g., which recording of "The Unanswered Question" by Ives do they use? It is impossible to recreate the dataset used in [48, 49] since they only state that they assemble 850 audio examples in 17 different genres. We find that about 51 % (224) of the works we survey having experimental components use datasets that are publicly available. Of these, over 79 % (177) only use public data.

Table 2 lists 18 publicly available datasets used in the work we survey. *GTZAN* appears in 23 % (100) of the work having an experimental component [6,12,15,17, 19,27,33,35–39,41,46,55,59–61,94,103,121,122,124,129,146,147,153,155,161,162, 182,183,195,200,206,214,222,223,226,227,230–232,234,235,238–240,243,244,249, 263–265,269,276,303,306,321–326,352,356,357,361,362,364,371,377,379–382,384, 387,388,405–407,409–412,416,419–422,426,427,450,451,454,456,457,461,466]. This dataset has only recently been analyzed and shown to have faults [408]. The second most-used publicly available dataset is that created for the 2004 Audio Description Contest of ISMIR [170], which appears in 76 works [15,32,45,50,75,94,114,116,117, 161,162,167,170,174,189,208,209,212–216,222,223,225,238–240,242,243,256,264–268,276,290,303,311,319–327,332,342,343,345,357,359,366,367,370,377,379,381, 382,384,395–397,402,403,412,417,424,436,450,457,460,461]. Datasets derived from Magnatune, e.g., Magnatagatune [485] but excepting *ISMIR2004* [170], appear in

Table 2. Datasets used in MGR, the type of data they contain, the references in which they are used, and the percentage of experimental work (435) that use them. All datasets listed after *Private* are public.

Dataset	Description	%
Private	Constructed for research but not made available; used in: *see text*	58.2
GTZAN	Audio; http://marsyas.info/download/data_sets; used in: *see text*	23.0
ISMIR2004	Audio; http://ismir2004.ismir.net/genre_contest; used in: *see text*	17.4
Latin [394]	Features; http://www.ppgia.pucpr.br/~silla/lmd/; used in [74–76, 97, 102, 242, 254, 267, 268, 295–299, 377, 391–397]	5.1
Ballroom	Audio; http://mtg.upf.edu/ismir2004/contest/ tempoContest/; used in [115, 139–141, 163, 164, 333, 345, 378, 381, 382, 384, 419–421]	3.4
Homburg [165]	Audio; http://www-ai.cs.uni-dortmund.de/audio.html; used in [20, 21, 46, 108, 165, 302, 303, 307, 345, 353, 355, 378, 381, 382, 384]	3.4
Bodhidharma	Symbolic; http://jmir.sourceforge.net/Codaich.html; used in [52, 86, 128, 192, 279–281, 284, 285, 293, 399]	2.5
USPOP2002 [482]	Audio; http://labrosa.ee.columbia.edu/projects/ musicsim/uspop2002.html; used in [38, 42, 239, 262, 290, 293, 349, 354]	1.8
1517-artists	Audio; http://www.seyerlehner.info; used in [378, 381–384]	1.1
RWC [483]	Audio; http://staff.aist.go.jp/m.goto/RWC-MDB/; used in [106, 107, 153, 353]	0.9
SOMeJB	Features; http://www.ifs.tuwien.ac.at/~andi/somejb/; used in [177, 236, 237, 351]	0.9
SLAC	Audio & symbols; http://jmir.sourceforge.net/Codaich. html; used in [283–286]	0.9
SALAMI [400]	Features; http://ddmal.music.mcgill.ca/research/salami; used in [309, 310, 400]	0.7
Unique	Features; http://www.seyerlehner.info; used in [381, 382, 384]	0.7
Million Song [484]	Features; http://labrosa.ee.columbia.edu/millionsong/; used in [90, 168, 376]	0.7
ISMIS2011	Features; http://tunedit.org/challenge/music-retrieval; used in [171, 194, 375]	0.4

at least 5.7 % (25) of the references having an experimental component [16,28–31, 38,112,113,146,225,269,290,293,319,320,342,349,353,355,367–369,414,446,447].

Over 79 % (344) of the experimental work we survey approaches MGR using audio data or features of audio [1–3, 6–10, 12, 13, 15–22, 27–32, 35–39, 42, 44–50, 52,55–58,60–64,68,73–76,79,88–93,96,97,99,102–119,121,122,124,127,129–141, 143–151,153–155,161–165,167–179,182,183,186–191,193–195,200,201,204–217, 222,223,225–228,230–232,234–240,244,246–250,252,254–256,258,260–269,271, 273–277,283,285–307,309–313,317–320,327–334,341–345,347–353,355–362,364– 372,375,376,378–384,386–397,401–403,405–407,409–412,414,416–422,424–433, 436–440,443–448,450–457,459–467]. The use of symbolic data, e.g., MIDI and humdrum, appears in over 18 % (81) of these references [1–5,11,13–15,26,34,43, 52,53,65–67,69,70,72,80–83,86,87,94,95,120,128,156–160,166,180,184,185,192, 196,197,199,202,203,218–221,229,240,243,245,251,253,257,259,261,279–281, 283–286,293,335–341,346,363,385,399,413,428,434,435]. We find about 6 % (27) of the work having an experimental component approaches MGR using other kinds of data, e.g., lyrics, co-occurrences on the WWW, album covers, and so on [25,40,62,98,125,142,272–277,283–286,308,309,354,374,381,415,438,448,458, 464,465].

Figure 3 shows the number of experiments in the evaluative work we survey using datasets with specific numbers of labels. We can clearly see the influence of the *GTZAN* (10 genre labels) and *ISMIR2004* (6 genre labels) datasets. We find 16 works using datasets having 25 or more labels [25,30,31,40,106,107,153, 199,228,232,280,309,353,376,434,437], and only two using datasets having more than 100 labels [40, 437]. Over 72 % (316) of the papers with an experimental component uses only a single dataset, at least 20 % (90) use two datasets, and 6.2 % (27) use more than two datasets. Three references provide no details about the dataset used [54,145,331].

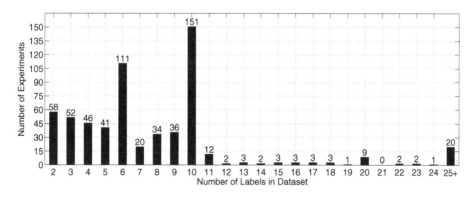

Fig. 3. The number of experiments in this survey employing datasets with specific numbers of labels.

We find a majority of the works with experimental components involves datasets that consist primarily of Western music. For instance, the label "classical" is part of *GTZAN, ISMIR2004,* and *RWC,* and exists in the private datasets used in [22, 125, 144, 146, 173, 209, 225, 246, 266, 313, 341, 342, 344, 360, 363, 387]. The label "blues" is in *GTZAN, ISMIR2004,* and *Homburg,* and exists in the private datasets used by [4, 5, 19, 22, 125, 313, 342, 344, 358, 387]. And the label "jazz" is in *GTZAN, ISMIR2004, Homburg,* and *RWC,* and exists in the private datasets used by [19, 22, 57, 125, 144, 146, 173, 209, 225, 229, 266, 303, 313, 341, 342, 344, 358, 360, 387, 389]. However, we find that only about 10 % (48) of the private datasets used include music from around the world, such as Asia, Africa, and South America [3–5, 11, 19, 22, 57, 96–98, 125, 133–135, 137, 163, 164, 173, 179, 187, 202, 205, 209, 225, 229, 242, 243, 247–250, 255, 257, 266, 303, 308, 312, 313, 342, 344, 358, 360, 363, 385, 389, 430, 462, 463]. Finally, we find only 5 % (22) of the work with experimental components perform human validation of the "ground truth" labels in the public and/or private datasets used [8, 9, 34, 45, 79, 246, 247, 287, 289–291, 301, 346, 366, 367, 370, 383, 387, 394, 401, 408, 434]. For instance, Soltau et al. [401] validate the labels in their private four-class dataset with a human listening experiment.

2.3 Figures of Merit (FoMs)

Table 3 defines several FoMs we find in the work we surveyed. The FoMs most often reported in the work we survey here are those that accompany the *Classify* experimental design: *Mean accuracy, Recall, Precision, F-measure, Receiver Operating Characteristic (ROC),* and the *Contingency table.* We find *Mean accuracy* in over 82 % (385) of the references. For instance, Fu et al. [123] compare the reported mean accuracies of 16 MGR algorithms using *Classify* in *GTZAN.* This computation can also involve taking into consideration "partial credit" for labelings in the correct hierarchical branch, e.g., [293,294,296]. When it appears, *Mean accuracy* is accompanied by a standard deviation (SD), or standard error of the mean (SEM), about 25 % (96) of the time. For instance, [116] uses these statistics to test the null hypothesis that the *Mean accuracy* of two MGR systems are not significantly different.

We find *Recall* in over 25 % (119) of the references. For instance, this FoM appears in the MIREX evaluations of MGR algorithms [295, 297–299]. When it appears, *Recall* is accompanied by the standard deviation or standard error of the mean in about 10 % (12) of the references. *Precision* appears in over 10 % (47) of the references. Together, *Mean accuracy, Recall* and *Precision* appear in over 6 % (31) of the work we survey. The *F-measure* can be computed in "Micro form" and "Macro form" [437], but we make no distinction here. This FoM appears in at least 17 works. For instance, Burred and Peeters [50] cite the *F-measure* of their MGR system, as well as its *Mean accuracy, Recall,* and *Precision.* We find the *ROC* in only 7 references [105, 121, 245, 349, 432, 440, 466]. For instance, Watanabe and Sato [440] plot the ROC of their sparrows trained to discriminate Baroque and Modern music.

We find a *Contingency table* reported in over 32 % (150) of the work we survey. For instance, Soltau et al. [401] show their MGR system often confuses the music in their private dataset having the labels "rock" or "pop," and rarely confuses

Table 3. Figures of merit (FoMs) of MGR, their description, and the percentage of work (467) that cite them

FoM	Description	%
Mean accuracy	Proportion of the number of correct trials to the total number of trials of the system	82
Contingency table	Counts of labeling outcomes of the system for each labeled input	32
Recall	Proportion of the number of correct trials of the system to the total number of a specific input label	25
Confusions	Discussion of confusions of the system in general or with specifics	24
Precision	Proportion of the number of correct trials of the system to the total number of a specific output label	10
F-measure	Twice the product of *Recall* and *Precision* divided by their sum	4
Composition	Observations of the composition of clusters created by the system, and distances within and between	4
Precision@ k	Proportion of the number of correct items of a specific label in the k items retrieved by the system	3
ROC	*Precision* vs. *Recall* for several systems, parameters, etc.	1

music labeled "classic" with music labeled "techno" or "rock." Of those works that present contingency tables, only 52 % (78) of them are accompanied by some musical reflection of the results. For instance, in the analysis of their *Contingency table*, Dixon et al. [93] reason that the high number of confusions produced between three of eight classes come from the fact that they are indistinguishable using meter- and tempo-sensitive features employed in their system. When they expand their feature set, the new *Contingency table* confirms this hypothesis.

General discussions about observed confusions without reference to a *Contingency table* are reported in over 8 % (39) of the references. For instance, Matityaho and Furst [271] note that their MGR system trained to discriminate between music labeled "classic" and "pop" classifies as "classic" a signal of complete silence and a "complex tone," and as "pop" a signal of white noise. Using *Eyeball*, Bigerelle and Iost [44] argue that "Music classification becomes very logical [by comparing the fractal dimension]. ... Progressive music has the same fractal dimension as the electronic one: we could explain this fact by the abundance of synthesizers used in progressive music." Only 15 works mention confusions in detail, e.g., a specific piece of misclassified music [3,68,98,210,228, 301,342,366,367,370,407–410,412]. For instance, [410] notices that one MGR system persistently misclassifies as Hip hop "Kung Fu Fighting" by Carl Douglas, and as Classical "Why?" by Bronski Beat.

We find over 44 % (175) of the 397 works employing *Classify* report only one FoM and over 53 % (214) report more than one FoM. Only 21 present four or more FoM [6,34,50,56,93,156,188,200,208,238–240,333,353,355,367,368,410,412, 418, 433]. For instance, Lidy [239] reports *mean accuracies, recalls, precisions, F-measures,* and *contingency tables* of the systems he tests.

The FoM most often reported in the case of the *Retrieve* experimental design is *Precision@k*. This FoM is reported in 12 of the 19 works using *Retrieve* [10, 57,61,86,203,222,262,320,384,446,447,466]; and [388] reports "normalized precision" and "normalized recall," which takes into account the ranking of retrieved elements. Of the references using *Retrieval,* the *ROC* is reported in [121, 466]. For instance, Fu et al. [121] plot the ROC of four systems to show their retrieval approach provides statistically significant improvement.

The FoMs most often reported in the case of *Cluster* experimental design are based on observations of the cluster compositions. The contents of clusters are analyzed in over 62 % (18) of the *Clustering* experiments [33, 107, 136, 196, 236, 237,253,301,302,304,318,320,334,350,351,365,415,430]. For instance, Rauber and Frühwirth [350] show that one cluster created by the self-organizing map method consists mainly of music labeled "classical." Comparisons of cluster distances, e.g., that within classes to that between classes, are reported in five works [22, 72, 302, 318, 334]. For instance, Aucouturier and Pachet [22] compare average distances between neighbors of the same class to those between neighbors of different classes. Visualizations of the clusters, e.g., using self-organizing maps, are presented in seven works [189, 218, 241, 242, 350, 351, 473]. Both [233, 417] report the "purity" of a collection of clusters, which measures the mean class homogeneity of the clusters.

Human-weighted ratings of classification and/or clustering results are reported in at least six works [22, 154, 203, 246, 366, 370]. Other FoMs include, "staying time" [278, 439] (measuring the time during which the subject stayed in the presence of musical stimuli for particular classes), "stability measure" [161,162] (essentially inter-intra class distance), "Hamming loss" [367–369] (describing instance-label pair misclassifications in a multilabel scenario), and "persistent misclassifications" [65,342,409,410] (noting instances that a system always mislabels).

3   Conclusion

While genre is an inevitable condition of human communication in general [469], a way to automatically identify it in music remains elusive. In this paper, we have attempted to present an exhaustive survey of evaluation in MGR, and to organize it along three dimensions: experimental design, datasets, and figures of merit. By the sheer size of this task, it is certain that we have missed some relevant work, misunderstood aspects of evaluation in some of the works we cite, and committed errors in the bibliography. We will thus continue to maintain this bibliography, and expand it when new work is published. The bibliography file we assembled for this work is available here: http://imi.aau. dk/~bst/research/MGRbibliography.bib; and the spreadsheet we created identifying the characteristics of the references is available here: http://imi.aau.dk/ ~bst/research/MGRspreadsheet.xlsx.

Acknowledgments. For Pepi [58]. Thank you to Carla Sturm for her bibliographic entry prowess. This work is supported in part by: Independent Postdoc Grant 11-105218 from Det Frie Forskningsråd; and the Danish Council for Strategic Research of the Danish Agency for Science Technology and Innovation in project CoSound, case no. 11-115328.

References

1. Abeßer, J., Dittmar, C., Großmann, H.: Automatic genre and artist classification by analyzing improvised solo parts from musical recordings. In: Proceedings of the Audio Mostly Conference, Piteå, Sweden, pp. 127–131 (2008)
2. Abeßer, J., Lukashevich, H.M., Dittmar, C., Schuller, G.: Genre classification using bass-related high-level features and playing styles. In: Proceedings of the ISMIR, pp. 453–458 (2009)
3. Abeßer, J., Lukashevich, H., Dittmar, C., Bräuer, P., Karuse, F.: Rule-based classification of musical genres from a global cultural background. In: Proceedings of the CMMR, pp. 317–336 (2010)
4. Abeßer, J., Bräuer, P., Lukashevich, H.M., Schuller, G.: Bass playing style detection based on high-level features and pattern similarity. In: Proceedings of the ISMIR, pp. 93–98 (2010)
5. Abeßer, J., Lukashevich, H., Bräuer, P.: Classification of music genres based on repetitive basslines. J. New Music Res. **41**(3), 239–257 (2012)
6. Ahonen, T.E.: Compressing lists for audio classification. In: Proceedings of the International Workshop on Machine Learning and Music. MML '10, pp. 45–48. ACM, New York (2010)
7. Ahrendt, P., Meng, A., Larsen, J.: Decision time horizon for music genre classification using short-time features. In: Proceedings of the EUSIPCO (2004)
8. Ahrendt, P., Larsen, J., Goutte, C.: Co-occurrence models in music genre classification. In: Proceedings of the IEEE Workshop Machine Learning Signal Process, Sept 2005
9. Ahrendt, P.: Music genre classification systems - a computational approach. Ph.D. thesis, Technical University of Denmark (2006)
10. Almoosa, N., Bae, S.H., Juang, B.H.: Feature extraction by incremental parsing for music indexing. In: Proceedings of the ICASSP, pp. 2410–2413, Mar 2010
11. Anan, Y., Hatano, K., Bannai, H., Takeda, M.: Music genre classification using similarity functions. In: Proceedings of the ISMIR, pp. 693–698 (2011)
12. Andén, J., Mallat, S.: Multiscale scattering for audio classification. In: Proceedings of the ISMIR, pp. 657–662 (2011)
13. Anglade, A., Ramirez, R., Dixon, S.: Genre classification using harmony rules induced from automatic chord transcriptions. In: Proceedings of the ISMIR (2009)
14. Anglade, A., Ramirez, R., Dixon, S.: First-order logic classification models of musical genres based on harmony. In: Proceedings of the SMC (2009)
15. Anglade, A., Benetos, E., Mauch, M., Dixon, S.: Improving music genre classification using automatically induced harmony rules. J. New Music Res. **39**(4), 349–361 (2010)
16. Annesi, P., Basili, R., Gitto, R., Moschitti, A., Petitti, R.: Audio feature engineering for automatic music genre classification. In: Proceedings of the Recherche d'Information Assistée par Ordinateur, Pittsburgh, Pennsylvania, pp. 702–711 (2007)

17. Arabi, A.F., Lu, G.: Enhanced polyphonic music genre classification using high level features. In: IEEE International Conference on Signal and Image Processing Applications (2009)
18. Arenas, J., Larsen, J., Hansen, L., Meng, A.: Optimal filtering of dynamics in short-time features for music organization. In: Proceedings of the ISMIR (2006)
19. Ariyaratne, H., Zhang, D.: A novel automatic hierarchical approach to music genre classification. In: Proceedings of the ICME, pp. 564–569, July 2012
20. Aryafar, K., Shokoufandeh, A.: Music genre classification using explicit semantic analysis. In: Proceedings of the ACM MIRUM Workshop, Scottsdale, AZ, USA, pp. 33–38, Nov 2011
21. Aryafar, K., Jafarpour, S., Shokoufandeh, A.: Music genre classification using sparsity-eager support vector machines. Technical report, Drexel University (2012)
22. Aucouturier, J.J., Pachet, F.: Music similarity measures: what's the use? In: Proceedings of the ISMIR, Paris, France, Oct 2002
23. Aucouturier, J.J., Pachet, F.: Representing music genre: a state of the art. J. New Music Res. 32(1), 83–93 (2003)
24. Aucouturier, J.J., Pampalk, E.: Introduction - from genres to tags: a little epistemology of music information retrieval research. J. New Music Res. 37(2), 87–92 (2008)
25. Aucouturier, J.J.: Sounds like teen spirit: computational insights into the grounding of everyday musical terms. In: Minett, J., Wang, W. (eds.) Language, Evolution and the Brain. Frontiers in Linguistic Series. Academia Sinica Press, Taipei (2009)
26. Backer, E., van Kranenburg, P.: On musical stylometry - a pattern recognition approach. Pattern Recogn. Lett. 26, 299–309 (2005)
27. Bağci, U., Erzin, E.: Automatic classification of musical genres using inter-genre similarity. IEEE Signal Proc. Lett. 14(8), 521–524 (2007)
28. Balkema, W.: Variable-size gaussian mixture models for music similarity measures. In: Proceedings of the ISMIR, pp. 491–494 (2007)
29. Balkema, W., van der Heijden, F.: Music playlist generation by assimilating GMMs into SOMs. Pattern Recogn. Lett. 31(1), 1396–1402 (2010)
30. Barbedo, J.G.A., Lopes, A.: Automatic genre classification of musical signals. EURASIP J. Adv. Signal Process. 2007, 1–12 (2007)
31. Barbedo, J.G.A., Lopes, A.: Automatic musical genre classification using a flexible approach. J. Audio Eng. Soc. 56(7/8), 560–568 (2008)
32. Barbieri, G., Esposti, M.D., Pachet, F., Roy, P.: Is there a relation between the syntax and the fitness of an audio feature? In: Proceedings of the ISMIR (2010)
33. Barreira, L., Cavaco, S., da Silva, J.: Unsupervised music genre classification with a model-based approach. In: Proceedings of the Portuguese Conference on Progress in Artificial Intelligence, pp. 268–281 (2011)
34. Basili, R., Serafini, A., Stellato, A.: Classification of musical genre: a machine learning approach. In: Proceedings of the ISMIR (2004)
35. Behun, K.: Image features in music style recognition. In: Proceedings of the Central European Seminar on Computer Graphics (2012)
36. Benetos, E., Kotropoulos, C.: A tensor-based approach for automatic music genre classification. In: Proceedings of the EUSIPCO, Lausanne, Switzerland (2008)
37. Benetos, E., Kotropoulos, C.: Non-negative tensor factorization applied to music genre classification. IEEE Trans. Audio Speech Lang. Process. 18(8), 1955–1967 (2010)

38. Bergstra, J., Casagrande, N., Erhan, D., Eck, D., Kégl, B.: Aggregate features and Adaboost for music classification. Mach. Learn. **65**(2–3), 473–484 (2006)
39. Bergstra, J.: Algorithms for classifying recorded music by genre. Master's thesis, Université de Montréal, Montréal, Canada, Aug 2006
40. Bergstra, J., Lacoste, A., Eck, D.: Predicting genre labels for artist using FreeDB. In: Proceedings of the ISMIR, pp. 85–88 (2006)
41. Bergstra, J., Mandel, M., Eck, D.: Scalable genre and tag prediction with spectral covariance. In: Proceedings of the ISMIR (2010)
42. Bertin-Mahieux, T., Weiss, R.J., Ellis, D.P.W.: Clustering beat-chroma patterns in a large music database. In: Proceedings of the ISMIR, Utrecht, Netherlands, Aug 2010
43. Bickerstaffe, A.C., Makalic, E.: MML classification of music genres. In: Gedeon, T.T.D., Fung, L.C.C. (eds.) AI 2003. LNCS (LNAI), vol. 2903, pp. 1063–1071. Springer, Heidelberg (2003)
44. Bigerelle, M., Iost, A.: Fractal dimension and classification of music. Chaos Soliton. Fract. **11**(14), 2179–2192 (2000)
45. Blume, H., Haller, M., Botteck, M., Theimer, W.: Perceptual feature based music classification - a DSP perspective for a new type of application. In: International Conference on Embedded Computer Systems (2008)
46. Bogdanov, D., Serra, J., Wack, N., Herrera, P., Serra, X.: Unifying low-level and high-level music similarity measures. IEEE Trans. Multimed. **13**(4), 687–701 (2011)
47. Brecheisen, S., Kriegel, II.P., Kunath, P., Pryakhin, A.: Hierarchical genre classification for large music collections. In: Proceedings of the ICME, pp. 1385–1388, July 2006
48. Burred, J., Lerch, A.: A hierarchical approach to automatic musical genre classification. In: Proceedings of the DAFx, London, UK, Sept 2003
49. Burred, J.J., Lerch, A.: Hierarchical automatic audio signal classification. J. Audio Eng. Soc. **52**(7), 724–739 (2004)
50. Burred, J.J., Peeters, G.: An adaptive system for music classification and tagging. In: International Workshop on Learning Semantics of Audio Signals (2009)
51. Casey, M., Veltkamp, R., Goto, M., Leman, M., Rhodes, C., Slaney, M.: Content-based music information retrieval: current directions and future challenges. Proc. IEEE **96**(4), 668–696 (2008)
52. Cataltepe, Z., Yaslan, Y., Sonmez, A.: Music genre classification using MIDI and audio features. EURASIP J. Adv. Signal Process. **2007**, 1–8 (2007)
53. Chai, W., Vercoe, B.: Folk music classification using hidden Markov models. In: International Conference on Artificial Intelligence (2001)
54. Chang, L., Yu, X., Wan, W., Yao, J.: Research on fast music classification based on SVM in compressed domain. In: Proceedings of the ICALIP, pp. 638–642, July 2008
55. Chang, K., Jang, J.S.R., Iliopoulos, C.S.: Music genre classification via compressive sampling. In: Proceedings of the ISMIR, Amsterdam, The Netherlands, pp. 387–392, Aug 2010
56. Charami, M., Halloush, R., Tsekeridou, S.: Performance evaluation of TreeQ and LVQ classifiers for music information retrieval. In: Boukis, C., Pnevmatikakis, L., Polymenakos, L., et al. (eds.) Artificial Intelligence and Innovations 2007: From Theory to Applications. IFIP, vol. 247, pp. 331–338. Springer, Boston (2007)
57. Charbuillet, C., Tardieu, D., Peeters, G.: GMM supervector for content based music similarity. In: Proceedings of the DAFx, Paris, France, Sept 2011

58. Chase, A.: Music discriminations by carp "Cyprinus carpio". Learn. Behav. **29**, 336–353 (2001)
59. Chathuranga, D., Jayaratne, L.: Musical genre classification using ensemble of classifiers. In: Proceedings of the International Conference on Computational Intelligence, Modelling and Simulation, pp. 237–242, Sept 2012
60. Chen, K., Gao, S., Zhu, Y., Sun, Q.: Music genres classification using text categorization method. In: Proceedings of the IEEE Workshop on Multimedia Signal Processing, pp. 221–224, Oct 2006
61. Chen, G., Wang, T., Herrera, P.: Relevance feedback in an adaptive space with one-class SVM for content-based music retrieval. In: Proceedings of the ICALIP, pp. 1153–1158, July 2008
62. Chen, L., Wright, P., Nejdl, W.: Improving music genre classification using collaborative tagging data. In: International Conference on Web Search and Data Mining, Barcelona, Spain, Feb 2009
63. Chen, S.H., Chen, S.H.: Content-based music genre classification using timbral feature vectors and support vector machine. In: Proceedings of the International Conference on Interaction Sciences, pp. 1095–1101, Nov 2009
64. Chen, S.H., Chen, S.H., Guido, R.C.: Music genre classification algorithm based on dynamic frame analysis and support vector machine. In: IEEE International Symposium on Multimedia (2010)
65. Chew, E., Volk, A., Lee, C.Y.: Dance music classification using inner metric analysis. In: Golden, B., Raghavan, S., Wasil, E. (eds.) The Next Wave in Computing, Optimization, and Decision Technologies. Proceedings of the INFORMS Computing Society Conference, pp. 355–370. Kluwer, Dordrecht (2005)
66. Cilibrasi, R., Vitányi, P., de Wolf, R.: Algorithmic clustering of music based on string compression. Comput. Music J. **28**(4), 49–67 (2004)
67. Cilibrasi, R., Vitanyi, P.: Clustering by compression. IEEE Trans. Inf. Theory **51**(4), 1523–1545 (2005)
68. Collins, N.: Influence in early electronic dance music: an audio content analysis investigation. In: Proceedings of the ISMIR (2012)
69. Conklin, D.: Melodic analysis with segment classes. Mach. Learn. **65**, 349–360 (2006)
70. Conklin, D.: Melody classification using patterns. In: Proceedings of the International Workshop on Machine Learning and Music, pp. 37–41 (2009)
71. Cornelis, O., Lesaffre, M., Moelants, D., Leman, M.: Access to ethnic music: advances and perspectives in content-based music information retrieval. Signal Process. **90**(4), 1008–1031 (2010)
72. Correa, D.C., Saito, J.H., da Costa, L.F.: Musical genres: beating to the rhythms of different drums. New. J. Phys. **12**(5), 053030 (2010)
73. Costa, C.H.L., Valle Jr., J.D., Koerich, A.L.: Automatic classification of audio data. In: Proceedings of the IEEE International Conference on Systems, Man and Cybernetics, pp. 562–567 (2004)
74. Costa, Y.M.G., Oliveira, L.S., Koerich, A.L., Gouyon, F.: Music genre recognition using spectrograms. In: Proceedings of the International Conference on Systems, Signals and Image Processing (2011)
75. Costa, Y., Oliveira, L., Koerich, A., Gouyon, F., Martins, J.: Music genre classification using LBP textural features. Signal Process. **92**(11), 2723–2737 (2012)
76. Costa, Y.M.G., Oliveira, L.S., Koerich, A.L., Gouyon, F.: Comparing textural features for music genre classification. In: Proceedings of the IEEE World Congress on Computational Intelligence, June 2012

77. Craft, A., Wiggins, G.A., Crawform, T.: How many beans make five? The consensus problem in music-genre classification and a new evaluation method for single-genre categorisation systems. In: Proceedings of the ISMIR (2007)
78. Craft, A.: The role of culture in the music genre classification task: human behaviour and its effect on methodology and evaluation. Technical report, Queen Mary University of London, Nov 2007
79. Crump, M.: A principal components approach to the perception of musical style. Master's thesis, University of Lethbridge (2002)
80. Cruz-Alcáza, P.P., Vidal-Ruiz, E.: Modeling musical style using grammatical inference techniques: a tool for classifying and generating melodies. In: Proceedings of the WEDELMUSIC, pp. 77–84, Sept 2003
81. Cruz-Alcázar, P.P., Vidal-Ruiz, E., Pérez-Cortés, J.C.: Musical style identification using grammatical inference: the encoding problem. In: Sanfeliu, A., Ruiz-Shulcloper, J. (eds.) CIARP 2003. LNCS, vol. 2905, pp. 375–382. Springer, Heidelberg (2003)
82. Cruz-Alcáza, P.P., Vidal-Ruiz, E.: Two grammatical inference applications in music processing. Appl. Artif. Intell. **22**(1/2), 53–76 (2008)
83. Dannenberg, R.B., Thom, B., Watson, D.: A machine learning approach to musical style recognition. In: Proceedings of the ICMC, pp. 344–347 (1997)
84. Dannenberg, R., Foote, J., Tzanetakis, G., Weare, C.: Panel: new directions in music information retrieval. In: Proceedings of the ICMC (2001)
85. Dannenberg, R.B.: Style in music. In: Argamon, S., Burns, K., Dubnov, S. (eds.) The Structure of Style, pp. 45–57. Springer, Heidelberg (2010)
86. DeCoro, C., Barutcuoglu, S., Fiebrink, R.: Bayesian aggregation for hierarchical genre classification. In: Proceedings of the ISMIR (2007)
87. Dehghani, M., Lovett, A.M.: Efficient genre classification using qualitative representations. In: Proceedings of the ISMIR, pp. 353–354 (2006)
88. Dellandrea, E., Harb, H., Chen, L.: Zipf, neural networks and SVM for musical genre classification. In: Proceedings of the IEEE International Symposium on Signal Processing and Information Technology, pp. 57–62, Dec 2005
89. Deshpande, H., Singh, R., Nam, U.: Classification of music signals in the visual domain. In: Proceedings of the DAFx, Limerick, Ireland, Dec 2001
90. Dieleman, S., Brakel, P., Schrauwen, B.: Audio-based music classification with a pretrained convolutional network. In: Proceedings of the ISMIR (2011)
91. Diodati, P., Piazza, S.: Different amplitude and time distribution of the sound of light and classical music. Eur. Phys. J. B - Condens. Matter Complex Syst. **17**, 143–145 (2000)
92. Dixon, S., Pampalk, E., Widmer, G.: Classification of dance music by periodicity patterns. In: Proceedings of the ISMIR (2003)
93. Dixon, S., Gouyon, F., Widmer, G.: Towards characterisation of music via rhythmic patterns. In: Proceedings of the ISMIR, Barcelona, Spain, pp. 509–517 (2004)
94. Dixon, S., Mauch, M., Anglade, A.: Probabilistic and logic-based modelling of harmony. In: Ystad, S., Aramaki, M., Kronland-Martinet, R., Jensen, K. (eds.) CMMR 2010. LNCS, vol. 6684, pp. 1–19. Springer, Heidelberg (2011)
95. Dor, O., Reich, Y.: An evaluation of musical score characteristics for automatic classification of composers. Comput. Music J. **35**(3), 86–97 (2011)
96. Doraisamy, S., Golzari, S., Norowi, N.M., Sulaiman, M.N.B., Udzir, N.I.: A study on feature selection and classification techniques for automatic genre classification of traditional Malay music. In: Proceedings of the ISMIR, Philadelphia, PA (2008)

97. Doraisamy, S., Golzari, S.: Automatic musical genre classification and artificial immune recognition system. In: Raś, Z.W., Wieczorkowska, A.A. (eds.) Advances in Music Information Retrieval. SCI, vol. 274, pp. 390–402. Springer, Heidelberg (2010)
98. Doudpota, S.M., Guha, S.: Mining movies for song sequences with video based music genre identification system. Int. J. Info. Process. Manag. **49**(2), 529–544 (2013)
99. Downie, J., Ehmann, A., Tcheng, D.: Real-time genre classification for music digital libraries. In: Proceedings of the Joint ACM/IEEE Conference on Digital Libraries, p. 377, June 2005
100. Downie, J.S.: The music information retrieval evaluation exchange (2005–2007): a window into music information retrieval research. Acoust. Sci. Technol. **29**(4), 247–255 (2008)
101. Downie, J.S., Ehmann, A.F., Bay, M., Jones, M.C.: The music information retrieval evaluation exchange: some observations and insights. In: Raś, Z.W., Wieczorkowska, A.A. (eds.) Advances in Music Information Retrieval. SCI, vol. 274, pp. 93–115. Springer, Heidelberg (2010)
102. Draman, N.A., Wilson, C., Ling, S.: Modified AIS-based classifier for music genre classification. In: Proceedings of the ISMIR, pp. 369–374 (2010)
103. Draman, N.A., Ahmad, S., Muda, A.K.: Recognizing patterns of music signals to songs classification using modified AIS-based classifier. In: Zain, J.M., Wan Mohd, W.M., El-Qawasmeh, E. (eds.) ICSECS 2011, Part II. CCIS, vol. 180, pp. 724–737. Springer, Heidelberg (2011)
104. Dunker, P., Dittmar, C., Begau, A., Nowak, S., Gruhne, M.: Semantic high-level features for automated cross-modal slideshow generation. In: Proceedings of the Content-Based Multimedia Indexing, pp. 144–149 (2009)
105. Esmaili, S., Krishnan, S., Raahemifar, K.: Content based audio classification and retrieval using joint time-frequency analysis. In: Proceedings of the ICASSP, vol. 5, pp. 665–668 (2004)
106. Ezzaidi, H., Rouat, J.: Comparison of the statistical and information theory measures: Application to automatic musical genre classification. In: Proceedings of the IEEE Workshop on Machine Learning for Signal Processing, pp. 241–246, Aug 2007
107. Ezzaidi, H., Bahoura, M., Rouat, J.: Taxonomy of musical genres. In: International Conference on Signal Image Technology and Internet Based Systems (2009)
108. Fadeev, A., Missaoui, O., Frigui, H.: Dominant audio descriptors for audio classification and retrieval. In: Proceedings of the ICMLA, Louisville, KY, USA, pp. 75–78, Dec 2009
109. Feng, Y., Dou, H., Qian, Y.: A study of audio classification on using different feature schemes with three classifiers. In: Proceedings of the International Conference on Information, Networking, Automation, pp. 298–302 (2010)
110. Fernández, F., Chávez, F., Alcala, R., Herrera, F.: Musical genre classification by means of fuzzy rule-based systems: a preliminary approach. In: IEEE Congress on Evolutionary Computation (2011)
111. Fernández, F., Chávez, F.: Fuzzy rule based system ensemble for music genre classification. In: Machado, P., Romero, J., Carballal, A. (eds.) EvoMUSART 2012. LNCS, vol. 7247, pp. 84–95. Springer, Heidelberg (2012)
112. Fiebrink, R., Fujinaga, I.: Feature selection pitfalls and music classification. In: Proceedings of the ISMIR, Victoria, BC, Canada, pp. 340–341 (2006)
113. Fiebrink, R.: An exploration of feature selection as a tool for optimizing musical genre classification. Master's thesis, McGill University, June 2006

114. Flexer, A., Pampalk, E., Widmer, G.: Hidden Markov models for spectral similarity of songs. In: Proceedings of the DAFx, Madrid, Spain, Sept 2005
115. Flexer, A., Gouyon, F., Dixon, S., Widmer, G.: Probabilistic combination of features for music classification. In: Proceedings of the ISMIR, Victoria, BC, Canada, pp. 111–114, Oct 2006
116. Flexer, A.: Statistical evaluation of music information retrieval experiments. J. New Music Res. **35**(2), 113–120 (2006)
117. Flexer, A.: A closer look on artist filters for musical genre classification. In: Proceedings of the ISMIR, Vienna, Austria, Sept 2007
118. Flexer, A., Schnitzer, D.: Album and artist effects for audio similarity at the scale of the web. In: Proceedings of the SMC, Porto, Portugal, pp. 59–64, July 2009
119. Flexer, A., Schnitzer, D.: Effects of album and artist filters in audio similarity computed for very large music databases. Comput. Music J. **34**(3), 20–28 (2010)
120. Frederico, G.: Classification into musical genres using a rhythmic kernel. In: Proceedings of the SMC (2004)
121. Fu, Z., Lu, G., Ting, K.M., Zhang, D.: Learning naive Bayes classifiers for music classification and retrieval. In: Proceedings of the ICPR, pp. 4589–4592 (2010)
122. Fu, Z., Lu, G., Ting, K.M., Zhang, D.: On feature combination for music classification. In: Proceedings of the International Workshop on Structural and Syntactic Pattern Recognition, pp. 453–462 (2010)
123. Fu, Z., Lu, G., Ting, K.M., Zhang, D.: A survey of audio-based music classification and annotation. IEEE Trans. Multimed. **13**(2), 303–319 (2011)
124. Fu, Z., Lu, G., Ting, K.M., Zhang, D.: Music classification via the bag-of-features approach. Pattern Recogn. Lett. **32**(14), 1768–1777 (2011)
125. García, J., Hernández, E., Meng, A., Hansen, L.K., Larsen, J.: Discovering music structure via similarity fusion. In: Proceedings of the Music, Brain and Cognition Workshop (2007)
126. García-García, D., Arenas-García, J., Parrado-Hernandez, E., de Maria, F.D.: Music genre classification using the temporal structure of songs. In: IEEE International Workshop on Machine Learning for Signal Processing, Kittilä, Finland, Aug–Sept 2010
127. García, A., Arenas, J., García, D., Parrado, E.: Music genre classification based on dynamical models. In: International Conference on Pattern Recognition Applications and Methods, pp. 250–256 (2012)
128. Gedik, A.C., Alpkocak, A.: Instrument independent musical genre classification using random 3000 ms segment. In: Savacı, F.A. (ed.) TAINN 2005. LNCS (LNAI), vol. 3949, pp. 149–157. Springer, Heidelberg (2006)
129. Genussov, M., Cohen, I.: Musical genre classification of audio signals using geometric methods. In: Proceedings of the EUSIPCO, Aalborg, Denmark, pp. 497–501, Aug 2010
130. Ghosal, A., Chakraborty, R., Dhara, B., Saha, S.: Instrumental/song classification of music signal using RANSAC. In: Proceedings of the International Conference on Electronics Computer Technology, pp. 269–272, Apr 2011
131. Gjerdingen, R.O., Perrott, D.: Scanning the dial: the rapid recognition of music genres. J. New Music Res. **37**(2), 93–100 (2008)
132. Golub, S.: Classifying recorded music. Master's thesis, University of Edinburgh, Edinburgh, Scotland, UK (2000)
133. Golzari, S., Doraisamy, S., Sulaiman, N., Udzir, N.I.: A hybrid approach to traditional Malay music genre classification: combining feature selection and artificial immune recognition system. In: Proceedings of the International Symposium on Information Technology, Aug 2008

134. Golzari, S., Doraisamy, S., Sulaiman, M.N.B., Udzir, N.I., Norowi, N.M.: Artificial immune recognition system with nonlinear resource allocation method and application to traditional Malay music genre classification. In: Bentley, P.J., Lee, D., Jung, S. (eds.) ICARIS 2008. LNCS, vol. 5132, pp. 132–141. Springer, Heidelberg (2008)

135. Golzari, S., Doraisamy, S., Norowi, N.M., Sulaiman, M.N., Udzir, N.I.: A comprehensive study in benchmarking feature selection and classification approaches for traditional Malay music genre classification. In: Proceedings of Data Mining, pp. 71–77, July 2008

136. González, A., Granados, A., Camacho, D., de Borja Rodríguez, F.: Influence of music representation on compression-based clustering. In: Proceedings of the IEEE Congress on Evolutionary Computation (2010)

137. Goulart, A.J.H., Maciel, C.D., Guido, R.C., Paulo, K.C.S., da Silva, I.N.: Music genre classification based on entropy and fractal lacunarity. In: IEEE International Symposium on Multimedia (2011)

138. Goulart, A., Guido, R., Maciel, C.: Exploring different approaches for music genre classification. Egypt. Inf. J. **13**(2), 59–63 (2012)

139. Gouyon, F., Dixon, S., Pampalk, E., Widmer, G.: Evaluating rhythmic descriptors for musical genre classification. In: Proceedings of the International Audio Engineering Society Conference, pp. 196–204 (2004)

140. Gouyon, F., Dixon, S.: Dance music classification: a tempo-based approach. In: Proceedings of the ISMIR, pp. 501–504 (2004)

141. Gouyon, F.: A computational approach to rhythm description – audio features for the computation of rhythm periodicity functions and their use in tempo induction and music content processing. Ph.D. thesis, Universitat Pompeu Fabra (2005)

142. Govaerts, S., Corthaut, N., Duval, E.: Using search engine for classification: does it still work? In: Proceedings of the IEEE International Symposium on Multimedia, pp. 483–488, Dec 2009

143. Grimaldi, M., Cunningham, P., Kokaram, A.: A wavelet packet representation of audio signals for music genre classification using different ensemble and feature selection techniques. In: Proceedings of the ACM Multimedia, pp. 102–108 (2003)

144. Grimaldi, M., Cunningham, P., Kokaram, A.: Discrete wavelet packet transform and ensembles of lazy and eager learners for music genre classification. Multimed. Syst. **11**, 422–437 (2006)

145. Grosse, R., Raina, R., Kwong, H., Ng, A.Y.: Shift-invariant sparse coding for audio classification. In: Proceedings of the Twenty-third Conference on Uncertainty in Artificial Intelligence (2007)

146. Guaus, E.: Audio content processing for automatic music genre classification: descriptors, databases, and classifiers. Ph.D. thesis, Universitat Pompeu Fabra, Barcelona, Spain (2009)

147. Hamel, P., Eck, D.: Learning features from music audio with deep belief networks. In: Proceedings of the ISMIR (2010)

148. Han, K.P., Park, Y.S., Jeon, S.G., Lee, G.C., Ha, Y.H.: Genre classification system of TV sound signals based on a spectrogram analysis. IEEE Trans. Consumer Elect. **44**(1), 33–42 (1998)

149. Hansen, L.K., Ahrendt, P., Larsen, J.: Towards cognitive component analysis. In: Proceedings of the International and Interdisciplinary Conference on Adaptive Knowledge Representation and Reasoning, Espoo, Finland, pp. 148–153, June 2005

150. Harb, H., Chen, L., Auloge, J.Y.: Mixture of experts for audio classification: an application to male female classification and musical genre recognition. In: Proceedings of the ICME (2004)
151. Harb, H., Chen, L.: A general audio classifier based on human perception motivated model. Multimed. Tools Appl. **34**, 375–395 (2007)
152. Hartmann, K., Büchner, D., Berndt, A., Nürnberger, A., Lange, C.: Interactive data mining and machine learning techniques for musicology. In: Proceedings of the Conference on Interdisciplinary Musicology, pp. 1–8 (2007)
153. Hartmann, M.A.: Testing a spectral-based feature set for audio genre classification. Master's thesis, University of Jyväskylä, June 2011
154. Heittola, T.: Automatic classification of music signals. Master's thesis, Tampere University of Technology, Feb 2003
155. Henaff, M., Jarrett, K., Kavukcuoglu, K., LeCun, Y.: Unsupervised learning of sparse features for scalable audio classification. In: Proceedings of the ISMIR, Miami, FL, Oct 2011
156. de la Higuera, C., Piat, F., Tantini, F.: Learning stochastic finite automata for musical style recognition. In: Farré, J., Litovsky, I., Schmitz, S. (eds.) CIAA 2005. LNCS, vol. 3845, pp. 345–346. Springer, Heidelberg (2006)
157. Hillewaere, R., Manderick, B., Conklin, D.: Global feature versus event models for folk song classification. In: Proceedings of the ISMIR, pp. 729–733 (2009)
158. Hillewaere, R., Manderick, B., Conklin, D.: Melodic models for polyphonic music classification. In: Proceedings of the International Workshop on Machine Learning and Music (2009)
159. Hillewaere, R., Manderick, B., Conklin, D.: String quartet classification with monophonic models. In: Proceedings of the ISMIR, pp. 537–542 (2010)
160. Hillewaere, R., Manderick, B., Conklin, D.: String methods for folk tune genre classification. In: Proceedings of the ISMIR (2012)
161. Holzapfel, A., Stylianou, Y.: A statistical approach to musical genre classification using non-negative matrix factorization. In: Proceedings of the ICASSP, pp. 693–696, Apr 2007
162. Holzapfel, A., Stylianou, Y.: Musical genre classification using nonnegative matrix factorization-based features. IEEE Trans. Audio Speech Lang. Process. **16**(2), 424–434 (2008)
163. Holzapfel, A., Stylianou, Y.: Rhythmic similarity of music based on dynamic periodicity warping. In: Proceedings of the ICASSP, pp. 2217–2220 (2008)
164. Holzapfel, A., Stylianou, Y.: A scale based method for rhythmic similarity of music. In: Proceedings of the ICASSP, Taipei, Taiwan, pp. 317–320, Apr 2009
165. Homburg, H., Mierswa, I., Möller, B., Morik, K., Wurst, M.: A benchmark dataset for audio classification and clustering. In: Proceedings of the ISMIR, London, UK (2005)
166. Honingh, A., Bod, R.: Clustering and classification of music by interval categories. In: Agon, C., Andreatta, M., Assayag, G., Amiot, E., Bresson, J., Mandereau, J. (eds.) MCM 2011. LNCS, vol. 6726, pp. 346–349. Springer, Heidelberg (2011)
167. Hsieh, C.T., Han, C.C., Lee, C.H., Fan, K.C.: Pattern classification using eigenspace projection. In: Proceedings of the International Conference on Intelligent Information Hiding and Multimedia Signal Processing, pp. 154–157, July 2012
168. Hu, Y., Ogihara, M.: Genre classification for million song dataset using confidence-based classifiers combination. In: Proceedings of the ACM SIGIR, pp. 1083–1084. ACM, New York (2012)

169. Iñesta, J.M., Ponce de León, P.J., Heredia, J.L.: A ground-truth experiment on melody genre recognition in absence of timbre. In: Proceedings of the International Conference on Music Perception and Cognition, pp. 758–761 (2009)
170. ISMIR: Genre results (2004). http://ismir2004.ismir.net/genre_contest/index.htm
171. ISMIS: Genre results, Mar 2011. http://tunedit.org/challenge/music-retrieval
172. Jang, D., Jin, M., Yoo, C.D.: Music genre classification using novel features and a weighted voting method. In: Proceedings of the ICME, pp. 1377–1380 (2008)
173. Jennings, H., Ivanov, P., Martins, A., da Silva, P., Viswanathan, G.: Variance fluctuations in nonstationary time series: a comparative study of music genres. Phys. A: Stat. Theor. Phys. **336**(3–4), 585–594 (2004)
174. Jensen, J., Christensen, M., Murthi, M., Jensen, S.: Evaluation of MFCC estimation techniques for music similarity. In: Proceedings of the EUSIPCO (2006)
175. Jensen, K.: Music genre classification using an auditory memory model. In: Ystad, S., Aramaki, M., Kronland-Martinet, R., Jensen, K., Mohanty, S. (eds.) CMMR and FRSM 2011. LNCS, vol. 7172, pp. 79–88. Springer, Heidelberg (2012)
176. Jiang, D.N., L.-Lu, Zhang, H.J., Tao, J.H., Cai, L.H.: Music type classification by spectral contrast features. In: Proceedings of the ICME (2002)
177. Jin, X., Bie, R.: Random forest and PCA for self-organizing maps based automatic music genre discrimination. In: Proceedings of the Data Mining, pp. 414–417 (2006)
178. Lu, J., Wan, W., Yu, X., Li, C.: Music style classification using support vector machine. In: Proceedings of the International Conference on Wireless Communication and Mobile Computing, pp. 452–455, Dec 2009
179. Jothilakshmi, S., Kathiresan, N.: Automatic music genre classification for Indian music. In: Proceedings of the International Conference on Software and Computer Applications (2012)
180. Ju, H., Xu, J.X., VanDongen, A.M.J.: Classification of musical styles using liquid state machines. In: Proceedings of the International Joint Conference on Neural Networks, pp. 1–7 (2010)
181. Kaminskas, M., Ricci, F.: Contextual music information retrieval and recommendation: state of the art and challenges. Comput. Sci. Rev. **6**(2–3), 89–119 (2012)
182. Karkavitsas, G.V., Tsihrintzis, G.A.: Automatic music genre classification using hybrid genetic algorithms. In: Tsihrintzis, G.A., Virvou, M., Jain, L.C., Howlett, R.J. (eds.) IIMSS 2011. SIST, vol. 11, pp. 323–335. Springer, Heidelberg (2011)
183. Karkavitsas, G.V., Tsihrintzis, G.A.: Optimization of an automatic music genre classification system via hyper-entities. In: Proceedings of the International Conference on Intelligent Information Hiding and Multimedia Signal Processing, pp. 449–452 (2012)
184. Karydis, I.: Symbolic music genre classification based on note pitch and duration. In: Manolopoulos, Y., Pokorný, J., Sellis, T.K. (eds.) ADBIS 2006. LNCS, vol. 4152, pp. 329–338. Springer, Heidelberg (2006)
185. Karydis, I., Nanopoulos, A., Manolopoulos, Y.: Symbolic musical genre classification based on repeating patterns. In: Proceedings of the ACM Workshop on Audio and Music Computing Multimedia, pp. 53–58 (2006)
186. Kim, H.G., Cho, J.M.: Car audio equalizer system using music classification and loudness compensation. In: Proceedings of the International Conference on ICT Convergence (2011)
187. Kini, S., Gulati, S., Rao, P.: Automatic genre classification of North Indian devotional music. In: National Conference on Communications (2011)

188. Kirss, P.: Audio based genre classification of electronic music. Master's thesis, University of Jyväskylä, June 2007
189. Kitahara, T., Tsuchihashi, Y., Katayose, H.: Music genre classification and similarity calculation using bass-line features. In: Proceedings of the IEEE International Symposium on Multimedia, pp. 574–579, Dec 2008
190. Kobayakawa, M., Hoshi, M.: Musical genre classification of MPEG-4 Twin VQ audio data. In: Proceedings of the ICME (2011)
191. Koerich, A., Poitevin, C.: Combination of homogeneous classifiers for musical genre classification. In: IEEE International Conference on Systems, Man and Cybernetics, Oct 2005
192. Kofod, C., Ortiz-Arroyo, D.: Exploring the design space of symbolic music genre classification using data mining techniques. In: Proceedings of the International Conference on Computational Intelligence for Modelling, Control and Automation, pp. 432–448, Dec 2008
193. Kosina, K.: Music genre recognition. Master's thesis, Hagenberg Technical University, Hagenberg, Germany, June 2002
194. Kostek, B., Kupryjanow, A., Zwan, P., Jiang, W., Raś, Z.W., Wojnarski, M., Swietlicka, J.: Report of the ISMIS 2011 contest: music information retrieval. In: Kryszkiewicz, M., Rybinski, H., Skowron, A., Raś, Z.W. (eds.) ISMIS 2011. LNCS, vol. 6804, pp. 715–724. Springer, Heidelberg (2011)
195. Kotropoulos, C., Arce, G.R., Panagakis, Y.: Ensemble discriminant sparse projections applied to music genre classification. In: Proceedings of the ICPR, pp. 823–825, Aug 2010
196. van Kranenburg, P., Baker, W.: Musical style recognition - a quantitative approach. In: Proceedings of the Conference on Interdisciplinary Musicology (2004)
197. van Kranenburg, P.: On measuring musical style - the case of some disputed organ fugues in the J.S. Bach (BWV)catalogue. Computing In Musicology (2007-8)
198. van Kranenburg, P., Garbers, J., Volk, A., Wiering, F., Grijp, L., Veltkamp, R.: Collaboration perspectives for folk song research and music information retrieval: the indispensable role of computational musicology. J. Interdiscipl. Music Stud. **4**(1), 17–43 (2010)
199. van Kranenburg, P., Volk, A., Wiering, F.: A comparison between global and local features for computational classification of folk song melodies. J. New Music Res. **42**, 1–18 (2012)
200. Krasser, J., Abeßer, J., Großmann, H., Dittmar, C., Cano, E.: Improved music similarity computation based on tone objects. In: Proceedings of the Audio Mostly Conference, pp. 47–54 (2012)
201. Krumhansl, C.L.: Plink: "thin slices" of music. Music Percept.: Interdiscipl. J. **27**(5), 337–354 (2010)
202. Kuo, F.F., Shan, M.K.: A personalized music filtering system based on melody style classification. In: Proceedings of the IEEE International Conference on Data Mining, pp. 649–652 (2002)
203. Kuo, F.F., Shan, M.K.: Looking for new, not known music only: music retrieval by melody style. In: Proceedings of the Joint ACM/IEEE Conference on Digital Libraries, pp. 243–251, June 2004
204. Lambrou, T., Kudumakis, P., Speller, R., Sandler, M., Linney, A.: Classification of audio signals using statistical features on time and wavelet transform domains. In: Proceedings of the ICASSP, pp. 3621–3624, May 1998
205. Lampropoulos, A.S., Lampropoulou, P.S., Tsihrintzis, G.A.: Musical genre classification enhanced by improved source separation techniques. In: Proceedings of the ISMIR (2005)

206. Lampropoulos, A.S., Lampropoulou, P.S., Tsihrintzis, G.A.: Music genre classification based on ensemble of signals produced by source separation methods. Intell. Dec. Technol. **4**(3), 229–237 (2010)

207. Lampropoulos, A., Lampropoulou, P., Tsihrintzis, G.: A cascade-hybrid music recommender system for mobile services based on musical genre classification and personality diagnosis. Multimed. Tools Appl. **59**, 241–258 (2012)

208. Langlois, T., Marques, G.: A music classification method based on timbral features. In: Proceedings of the ISMIR (2009)

209. Langlois, T., Marques, G.: Automatic music genre classification using a hierarchical clustering and a language model approach. In: Proceedings of the International Conference on Advances in Multimedia (2009)

210. Lee, J.-W., Park, S.-B., Kim, S.-K.: Music genre classification using a time-delay neural network. In: Wang, J., Yi, Z., Żurada, J.M., Lu, B.-L., Yin, H. (eds.) ISNN 2006. LNCS, vol. 3972, pp. 178–187. Springer, Heidelberg (2006)

211. Lee, C.H., Shih, J.L., Yu, K.M., Su, J.M.: Automatic music genre classification using modulation spectral contrast feature. In: Proceedings of the ICME (2007)

212. Lee, C.H., Shih, J.L., Yu, K.M., Lin, H.S., Wei, M.H.: Fusion of static and transitional information of cepstral and spectral features for music genre classification. In: IEEE Asia-Pacific Service Computing Conference (2008)

213. Lee, C.H., Lin, H.S., Chou, C.H., Shih, J.L.: Modulation spectral analysis of static and transitional information of cepstral and spectral features for music genre classification. In: Proceedings of the International Conference on Intelligent Information Hiding and Multimedia Signal Processing (2009)

214. Lee, C., Shih, J., Yu, K., Lin, H.: Automatic music genre classification based on modulation spectral analysis of spectral and cepstral features. IEEE Trans. Multimed. **11**(4), 670–682 (2009)

215. Lee, H., Largman, Y., Pham, P., Ng, A.Y.: Unsupervised feature learning for audio classification using convolutional deep belief networks. In: Proceedings of the Neural Information Processing Systems, Vancouver, BC, Canada, Dec 2009

216. Lee, C.H., Chou, C.H., Lien, C.C., Fang, J.C.: Music genre classification using modulation spectral features and multiple prototype vectors representation. In: International Congress on Image and Signal Processing (2011)

217. Lehn-Schioler, T., Arenas-García, J., Petersen, K.B., Hansen, L.: A genre classification plug-in for data collection. In: Proceedings of the ISMIR (2006)

218. de León, P., Iñesta, J.: Musical style identification using self-organising maps. In: Proceedings of the WEDELMUSIC, pp. 82–89 (2002)

219. de León, P., Iñesta, J.: Feature-driven recognition of music styles. In: Perales, F.J., Campilho, A.C., Pérez, N., Sanfeliu, A. (eds.) IbPRIA 2003. LNCS, vol. 2652, pp. 773–781. Springer, Heidelberg (2003)

220. de León, P.J.P., Iñesta, J.M.: Musical style classification from symbolic data: a two-styles case study. In: Wiil, U.K. (ed.) CMMR 2003. LNCS, vol. 2771, pp. 167–178. Springer, Heidelberg (2004)

221. de León, P.P., Iñesta, J.: Pattern recognition approach for music style identification using shallow statistical descriptors. IEEE Trans. Syst. Man Cybern.: Part C: Appl. Rev. **37**(2), 248–257 (2007).

222. de León, F., Martinez, K.: Enhancing timbre model using MFCC and its time derivatives for music similarity estimation. In: Proceedings of the EUSIPCO, Bucharest, Romania, pp. 2005–2009, Aug 2012

223. de León, F., Martinez, K.: Towards efficient music genre classification using FastMap. In: Proceedings of the DAFx (2012)

224. Lerch, A.: An Introduction to Audio Content Analysis: Applications in Signal Processing and Music Informatics. Wiley/IEEE Press, Hoboken (2012)
225. Levy, M., Sandler, M.: Lightweight measures for timbral similarity of musical audio. In: Proceedings of the ACM Workshop on Audio and Music Computing Multimedia, pp. 27–36 (2006)
226. Li, T., Ogihara, M., Li, Q.: A comparative study on content-based music genre classification. In: Proceedings of the International ACM SIGIR Conference on Research and Development in Information Retrieval (2003)
227. Li, T., Tzanetakis, G.: Factors in automatic musical genre classification of audio signals. In: Proceedings of the IEEE Workshop on Applications of the Signal Processing to Audio and Acoustics (2003)
228. Li, T., Ogihara, M.: Music artist style identification by semi-supervised learning from both lyrics and contents. In: Proceedings of the ACM Multimedia (2004)
229. Li, M., Sleep, R.: Melody classification using a similarity metric based on Kolmogorov complexity. In: Proceedings of the SMC (2004)
230. Li, M., Sleep, R.: Genre classification via an LZ78-based string kernel. In: Proceedings of the ISMIR (2005)
231. Li, T., Ogihara, M.: Music genre classification with taxonomy. In: Proceedings of the ICASSP, Philadelphia, PA, pp. 197–200, Mar 2005
232. Li, T., Ogihara, M.: Toward intelligent music information retrieval. IEEE Trans. Multimed. **8**(3), 564–574 (2006)
233. Li, T., Ogihara, M., Shao, B., Wang, D.: Machine learning approaches for music information retrieval. In: Theory and Novel Applications of Machine Learning. I-Tech, Austria (2009)
234. Li, T.L., Chan, A.B., Chun, A.H.: Automatic musical pattern feature extraction using convolutional neural network. In: Proceedings of the International Conference on Data Mining and Applications (2010)
235. Li, T., Chan, A.: Genre classification and the invariance of MFCC features to key and tempo. In: Proceedings of the International Conference on Multimedia Modeling, Taipei, China, Jan 2011
236. Lidy, T., Rauber, A.: Genre-oriented organization of music collections using the SOMeJB system: an analysis of rhythm patterns and other features. In: Proceedings of the DELOS Workshop Multimedia Contents in Digital Libraries (2003)
237. Lidy, T.: Marsyas and rhythm patterns: evaluation of two music genre classification systems. In: Proceedings of the Workshop Data Analysis, June 2003
238. Lidy, T., Rauber, A.: Evaluation of feature extractors and psycho-acoustic transformations for music genre classification. In: Proceedings of the ISMIR (2005)
239. Lidy, T.: Evaluation of new audio features and their utilization in novel music retrieval applications. Master's thesis, Vienna University of Technology, Dec 2006
240. Lidy, T., Rauber, A., Pertusa, A., Iñesta, J.M.: Improving genre classification by combination of audio and symbolic descriptors using a transcription system. In: Proceedings of the ISMIR, Vienna, Austria, pp. 61–66, Sept 2007
241. Lidy, T., Rauber, A.: Classification and clustering of music for novel music access applications. In: Cord, M., Cunningham, P. (eds.) Machine Learning Techniques for Multimedia, pp. 249–285. Springer, Heidelberg (2008)
242. Lidy, T., Silla, C., Cornelis, O., Gouyon, F., Rauber, A., Kaestner, C.A., Koerich, A.L.: On the suitability of state-of-the-art music information retrieval methods for analyzing, categorizing and accessing non-western and ethnic music collections. Signal Process. **90**(4), 1032–1048 (2010)

243. Lidy, T., Mayer, R., Rauber, A., de León, P.P., Pertusa, A., Quereda, J.: A carte-sian ensemble of feature subspace classifiers for music categorization. In: Proceedings of the ISMIR, pp. 279–284 (2010)

244. Lim, S.C., Jang, S.J., Lee, S.P., Kim, M.Y.: Music genre/mood classification using a feature-based modulation spectrum. In: Proceedings of the International Conference on Modelling, Identification and Control (2011)

245. Lin, C.-R., Liu, N.-H., Wu, Y.-H., Chen, A.L.P.: Music classification using significant repeating patterns. In: Lee, Y.J., Li, J., Whang, K.-Y., Lee, D. (eds.) DASFAA 2004. LNCS, vol. 2973, pp. 506–518. Springer, Heidelberg (2004)

246. Lippens, S., Martens, J., De Mulder, T.: A comparison of human and automatic musical genre classification. In: Proceedings of the ICASSP, pp. 233–236, May 2004

247. Liu, Y., Xu, J., Wei, L., Tian, Y.: The study of the classification of Chinese folk songs by regional style. In: Proceedings of the International Conference on Semantic Computing, pp. 657–662, Sept 2007

248. Liu, X., Yang, D., Chen, X.: New approach to classification of Chinese folk music based on extension of hmm. In: Proceedings of the ICALIP, pp. 1172–1179, July 2008

249. Liu, Y., Wei, L., Wang, P.: Regional style automatic identification for Chinese folk songs. In: World Congress on Computer Science and Information Engineering (2009)

250. Liu, Y., Xiang, Q., Wang, Y., Cai, L.: Cultural style based music classification of audio signals. In: Proceedings of the ICASSP, Taipei, Taiwan, Apr 2009

251. Lo, Y.L., Lin, Y.C.: Content-based music classification. In: Proceedings of the International Conference on Computer Science Information Technology, pp. 112–116 (2010)

252. Loh, Q.J.B., Emmanuel, S.: ELM for the classification of music genres. In: Proceedings of the International Conference on Control, Automation, Robotics and Vision, pp. 1–6 (2006)

253. Londei, A., Loreto, V., Belardinelli, M.O.: Musical style and authorship categorization by informative compressors. In: Proceedings of the ESCOM Conference on Hanover, Germany, pp. 200–203, Sept 2003

254. Lopes, M., Gouyon, F., Koerich, A., Oliveira, L.E.S.: Selection of training instances for music genre classification. In: Proceedings of the ICPR, Istanbul, Turkey (2010)

255. Lukashevich, H., Abeßer, J., Dittmar, C., Großmann, H.: From multi-labeling to multi-domain-labeling: a novel two-dimensional approach to music genre classification. In: Proceedings of the ISMIR (2009)

256. Lukashevich, H.: Applying multiple kernel learning to automatic genre classification. In: Gaul, W.A., Geyer-Schulz, A., Schmidt-Thieme, L., Kunze, J. (eds.) Challenges at the Interface of Data Analysis, Computer Science, and Optimization, pp. 393–400. Springer, Berlin (2012)

257. Christanti, M.V., Kurniawan, F., Tony: Automatic music classification for Dangdut and Campursari using Naïve Bayes. In: International Conference on Electrical Engineering and Informatics (2011)

258. Mace, S.T., Wagoner, C.L., Teachout, D.J., Hodges, D.A.: Genre identification of very brief musical excerpts. Psychol. Music 40(1), 112–128 (2011)

259. Manaris, B., Romero, J., Machado, P., Krehbiel, D., Hirzel, T., Pharr, W., Davis, R.B.: Zipf's law, music classification, and aesthetics. Comput. Music J. 29(1), 55–69 (2005)

260. Manaris, B., Krehbiel, D., Roos, P., Zalonis, T.: Armonique: experiments in content-based similarity retrieval using power-law melodic and timbre metrics. In: Proceedings of the ISMIR, pp. 343–348 (2008)
261. Manaris, B., Roos, P., Krehbiel, D., Zalonis, T., Armstrong, J.: Zipf's law, power laws and music aesthetics. In: Li, T., Ogihara, M., Tzanetakis, G. (eds.) Music Data Mining, pp. 169–216. CRC Press, Boca Raton (2011)
262. Mandel, M.I., Poliner, G.E., Ellis, D.P.W.: Support vector machine active learning for music retrieval. Multimed. Syst. **12**, 3–13 (2006)
263. Manzagol, P.A., Bertin-Mahieux, T., Eck, D.: On the use of sparse time-relative auditory codes for music. In: Proceedings of the ISMIR, Philadelphia, PA, pp. 603–608, Sept 2008
264. Markov, K., Matsui, T.: Music genre classification using self-taught learning via sparse coding. In: Proceedings of the ICASSP, pp. 1929–1932, Mar 2012
265. Markov, K., Matsui, T.: Nonnegative matrix factorization based self-taught learning with application to music genre classification. In: Proceedings of the IEEE International Workshop on Machine Learning for Signal Processing, pp. 1–5, Sept 2012
266. Marques, G., Langlois, T.: A language modeling approach for the classification of music pieces. In: Proceedings of the Data Mining, pp. 193–198 (2009)
267. Marques, G., Lopes, M., Sordo, M., Langlois, T., Gouyon, F.: Additional evidence that common low-level features of individual audio frames are not representative of music genres. In: Proceedings of the SMC, Barcelona, Spain, July 2010
268. Marques, G., Langlois, T., Gouyon, F., Lopes, M., Sordo, M.: Short-term feature space and music genre classification. J. New Music Res. **40**(2), 127–137 (2011)
269. Marques, C., Guiherme, I.R., Nakamura, R.Y.M., Papa, J.P.: New trends in musical genre classification using optimum-path forest. In: Proceedings of the ISMIR (2011)
270. Martin, K.D., Scheirer, E.D., Vercoe, B.L.: Music content analysis through models of audition. In: Proceedings of the ACM Multimedia Workshop on Content Processing of Music for Multimedia Applications, Sept 1998
271. Matityaho, B., Furst, M.: Neural network based model for classification of music type. In: Proceedings of the Convention of Electrical and Electronics Engineers in Israel, pp. 1–5, Mar 1995
272. Mayer, R., Neumayer, R., Rauber, A.: Rhyme and style features for musical genre classification by song lyrics. In: Proceedings of the ISMIR (2008)
273. Mayer, R., Neumayer, R., Rauber, A.: Combination of audio and lyrics features for genre classification in digital audio collections. In: Proceedings of the ACM Multimedia, pp. 159–168, Oct 2008
274. Mayer, R., Rauber, A.: Building ensembles of audio and lyrics features to improve musical genre classification. In: International Conference on Distributed Frameworks for Multimedia Applications (2010)
275. Mayer, R., Rauber, A.: Multimodal aspects of music retrieval: audio, song lyrics - and beyond? Stud. Comput. Intell. **274**, 333–363 (2010)
276. Mayer, R., Rauber, A., Ponce de León, P.J., Pérez-Sancho, C., Iñesta, J.M.: Feature selection in a cartesian ensemble of feature subspace classifiers for music categorisation. In: Proceedings of the ACM International Workshop on Machine Learning and Music, pp. 53–56 (2010)
277. Mayer, R., Rauber, A.: Music genre classification by ensembles of audio and lyrics features. In: Proceedings of the ISMIR, pp. 675–680 (2011)
278. McDermott, J., Hauser, M.D.: Nonhuman primates prefer slow tempos but dislike music overall. Cognition **104**(3), 654–668 (2007)

279. McKay, C., Fujinaga, I.: Automatic genre classification using large high-level musical feature sets. In: Proceedings of the ISMIR (2004)
280. McKay, C.: Automatic genre classification of MIDI recordings. Ph.D. thesis, McGill University, Montréal, Canada, June 2004
281. McKay, C., Fujinaga, I.: Automatic music classification and the importance of instrument identification. In: Proceedings of the Conference on Interdisciplinary Musicology (2005)
282. McKay, C., Fujinaga, I.: Music genre classification: is it worth pursuing and how can it be improved? In: Proceedings of the ISMIR, Victoria, Canada, Oct 2006
283. McKay, C., Fujinaga, I.: Combining features extracted from audio, symbolic and cultural sources. In: Proceedings of the ISMIR, pp. 597–602 (2008)
284. McKay, C.: Automatic music classification with jMIR. Ph.D. thesis, McGill University, Montréal, Canada, Jan 2010
285. McKay, C., Fujinaga, I.: Improving automatic music classification performance by extracting features from different types of data. In: Multimedia Information Retrieval, pp. 257–266 (2010)
286. McKay, C., Burgoyne, J.A., Hockman, J., Smith, J.B.L., Vigliensoni, G., Fujinaga, I.: Evaluating the genre classification performance of lyrical features relative to audio, symbolic and cultural features. In: Proceedings of the ISMIR, pp. 213–218 (2010)
287. McKinney, M.F., Breebaart, J.: Features for audio and music classification. In: Proceedings of the ISMIR, Baltimore, MD, Oct 2003
288. Mendes, R.S., Ribeiro, H.V., Freire, F.C.M., Tateishi, A.A., Lenzi, E.K.: Universal patterns in sound amplitudes of songs and music genres. Phys. Rev. E **83**, 017101 (2011)
289. Meng, A., Ahrendt, P., Larsen, J.: Improving music genre classification by short-time feature integration. In: Proceedings of the ICASSP, Philadelphia, PA, pp. 497–500, Mar 2005
290. Meng, A.: Temporal feature integration for music organization. Ph.D. thesis, Technical University of Denmark (2006)
291. Meng, A., Shawe-Taylor, J.: An investigation of feature models for music genre classification using the support vector classifier. In: Proceedings of the ISMIR (2008)
292. Mierswa, I., Morik, K.: Automatic feature extraction for classifying audio data. Mach. Learn. **58**(2–3), 127–149 (2005)
293. MIREX: Genre results (2005). http://www.music-ir.org/mirex/wiki/2005: MIREX2005_Results
294. MIREX: Genre results (2007). http://www.music-ir.org/mirex/wiki/2007: MIREX2007_Results
295. MIREX: Genre results (2008). http://www.music-ir.org/mirex/wiki/2008: MIREX2008_Results
296. MIREX: Genre results (2009). http://www.music-ir.org/mirex/wiki/2009: MIREX2009_Results
297. MIREX: Genre results (2010). http://www.music-ir.org/mirex/wiki/2010: MIREX2010_Results
298. MIREX: Genre results (2011). http://www.music-ir.org/mirex/wiki/2011: MIREX2011_Results
299. MIREX: Genre results (2012). http://www.music-ir.org/mirex/wiki/2012: MIREX2012_Results
300. Mitra, V., Wang, C.J.: Content based audio classification: a neural network approach. Soft Comput. - A Fusion Found. Methodol. Appl. **12**, 639–646 (2008)

301. Mitri, G., Uitdenbogerd, A.L., Ciesielski, V.: Automatic music classification problems. In: Proceedings of the Autralasian Computer Science Conference (2004)
302. Moerchen, F., Ultsch, A., Nöcker, M., Stamm, C.: Databionic visualization of music collections according to perceptual distance. In: Proceedings of the ISMIR, London, UK, pp. 396–403, Sept 2005
303. Moerchen, F., Mierswa, I., Ultsch, A.: Understandable models of music collections based on exhaustive feature generation with temporal statistics. In: International Conference on Knowledge Discover and Data Mining (2006)
304. Mostafa, M.M., Billor, N.: Recognition of western style musical genres using machine learning techniques. Expert Syst. Appl. **36**(8), 11378–11389 (2009)
305. Nagathil, A., Gerkmann, T., Martin, R.: Musical genre classification based on highly-resolved cepstral modulation spectrum. In: Proceedings of the EUSIPCO, Aalborg, Denmark, pp. 462–466, Aug 2010
306. Nagathil, A., Göttel, P., Martin, R.: Hierarchical audio classification using cepstral modulation ratio regressions based on Legendre polynomials. In: Proceedings of the ICASSP, pp. 2216–2219, July 2011
307. Nayak, S., Bhutani, A.: Music genre classification using GA-induced minimal feature-set. In: Proceedings of the National Conference on Computer Vision, Pattern Recognition, Image Processing and Graphics (2011)
308. Neubarth, K., Goienetxea, I., Johnson, C., Conklin, D.: Association mining of folk music genres and toponyms. In: Proceedings of the ISMIR (2012)
309. Neumayer, R., Rauber, A.: Integration of text and audio features for genre classification in music information retrieval. In: Amati, G., Carpineto, C., Romano, G. (eds.) ECiR 2007. LNCS, vol. 4425, pp. 724–727. Springer, Heidelberg (2007)
310. Ni, Y., McVicar, M., Santos, R., Bie, T.D.: Using hyper-genre training to explore genre information for automatic chord estimation. In: Proceedings of the ISMIR (2012)
311. Nie, F., Xiang, S., Song, Y., Zhang, C.: Extracting the optimal dimensionality for local tensor discriminant analysis. Pattern Recogn. **42**(1), 105–114 (2009)
312. Nopthaisong, C., Hasan, M.M.: Automatic music classification and retrieval: experiments with Thai music collection. In: Proceedings of the International Conference on Information and Communication Technology, pp. 76–81, Mar 2007
313. Norowi, N.M., Doraisamy, S., Wirza, R.: Factors affecting automatic genre classification: an investigation incorporating non-western musical forms. In: Proceedings of the ISMIR (2005)
314. Novello, A., McKinney, M.F., Kohlrausch, A.: Perceptual evaluation of music similarity. In: Proceedings of the ISMIR, pp. 246–249 (2006)
315. Orio, N.: Music retrieval: a tutorial and review. Found. Trends Inf. Retr. **1**(1), 1–90 (2006)
316. Orio, N., Rizo, D., Miotto, R., Schedl, M., Montecchio, N., Lartillot, O.: MusiClef: a benchmark activity in multimodal music information retrieval. In: Proceedings of the ISMIR, pp. 603–608 (2011)
317. Otsuka, Y., Yanagi, J., Watanabe, S.: Discriminative and reinforcing stimulus properties of music for rats. Behav. Process. **80**(2), 121–127 (2009)
318. Pampalk, E., Dixon, S., Widmer, G.: On the evaluation of perceptual similarity measures for music. In: Proceedings of the DAFx, London, UK, pp. 7–12, Sept 2003
319. Pampalk, E., Flexer, A., Widmer, G.: Improvements of audio-based music similarity and genre classification. In: Proceedings of the ISMIR, London, UK, pp. 628–233, Sept 2005

320. Pampalk, E.: Computational models of music similarity and their application in music information retrieval. Ph.D. thesis, Vienna University of Technology, Vienna, Austria, Mar 2006

321. Panagakis, Y., Benetos, E., Kotropoulos, C.: Music genre classification: a multilinear approach. In: Proceedings of the ISMIR, Philadelphia, PA, pp. 583–588, Sept 2008

322. Panagakis, Y., Kotropoulos, C., Arce, G.R.: Music genre classification via sparse representations of auditory temporal modulations. In: Proceedings of the EUSIPCO, Glasgow, Scotland, Aug 2009

323. Panagakis, Y., Kotropoulos, C., Arce, G.R.: Music genre classification using locality preserving non-negative tensor factorization and sparse representations. In: Proceedings of the ISMIR, Kobe, Japan, pp. 249–254, Oct 2009

324. Panagakis, Y., Kotropoulos, C., Arce, G.R.: Non-negative multilinear principal component analysis of auditory temporal modulations for music genre classification. IEEE Trans. Acoust. Speech Lang. Process. **18**(3), 576–588 (2010)

325. Panagakis, Y., Kotropoulos, C., Arce, G.R.: Sparse multi-label linear embedding nonnegative tensor factorization for automatic music tagging. In: Proceedings of the EUSIPCO, pp. 492–496, Aug 2010

326. Panagakis, Y., Kotropoulos, C.: Music genre classification via topology preserving non-negative tensor factorization and sparse representations. In: Proceedings of the ICASSP, pp. 249–252, Mar 2010

327. Paradzinets, A., Harb, H., Chen, L.: Multiexpert system for automatic music genre classification. Technical report, Ecole Centrale de Lyon, Lyon, France, June 2009

328. Park, D.C.: Classification of audio signals using fuzzy c-means with divergence-based kernel. Pattern Recon. Lett. **30**(9), 794–798 (2009)

329. Park, D.C.: Partitioned feature-based classifier model. In: Proceedings of the IEEE International Symposium on Signal Processing and Information Technology, pp. 412–417 (2009)

330. Park, D.C.: Partitioned feature-based classifier model with expertise table. In: Proceedings of the IEEE International Conference on Bio-inspired Computing (2010)

331. Park, S., Park, J., Sim, K.: Optimization system of musical expression for the music genre classification. In: Proceedings of the International Conference on Control, Automation, and Systems, pp. 1644–1648, Oct 2011

332. Peeters, G.: A generic system for audio indexing: application to speech/music segmentation and music genre recognition. In: Proceedings of the DAFx, Bordeaux, France, Sept 2007

333. Peeters, G.: Spectral and temporal periodicity representations of rhythm for the automatic classification of music audio signal. IEEE Trans. Audio Speech Lang. Process. **19**(5), 1242–1252 (2011)

334. Peng, W., Li, T., Ogihara, M.: Music clustering with constraints. In: Proceedings of the ISMIR, pp. 27–32 (2007)

335. Pérez-Sancho, C., Iñesta, J.M., Calera-Rubio, J.: A text categorization approach for music style recognition. In: Marques, J.S., Pérez de la Blanca, N., Pina, P. (eds.) IbPRIA 2005. LNCS, vol. 3523, pp. 649–657. Springer, Heidelberg (2005)

336. Iñesta, T.P.J.M., Rizo, D.: metamidi: a tool for automatic metadata extraction from MIDI files. In: Proceedings of the Workshop on Exploring Musical Information Spaces, pp. 36–40, Oct 2009

337. Pérez-García, T., Pérez-Sancho, C., Iñesta, J.: Harmonic and instrumental information fusion for musical genre classification. In: Proceedings of the ACM International Workshop on Machine Learning and Music, pp. 49–52 (2010)
338. Pérez-Sancho, C., Iñesta, J.M., Calera-Rubio, J.: Style recognition through statistical event models. J. New Music Res. **34**(4), 331–340 (2005)
339. Pérez-Sancho, C., Rizo, D., Iñesta, J.: Stochastic text models for music categorization. In: da Vitoria Lobo, N., et al. (eds.) SSPR & SPR 2008. LNCS, vol. 5342, pp. 55–64. Springer, Heidelberg (2008)
340. Pérez-Sancho, C., Rizo, D., Iñesta, J.M.: Genre classification using chords and stochastic language models. Connect. Sci. **21**, 145–159 (2009)
341. Pérez-Sancho, C.: Stochastic language models for music information retrieval. Ph.D. thesis, Universidad de Alicante, Spain, June 2009
342. Pohle, T.: Extraction of audio descriptors and their evaluation in music classification tasks. Ph.D. thesis, Technischen Universität Kaiserslautern, Jan 2005
343. Pohle, T., Knees, P., Schedl, M., Widmer, G.: Independent component analysis for music similarity computation. In: Proceedings of ISMIR, pp. 228–233 (2006)
344. Pohle, T., Pampalk, E., Widmer, G.: Evaluation of frequently used audio features for classification of music into perceptual categories. In: International Workshop on Content-Based Multimedia Indexing (2008)
345. Pohle, T., Schnitzer, D., Schedl, M., Knees, P., Widmer, G.: On rhythm and general music similarity. In: Proceedings of the ISMIR (2009)
346. Pollastri, E., Simoncelli, G.: Classification of melodies by composer with hidden Markov models. In: Proceedings of the WEDELMUSIC, pp. 88–95, Nov 2001
347. Porter, D., Neuringer, A.: Music discriminations by pigeons. Exp. Psychol.: Animal Behav. Process. **10**(2), 138–148 (1984)
348. Pye, D.: Content-based methods for the management of digital music. In: Proceedings of the ICASSP (2000)
349. Rafailidis, D., Nanopoulos, A., Manolopoulos, Y.: Nonlinear dimensionality reduction for efficient and effective audio similarity searching. Multimed. Tools Appl. **42**, 273–293 (2009)
350. Rauber, A., Frühwirth, M.: Automatically analyzing and organizing music archives. In: Constantopoulos, P., Sølvberg, I.T. (eds.) ECDL 2001. LNCS, vol. 2163, p. 402. Springer, Heidelberg (2001)
351. Rauber, A., Pampalk, E., Merkl, D.: Using psycho-acoustic models and self-organizing maps to create a hierarchical structuring of music by musical styles. In: Proceedings of the ISMIR, pp. 71–80, Oct 2002
352. Ravelli, E., Richard, G., Daudet, L.: Audio signal representations for indexing in the transform domain. IEEE Trans. Audio Speech Lang. Process. **18**(3), 434–446 (2010)
353. Reed, J., Lee, C.H.: A study on music genre classification based on universal acoustic models. In: Proceedings of the ISMIR (2006)
354. Reed, J., Lee, C.H.: A study on attribute-based taxonomy for music information retrieval. In: Proceedings of the ISMIR, pp. 485–490 (2007)
355. Rin, J.M., Chen, Z.S., Jang, J.S.R.: On the use of sequential patterns mining as temporal features for music genre classification. In: Proceedings of the ICASSP (2010)
356. Ren, J.M., Jang, J.S.R.: Time-constrained sequential pattern discovery for music genre classification. In: Proceedings of the ICASSP, pp. 173–176 (2011)
357. Ren, J.M., Jang, J.S.R.: Discovering time-constrained sequential patterns for music genre classification. IEEE Trans. Audio Speech Lang. Process. **20**(4), 1134–1144 (2012)

358. Ribeiro, H., Zunino, L., Mendes, R., Lenzi, E.: Complexity-entropy causality plane: a useful approach for distinguishing songs. Phys. A: Stat. Mech. Its Appl. **391**(7), 2421–2428 (2012)

359. Rizzi, A., Buccino, N.M., Panella, M., Uncini, A.: Genre classification of compressed audio data. In: Proceedings of the International Workshop on Multimedia Signal Processing (2008)

360. Ro, W., Kwon, Y.: 1/f noise analysis of songs in various genre of music. Chaos Soliton. Fract. **42**(4), 2305–2311 (2009)

361. Rocha, B.: Genre classification based on predominant melodic pitch contours. Master's thesis, Universitat Pompeu Fabra, Barcelona, Spain, Sept 2011

362. Rump, H., Miyabe, S., Tsunoo, E., Ono, N., Sagayama, S.: Autoregressive MFCC models for genre classification improved by harmonic-percussion separation. In: Proceedings of the ISMIR, pp. 87–92 (2010)

363. Ruppin, A., Yeshurun, H.: Midi music genre classification by invariant features. In: Proceedings of the ISMIR, pp. 397–399 (2006)

364. Salamon, J., Rocha, B., Gomez, E.: Musical genre classification using melody features extracted from polyphonic music signals. In: Proceedings of the ICASSP, Kyoto, Japan, Mar 2012

365. Sanden, C., Befus, C., Zhang, J.Z.: Clustering-based genre prediction on music data. In: Proceedings of the International C* Conference on Computer Science and Software Engineering, pp. 117–119 (2008)

366. Sanden, C., Befus, C.R., Zhang, J.: Perception based multi-label genre classification on music data. In: Proceedings of the ICMC, pp. 9–15 (2010)

367. Sanden, C.: An empirical evaluation of computational and perceptual multi-label genre classification on music. Master's thesis, University of Lethbridge (2010)

368. Sanden, C., Zhang, J.Z.: Enhancing multi-label music genre classification through ensemble techniques. In: Proceedings of the International ACM SIGIR Conference on Research and Development in Information Retrieval, pp. 705–714 (2011)

369. Sanden, C., Zhang, J.Z.: Algorithmic multi-genre classification of music: an empirical study. In: Proceedings of the ICMC (2011)

370. Sanden, C., Befus, C.R., Zhang, J.Z.: A perceptual study on music segmentation and genre classification. J. New Music Res. **41**(3), 277–293 (2012)

371. de los Santos, C.A.: Nonlinear audio recurrence analysis with application to music genre classification. Master's thesis, Universitat Pompeu Fabra, Barcelona, Spain (2010)

372. Scaringella, N., Zoia, G.: On the modeling of time information for automatic genre recognition systems in audio signals. In: Proceedings of the ISMIR, pp. 666–671 (2005)

373. Scaringella, N., Zoia, G., Mlynek, D.: Automatic genre classification of music content: a survey. IEEE Signal Process. Mag. **23**(2), 133–141 (2006)

374. Schedl, M., Pohle, T., Knees, P., Widmer, G.: Assigning and visualizing music genres by web-based co-occurrence analysis. In: Proceedings of the ISMIR (2006)

375. Schierz, A., Budka, M.: High-performance music information retrieval system for song genre classification. In: Kryszkiewicz, M., Rybinski, H., Skowron, A., Raś, Z.W. (eds.) ISMIS 2011. LNCS (LNAI), vol. 6804, pp. 725–733. Springer, Heidelberg (2011)

376. Schindler, A., Mayer, R., Rauber, A.: Facilitating comprehensive benchmarking experiments on the million song dataset. In: Proceedings of the ISMIR, Oct 2012

377. Schindler, A., Rauber, A.: Capturing the temporal domain in echonest features for improved classification effectiveness. In: Proceedings of the Adaptive Multimedia Retrieval, Oct 2012

378. Schlüter, J., Osendorfer, C.: Music similarity estimation with the mean-covariance restricted Boltzmann machine. In: Proceedings of the ICMLA (2011)
379. Seo, J., Lee, S.: Higher-order moments for musical genre classification. Signal Process. **91**(8), 2154–2157 (2011)
380. Serra, J., de los Santos, C.A., Andrzejak, R.G.: Nonlinear audio recurrence analysis with application to genre classification. In: Proceedings of the ICASSP (2011)
381. Seyerlehner, K.: Content-based music recommender systems: beyond simple frame-level audio similarity. Ph.D. thesis, Johannes Kepler University, Linz, Austria, Dec 2010
382. Seyerlehner, K., Widmer, G., Pohle, T.: Fusing block-level features for music similarity estimation. In: Proceedings of the DAFx (2010)
383. Seyerlehner, K., Widmer, G., Knees, P.: A comparison of human, automatic and collaborative music genre classification and user centric evaluation of genre classification systems. In: Detyniecki, M., Knees, P., Nürnberger, A., Schedl, M., Stober, S. (eds.) AMR 2010. LNCS, vol. 6817, pp. 118–131. Springer, Heidelberg (2012)
384. Seyerlehner, K., Schedl, M., Sonnleitner, R., Hauger, D., Ionescu, B.: From improved auto-taggers to improved music similarity measures. In: Proceedings of the Adaptive Multimedia Retrieval, Copenhagen, Denmark, Oct 2012
385. Shan, M.K., Kuo, F.F., Chen, M.F.: Music style mining and classification by melody. In: Proceedings of the ICME, vol. 1, pp. 97–100 (2002)
386. Shao, X., Xu, C., Kankanhalli, M.S.: Unsupervised classification of music genre using hidden Markov model. In: Proceedings of the ICME, pp. 2023–2026 (2004)
387. Shen, J., Shepherd, J.A., Ngu, A.H.H.: On efficient music genre classification. In: Zhou, L., Ooi, B.-C., Meng, X. (eds.) DASFAA 2005. LNCS, vol. 3453, pp. 253–264. Springer, Heidelberg (2005)
388. Shen, J., Shepherd, J., Ngu, A.H.H.: Towards effective content-based music retrieval with multiple acoustic feature combination. IEEE Trans. Multimed. **8**(6), 1179–1189 (2006)
389. Shen, Y., Li, X., Ma, N.W., Krishnan, S.: Parametric time-frequency analysis and its applications in music classification. EURASIP J. Adv. Signal Process. **2010**, 1–9 (2010)
390. Shih, J.L., Lee, C.H., Lin, S.W.: Automatic classification of musical audio signals. J. Inf. Technol. Appl. **1**(2), 95–105 (2006)
391. Silla Jr., C.N., Kaestner, C.A.A., Koerich, A.L.: Time-space ensemble strategies for automatic music genre classification. In: Sichman, J.S., Coelho, H., Rezende, S.O. (eds.) IBERAMIA 2006 and SBIA 2006. LNCS (LNAI), vol. 4140, pp. 339–348. Springer, Heidelberg (2006)
392. Silla, C.N., Koerich, A., Kaestner, C.: Automatic music genre classification using ensembles of classifiers. In: Proceedings of the IEEE International Conference on Systems, Man, Cybernetics, pp. 1687–1692 (2007)
393. Silla, C.N., Koerich, A.L., Kaestner, C.A.A.: Feature selection in automatic music genre classification. In: Proceedings of the IEEE International Symposium on Multimedia, pp. 39–44 (2008)
394. Silla, C.N., Koerich, A.L., Kaestner, C.A.A.: The Latin music database. In: Proceedings of the ISMIR (2008)
395. Silla, C., Freitas, A.: Novel top-down approaches for hierarchical classification and their application to automatic music genre classification. In: IEEE International Conference on Systems, Man, and Cybernetics, San Antonio, USA, Oct 2009

396. Silla, C.N., Koerich, A.L., Kaestner, C.A.A.: A feature selection approach for automatic music genre classification. Int. J. Semantic Comput. **3**(2), 183–208 (2009)

397. Silla, C., Koerich, A., Kaestner, C.: Improving automatic music genre classification with hybrid content-based feature vectors. In: Proceedings of the Symposium on Applied Computer, Sierre, Switzerland, Mar 2010

398. Silla, C.N., Freitas, A.A.: A survey of hierarchical classification across different application domains. Data Mining Knowl. Disc. **22**, 31–72 (2011)

399. Simsekli, U.: Automatic music genre classification using bass lines. In: Proceedings of the ICPR (2010)

400. Smith, J.B.L., Burgoyne, J.A., Fujinaga, I., Roure, D.D., Downie, J.S.: Design and creation of a large-scale database of structural annotations. In: Proceedings of the ISMIR (2011)

401. Soltau, H., Schultz, T., Westphal, M., Waibel, A.: Recognition of music types. In: Proceedings of the ICASSP (1998)

402. Song, Y., Zhang, C., Xiang, S.: Semi-supervised music genre classification. In: Proceedings of the ICASSP, pp. 729–732 (2007)

403. Song, Y., Zhang, C.: Content-based information fusion for semi-supervised music genre classification. IEEE Trans. Multimed. **10**(1), 145–152 (2008)

404. Sordo, M., Celma, O., Blech, M., Guaus, E.: The quest for musical genres: do the experts and the wisdom of crowds agree? In: Proceedings of the ISMIR (2008)

405. Sotiropoulos, D., Lampropoulos, A., Tsihrintzis, G.: Artificial immune system-based music genre classification. In: Tsihrintzis, G., Virvou, M., Howlett, R., Jain, L. (eds.) New Directions in Intelligent Interactive Multimedia. SCI, vol. 142, pp. 191–200. Springer, Heidelberg (2008)

406. Srinivasan, H., Kankanhalli, M.: Harmonicity and dynamics-based features for audio. In: Proceedings of the ICASSP, vol. 4, pp. 321–324 (2004)

407. Sturm, B.L., Noorzad, P.: On automatic music genre recognition by sparse representation classification using auditory temporal modulations. In: Proceedings of the CMMR, London, UK, June 2012

408. Sturm, B.L.: An analysis of the GTZAN music genre dataset. In: Proceedings of the ACM MIRUM Workshop, Nara, Japan, Nov 2012

409. Sturm, B.L.: Two systems for automatic music genre recognition: what are they really recognizing? In: Proceedings of the ACM MIRUM Workshop, Nara, Japan, Nov 2012

410. Sturm, B.L.: Classification accuracy is not enough: on the analysis of music genre recognition systems. J. Intell. Inf. Syst. **41**, 371–406 (2013)

411. Sturm, B.L.: On music genre classification via compressive sampling. In: Proceedings of the ICME, July 2013, pp. 1–6 (2013)

412. Sturm, B.L.: Music genre recognition with risk and rejection. In: Proceedings of the ICME, July 2013, pp. 1–6 (2013)

413. Su, Z.Y., Wu, T.: Multifractal analyses of music sequences. Phys. D: Nonlin. Phen. **221**(2), 188–194 (2006)

414. Sundaram, S., Narayanan, S.: Experiments in automatic genre classification of full-length music tracks using audio activity rate. In: Proceedings of the IEEE Workshop Multimedia Signal Processing (2007)

415. Tacchini, E., Damiani, E.: What is a "musical world"? An affinity propagation approach. In: Proceedings of the ACM MIRUM Workshop, Scottsdale, AZ, USA, pp. 57–62, Nov 2011

416. Happi Tietche, B., Romain, O., Denby, B., Benaroya, L., Viateur, S.: FPGA-based radio-on-demand broadcast receiver with musical genre identification. In: Proceedings of the IEEE International Symposium on Industrial Electronics, pp. 1381–1385, May 2012
417. Tsai, W.H., Bao, D.F.: Clustering music recordings based on genres. In: Proceedings of the International Conference on Information Science and Applications (2010)
418. Tsatsishvili, V.: Automatic subgenre classification of heavy metal music. Master's thesis, University of Jyväskylä, Nov 2011
419. Tsunoo, E., Tzanetakis, G., Ono, N., Sagayama, S.: Audio genre classification by clustering percussive patterns. In: Proceedings of the Acoustical Society of Japan (2009)
420. Tsunoo, E., Tzanetakis, G., Ono, N., Sagayama, S.: Audio genre classification using percussive pattern clustering combined with timbral features. In: Proceedings of the ICME (2009)
421. Tsunoo, E., Tzanetakis, G., Ono, N., Sagayama, S.: Beyond timbral statistics: improving music classification using percussive patterns and bass lines. IEEE Trans. Audio Speech Lang. Process. **19**(4), 1003–1014 (2011)
422. Turnbull, D., Elkan, C.: Fast recognition of musical genres using RBF networks. IEEE Trans. Knowl. Data Eng. **17**(4), 580–584 (2005)
423. Typke, R., Wiering, F., Veltkamp, R.C.: A survey of music information retrieval systems. In: Proceedings of the ISMIR, London, UK, Sept 2005
424. Tzagkarakis, C., Mouchtaris, A., Tsakalides, P.: Musical genre classification via generalized Gaussian and alpha-stable modeling. In: Proceedings of the ICASSP, May 2006
425. Tzanetakis, G., Essl, G., Cook, P.: Automatic music genre classification of audio signals. In: Proceedings of the ISMIR (2001)
426. Tzanetakis, G., Cook, P.: Musical genre classification of audio signals. IEEE Trans. Speech Audio Process. **10**(5), 293–302 (2002)
427. Tzanetakis, G.: Manipulation, analysis and retrieval systems for audio signals. Ph.D. thesis, Princeton University, June 2002
428. Tzanetakis, G., Ermolinskyi, A., Cook, P.: Pitch histograms in audio and symbolic music information retrieval. J. New Music Res. **32**(2), 143–152 (2003)
429. Umapathy, K., Krishnan, S., Jimaa, S.: Multigroup classification of audio signals using time-frequency parameters. IEEE Trans. Multimed. **7**(2), 308–315 (2005)
430. Valdez, N., Guevara, R.: Feature set for philippine gong music classification by indigenous group. In: Proceedings of the IEEE Region 10 Conference, pp. 339–343, Nov 2011
431. Vatolkin, I., Theimer, W.M., Botteck, M.: Partition based feature processing for improved music classification. In: Proceedings of the Annual Conference of the German Classification Society, pp. 411–419 (2010)
432. Vatolkin, I., Preuß, M., Rudolph, G.: Multi-objective feature selection in music genre and style recognition tasks. In: Genetic and Evolutionary Computation Conference (2011)
433. Vatolkin, I.: Multi-objective evaluation of music classification. In: Gaul, W.A., Geyer-Schulz, A., Schmidt-Thieme, L., Kunze, J. (eds.) Challenges at the Interface of Data Analysis, Computer Science, and Optimization, pp. 401–410. Springer, Berlin (2012)
434. Volk, A., van Kranenburg, P.: Melodic similarity among folk songs: an annotation study on similarity-based categorization in music. Musicae Scientiae **16**(3), 317–339 (2012)

435. Völkel, T., Abeßer, J., Dittmar, C., Großmann, H.: Automatic genre classification of Latin music using characteristic rhythmic patterns. In: Proceedings of the Audio Mostly Conference, Piteå, Sweden (2010)
436. Wang, L., Huang, S., Wang, S., Liang, J., Xu, B.: Music genre classification based on multiple classifier fusion. In: Proceedings of the International Conference on Natural Computation (2008)
437. Wang, F., Wang, X., Shao, B., Li, T., Ogihara, M.: Tag integrated multi-label music style classification with hypergraph. In: Proceedings of the ISMIR (2009)
438. Wang, D., Li, T., Ogihara, M.: Are tags better than audio? The effect of joint use of tags and audio content features for artistic style clustering. In: Proceedings of the ISMIR, pp. 57–62 (2010)
439. Watanabe, S., Nemoto, M.: Reinforcing property of music in Java sparrows (Padda oryzivora). Behav. Process. **43**(2), 211–218 (1998)
440. Watanabe, S., Sato, K.: Discriminative stimulus properties of music in Java sparrows. Behav. Process. **47**(1), 53–57 (1999)
441. Watanabe, S.: How animals perceive music? Comparative study of discriminative and reinforcing properties of music for infrahuman animals. CARLS Series of Advanced Study of Logic and Sensibility vol. 2, pp. 5–16 (2008)
442. Weihs, C., Ligges, U., Morchen, F., Mullensiefen, D.: Classification in music research. Adv. Data Anal. Classif. **1**(3), 255–291 (2007)
443. Welsh, M., Borisov, N., Hill, J., von Behren, R., Woo, A.: Querying large collections of music for similarity. Technical report, University of California, Berkeley (1999)
444. West, K., Cox, S.: Features and classifiers for the automatic classification of musical audio signals. In: Proceedings of the ISMIR (2004)
445. West, K., Cox, S.: Finding an optimal segmentation for audio genre classification. In: Proceedings of the ISMIR, pp. 680–685 (2005)
446. West, K., Lamere, P.: A model-based approach to constructing music similarity functions. EURASIP J. Appl. Signal Process. **1**(1), 149 (2007)
447. West, K.: Novel techniques for audio music classification and search. Ph.D. thesis, University of East Anglia (2008)
448. Whitman, B., Smaragdis, P.: Combining musical and cultural features for intelligent style detection. In: Proceedings of the ISMIR, Paris, France, Oct 2002
449. Wiggins, G.A.: Semantic gap?? Schemantic schmap!! Methodological considerations in the scientific study of music. In: Proceedings of the IEEE International Symposium on Multimedia, pp. 477–482, Dec 2009
450. Wu, M.J., Chen, Z.S., Jang, J.S.R., Ren, J.M.: Combining visual and acoustic features for music genre classification. In: International Conference on Machine Learning and Applications (2011)
451. Wülfing, J., Riedmiller, M.: Unsupervised learning of local features for music classification. In: Proceedings of the ISMIR, Porto, Portugal, Oct 2012
452. Xu, C., Maddage, M., Shao, X., Cao, F., Tian, Q.: Musical genre classification using support vector machines. In: Proceedings of the ICASSP (2003)
453. Yang, W., Yu, X., Deng, J., Pan, X., Wang, Y.: Audio classification based on fuzzy-rough nearest neighbour clustering. In: Proceedings of the International Conference on Wireless Communications and Mobile Computation, pp. 320–324 (2011)
454. Yang, X., Chen, Q., Zhou, S., Wang, X.: Deep belief networks for automatic music genre classification. In: Proceedings of the INTERSPEECH, pp. 2433–2436 (2011)

455. Yao, Q., Li, H., Sun, J., Ma, L.: Visualized feature fusion and style evaluation for musical genre analysis. In: International Conference on Pervasive Computing, Signal Processing and Applications (2010)
456. Yaslan, Y., Cataltepe, Z.: Audio music genre classification using different classifiers and feature selection methods. In: Proceedings of the ICPR, pp. 573–576 (2006)
457. Yeh, C.C.M., Yang, Y.H.: Supervised dictionary learning for music genre classification. In: Proceedings of the ACM International Conference on Multimedia Retrieval, Hong Kong, China, June 2012
458. Ying, T.C., Doraisamy, S., Abdullah, L.N.: Genre and mood classification using lyric features. In: International Conference on Information Retrieval and Knowledge Management (2012)
459. Yoon, W.-J., Lee, K.-K., Park, K.-S., Yoo, H.-Y.: Automatic classification of western music in digital library. In: Fox, E.A., Neuhold, E.J., Premsmit, P., Wuwongse, V. (eds.) ICADL 2005. LNCS, vol. 3815, pp. 293–300. Springer, Heidelberg (2005)
460. Zanoni, M., Ciminieri, D., Sarti, A., Tubaro, S.: Searching for dominant high-level features for music information retrieval. In: Proceedings of the EUSIPCO, Bucharest, Romania, pp. 2025–2029, Aug 2012
461. Zeng, Z., Zhang, S., Li, H., Liang, W., Zheng, H.: A novel approach to musical genre classification using probabilistic latent semantic analysis model. In: Proceedings of the ICME, pp. 486–489 (2009)
462. Zhang, Y., Zhou, J.: A study on content-based music classification. In: Proceedings of the International Symposium on Signal Processing and Its Applications, pp. 113–116, July 2003
463. Zhang, Y.B., Zhou, J., Wang, X.: A study on Chinese traditional opera. In: Proceedings of the International Conference on Machine Learning and Cybernetics, pp. 2476–2480, July 2008
464. Zhen, C., Xu, J.: Solely tag-based music genre classification. In: Proceedings of the International Conference on Web Information Systems and Mining (2010)
465. Zhen, C., Xu, J.: Multi-modal music genre classification approach. In: Proceedings of the IEEE International Conference on Computer Science and Information Technology (2010)
466. Zhou, G.T., Ting, K.M., Liu, F.T., Yin, Y.: Relevance feature mapping for content-based multimedia information retrieval. Pattern Recogn. 45, 1707–1720 (2012)
467. Zhu, J., Xue, X., Lu, H.: Musical genre classification by instrumental features. In: Proceedings of the ICMC (2004)
468. Fabbri, F.: A theory of musical genres: two applications. In: Proceedings of the International Conference on Popular Music Studies, Amsterdam, The Netherlands (1980)
469. Frow, J.: Genre. Routledge, New York (2005)
470. Bertin-Mahieux, T., Eck, D., Mandel, M.: Automatic tagging of audio: the state-of-the-art. In: Wang, W. (ed.) Machine Audition: Principles, Algorithms and Systems. IGI Publishing, Hershey (2010)
471. Kim, Y., Schmidt, E., Migneco, R., Morton, B., Richardson, P., Scott, J., Speck, J., Turnbull, D.: Music emotion recognition: a state of the art review. In: Proceedings of the ISMIR, pp. 255–266 (2010)
472. Soltau, H.: Erkennung von Musikstilen. Ph.D. thesis, Universität Karlsruhe, Karlsruhe, Germany, May 1997
473. Kiernan, F.J.: Score-based style recognition using artificial neural networks. In: Proceedings of the ISMIR (2000)

474. Avcu, N., Kuntalp, D., Alpkocak, V.A.: Musical genre classification using higher-order statistics. In: Proceedings of the IEEE Signal Processing and Communication Applications Conference, pp. 1–4, June 2007

475. Bagci, U., Erzin, E.: Inter genre similarity modeling for automatic music genre classification. In: Proceedings of the IEEE Signal Processing and Communications Applications, pp. 1–4, Apr 2006

476. Herkiloglu, K., Gursoy, O., Gunsel, B.: Music genre determination using audio fingerprinting. In: Proceedings of the IEEE Signal Processing and Communications Applications, pp. 1–4, Apr 2006

477. Sonmez, A.: Music genre and composer identification by using Kolmogorov distance. Master's thesis, Istanbul Technical University, Istanbul, Turkey (2005)

478. Yaslan, Y., Cataltepe, Z.: Music genre classification using audio features, different classifiers and feature selection methods. In: Proceedings of the IEEE Signal Processing and Communications Applications, pp. 1–4, Apr 2006

479. Yaslan, Y., Cataltepe, Z.: Audio genre classification with co-MRMR. In: Proceedings of the IEEE Signal Processing and Communications Applications, pp. 408–411, Apr 2009

480. Allamanche, E., Kastner, T., Wistorf, R., Lefebvre, N., Herre, J.: Music genre estimation from low level audio features. In: Proceedings of the International Audio Engineering Society Conference (2004)

481. Seo, J.S.: An informative feature selection method for music genre classification. Trans. Japanese Eng. Tech. Org. **94–D**(6), 1362–1365 (2011)

482. Berenzweig, A., Logan, B., Ellis, D.P.W., Whitman, B.: A large-scale evaluation of acoustic and subjective music-similarity measures. Comput. Music J. **28**(2), 63–76 (2004)

483. Goto, M., Hashiguchi, H., Nishimura, T., Oka, R.: RWC music database: music genre database and musical instrument sound database. In: Proceedings of the ISMIR (2003)

484. Bertin-Mahieux, T., Ellis, D.P., Whitman, B., Lamere, P.: The million song dataset. In: Proceedings of the ISMIR (2011)

485. Law, E.: Human computation for music classification. In: Li, T., Ogihara, M., Tzanetakis, G. (eds.) Music Data Mining, pp. 281–301. CRC Press, Boca Raton (2011)

The Reason Why: A Survey of Explanations for Recommender Systems

Christian Scheel[1], Angel Castellanos[2],
Thebin Lee[3], and Ernesto William De Luca[3(✉)]

[1] DAI-Labor, Technische Universität Berlin, Berlin, Germany
scheel@dai-lab.de
[2] UNED, Madrid, Spain
acastellanos@lsi.uned.es
[3] Fachhochschule Potsdam, Potsdam, Germany
{lee,deluca}@fh-potsdam.de

Abstract. Recommender Systems refer to those applications that offer contents or items to the users, based on their previous activity. These systems are broadly used in several fields and applications, being common that an user interact with several recommender systems during his daily activities. However, most of these systems are black boxes which users really don't understand how to work. This lack of transparency often causes the distrust of the users. A suitable solution is to offer explanations to the user about why the system is offering such recommendations. This work deals with the problem of retrieving and evaluating explanations based on hybrid recommenders. These explanations are meant to improve the perceived recommendation quality from the user's perspective. Along with recommended items, explanations are presented to the user to underline the quality of the recommendation. Hybrid recommenders should express relevance by providing reasons speaking for a recommended item. In this work we present an attribute explanation retrieval approach to provide these reasons and show how to evaluate such approaches. Therefore, we set up an online user study where users were asked to provide movie feedback. For each rated movie we additionally retrieved feedback about the reasons this movie was liked or disliked. With this data, explanation retrieval can be studied in general, but it can also be used to evaluate such explanations.

Keywords: Hybrid recommender systems · Explanations · Evaluation · Persuasion · Satisfaction · Decision support

1 Introduction

In recent years, recommender systems have seen a steady increase in predictive accuracy in terms of quantifiable measures such as precision, recall, or the root-mean-square error [1]. Generally, the assumption by recommender systems researchers and developers is that the goal of a recommender system is to reach

A. Nürnberger et al. (Eds.): AMR 2012, LNCS 8382, pp. 67–84, 2014.
DOI: 10.1007/978-3-319-12093-5_3

higher levels of accuracy or lower levels of predictive errors. Given this assumption, the recent development indicates progress. We should however ask whether
we might be missing a bigger point? Predictive accuracy is only a measure of
the algorithmic quality of a system, it does not affect the perception of the user.
Keeping this in mind, it is possible that even a "perfect" recommender might
generate recommendations which are poorly perceived by the user [2], if the
recommendations are presented in an inappropriate way - or if the user fails to
see why the recommendation should be good [3,4]. If a system can motivate why
a recommendation is a good one, there is a higher chance of a higher perceived
quality of the recommendation and the system in general [5].

Recommender systems do not only provide recommendations, they help users
in decision making processes, they try to persuade users, filter out irrelevant data,
etc. Thus, a recommendation should be more than just the presentation of an
item. The context of the recommendation, i.e. why the item was recommended
is just as important in order to satisfy the different needs of users [3]. For this
purpose, this work presents a dataset collected through a user study where the
task users where asked to perform was to find and rate movies, and in addition
to this provide information on why a certain movie was liked or disliked.

Because explanations in the field of hybrid recommender systems are an
important factor for the perception of recommended items, we provide a benchmark dataset from a user study for evaluating these explanations. In this user
study users were asked to rate movies and provide reasons for and against
watching them.

We analyze the dataset's characteristics, provide baseline explanation retrieval approaches and show how to evaluate its performances.

2   Related Work

Searching for "interesting" news, "exciting" videos, or items to purchase are
common activities conducted everyday by millions of users on the web. Nevertheless, since its birth in the early 1990s, Internet has grown exponentially, even
more with the rise of Social Networks and user-generated contents. The huge
amount of data available on the web, far from being and advantage, is one of the
main problems in the information searching process. This is because the ability
of the users to discover new relevant information is seriously affected. In this
context the first Recommender Systems (RS) appeared to automatically explore
big data repositories and offer interesting contents to the users [6].

Since these first seed works, one aspect was pointed out by the RS community:
how to convince the users to use recommended items. Even though, some RS
could achieve a high theoretical performance, it will fail on their operation if the
users don't interact with its recommendations. It is for this that some research
lines in the RS field are focused on user confidence in recommendations, like for
example Explanations in Recommendations. Through these explanations, RS
are able to explain its decisions, hoping that in this way the users trust and use
these recommendations [7].

2.1 Recommender Systems

The operation of a RS can be basically defined as the estimation of the interest that a given user could have in a given content. To carry out this estimation, RS have to take into account some considerations and challenges. Some of them are set out in [8]: (*1*) scalability, to be able to manage very large amounts of data; (*2*) pro-activity, to automatically offer recommendations without the users have to ask about [9], but without interfere with the user activity [10]; and (*3*) user privacy, since in spite of RS have to collect as much information as it is possible, users must be able to set which information want to share and which don't [11].

According to the technique followed to offer recommendations, in the state of the art three RS types have been traditionally proposed [12]:

– **Collaborative Filtering RS (CFRS)** [13,14]. These systems based their operation on the interactions conducted by the users (readings, consults, purchases). Through these interactions the system is able to bring together users with similar tastes; for instance, two users who have purchased the same products in the past. User interactions can be collected "explicitly", the system ask to the user for them (e.g. please, rate this movie), or "implicitly", the system automatically collect the information (e.g. the user clicks on some link about some movie). According to the way in which this interactions are managed, there are three types of CFRS:
  - **User based** [15]: Users who have interacted in the same way with the same item set are grouped together in the same *user neighbourhood*. It is expected that if in the past users shared the same tastes, in the future they will continue sharing tastes. Then, if an user like/purchase/interact with a new item, it will be recommended to the rest of the user in his neighbourhood. Some aspects that should be addressed in this kind of systems are: estimation of the neighbourhood size [16], how to compute the similarity between users or rating normalization [12].
  - **Item based** [17]: Item based CFRS use the user interactions but instead of grouping together users, they group together, in the same *item neighbourhood*, items that has been liked/purchased/consulted for the same users. Recommendation will be conducted by, given an user who likes a set of items, offering him new contents in the same item neighbourhood that the already liked items.
  - **Model based** [18]: Both previous methods (item and user based) has the same problem: its complexity is very high (quadratic) to be applied in a real scenario. If U is the number of total users, I is the number of total items and K is the neighbourhood size, the complexity of item based systems will be $O(I^2 * U * K)$, while the complexity of the user based will be $O(U^2 * I * K)$. Due to this problem, model based CFRS was proposed. Basically these systems process in somehow the information to generate offline data models that simplifies the recommendation process. Simplification techniques proposed are, among others [19,20]: clustering, Latent Semantic Analysis, Latent Dirichlet Allocation or Support Vector Machines. Through

these techniques complexity of the systems is reduced to $O(I * U * K * L)$, being L the number of iterations needed to generate the data models.

– **Content-Based RS (CBRS)** [21,22]. CBRS are based on the content of the items to be recommended. These systems try to improve the recommendation process by taking advantage of the information contained on the items, and not only of the interactions with them. Instead of the user USER0001 likes the item ID0001 (as CFRS do), in these systems the interaction will be: USER0001 likes the item ID0001, which is a *movie* with *title, content, genres, actors, director,* etc.

CBRS have to cope with two aspects: how to model item contents and how to use these models to recommend new items. Regarding to the former, the work of Pazzani and Billsus in [23] discusses about different ways to represent and model item contents. Regarding to the later, also in [23] it is exposed that CBRS can be seen as Information Retrieval (IR) Systems, where the items to be recommended compose the IR Index and the user profiles are understood as the IR queries. Other method that has been proposed to carry out recommendations is the application of classification methods. It has been posed, for instance, in the work in [24] where items are classified in relevant/irrelevant by using as training the previous items liked by the users.

– **Hybrid RS (HRS)** [25,26]. CBRS and CFRS address the recommendation problem from different points of view. HRS try to exploit the benefits of both approaches by combining CBRS and CFRS in a single system.

The most basic approach to hybridize systems is to separately execute both systems and then merge its results. This approach is followed by the systems presented in [27,28], or [29]. Although these first works achieve a performance increment of the basis systems, because of their simplicity they don't really explode the potential of hybridizing systems. In this sense other more elaborated approaches has been recently proposed. Jack and Duclaye in [30] present a system that use the data generated for a CFRS to enrich the item description (with information of similar items and users). The item content plus the information added by CFRS are used to generate recommendations by applying a CBRS. Berkovsky et al. present other hybridization technique [31]: if there is enough information about some item content, a CBRS is applied; however, if there is no enough information a CFRS is applied.

Independently of the type, there is some general considerations about RS, which are addressed in this work: Semantic Based Recommendation, Context-Aware Recommendation and Trust-Based Recommendation.

Semantic Recommendation. Semantic RS refer to all those systems that use some knowledge base, containing semantic information, for their operation. Rationale behind these systems is the exploitation of semantic information (present in such knowledge bases) to mitigate the lacks in the item contents representation [32]. Several works have been developed in this field [33–35]. These works are mostly based on the use of the information and the relationships in knowledge databases (for example, FreeBase or DBPedia) to infer some knowledge with which enrich both user profiles and item descriptions.

Context-Aware Recommendation. Context-Aware Recommendation is related with the problem of *situated action*, posed by Mobasher [36]: the relevance of an item for an user is dependant on the user context. This problem has been addressed by means of different approaches, but one of the most common is to generate sub-profiles of the user profile according to the context information about the user. An example of this approach is the work of Said et al. [37], where it is proposed a movie recommender system in which different sub-profiles are generated according to *where* and *when* an user has watched a movie. These sub-profiles are subsequently used to personalized the recommendation step according to the current context of the user. A broad study about context-awareness can be consulted in [38], where the authors motivates the problem and recompiles some of the more novel approach in this field. The issues related with context-awareness are also exposed in the survey presented by Bettini et al. [39]. In this survey, they start presenting the most primitive techniques based on keywords or object-roles. They explain the utilization of spatial models and finally the application of ontology based models.

Trust-based Recommendation. Trust-based Recommendation (TBR) pursues to increment the user confidence in recommendations offered by the system, as it also intended in the work presented here. This kind of systems try to reproduce the natural process in which an user get recommendations in the real world. With this, it is intended to obtain more accurate and reliable recommendations. As it is posed before, user satisfaction is not only dependent on system performance; it is also related to psychological aspects, as trust. In this sense, an interesting user behaviour is explained by Shina and Swearingen in [40]. They show that users prefers recommendation made by their friends and acquaintances, even though these recommendations tend to be less novel and accurate. Another points of view in TBR are: the one proposed by Barman and Dabber in [15]. The authors base the recommendation in the "popularity" of the recommended items; or the social-based approaches that pretends to take advantage of the information in such platforms [41].

2.2 Explanations

Explanations of recommendations is not a new research topic, e.g. [42,43], the body of conducted work in this topic is by no means small.

Explanations in Recommender Systems. Tintarev and Masthoff present in [44] an extensive survey about this field. In this survey the authors present seven aspects that represent the facilities that explanation can offer to recommender systems operation and how the existing systems cope with these aspects.

One of the most important conclusion of their study is that the system has to be able to offer the explanations to the users "in their own terms". In this sense, the approach presented in this paper cover this aspect by explaining recommendations based on how the users have explained this preferences. The dataset especially developed for this work also offers the possibility to evaluate system explanations in the same way.

A similar approach to the one presented here is proposed by Symeonidis et al. in [45]. The authors propose the use of item attributes of the rated items to construct user profiles and to offer and explain recommendations. Unlike to the work presented here, the work in [45] doesn't offer the possibility that users individually rate the item attribute/s. It only allows the rating of the whole item, setting this rating for all of its attributes, doing that the system can infer wrong user preferences (i.e. I like The Godfather only because I like Marlon Brando, but I don't like Robert Duvall). The same limitation is shared by the explanation system presented in [3]. This system uses the tags with which an item (movies in MovieLens collection) is annotated to explain recommendations. But, it doesn't allow rating these tags one by one. The own authors expose this limitation in their conclusions.

Herlocker et al. present a model for *how* and *why* recommendation explanations should be implemented based on the conceptual model of the user's recommendation process and support this by empirical evidence [4]. They argue that explanations help detect, or estimate the likelihood of faulty recommendations (so-called recommendation errors). In a conceptually similar work, Bilgic and Mooney measure explanations in terms of *satisfaction* and *promotion* [46], where promotion refers to the explanation that most successfully convinces the user to pick an item and satisfaction refers to the to the explanation that best allows the user to assess the quality of an item.

In one of the earlier works on explanations in recommender systems-related work, Johnson and Johnson [42] attempt to identify what an explanation is and present both strengths and weaknesses in explanations in information systems. They list three limitations in explanation-related work: (*1*) the lack of a unifying theory of explanation, (*2*) the inability to identify and develop criteria for evaluating explanations, and (*3*) the lack of empirical studies in the field. It should be noted that the as the first and second limitations still apply to some extent, during the two decades since their work, a significant amount of empirical studies in the field of explanations has been undertaken [2,46].

Explanations in Persuasive Systems. Persuasive systems aim at heightening the user's experience of a system by trying to persuade the user to change her attitude or behavior [47]. Fogg lists several persuasion techniques and how they apply to different contexts and situations [48]. Physical persuasion, for instance, is the way a person acts while trying to persuade someone else. Mostly related to the type of persuasion explanations create is the concept of language-based persuasion, which according to Fogg's examples is the "interactive language use", either by a person or a system.

In their overview of the last few years of state-of-the-art in persuasive systems, Torning and Oinas-Kukkonen [49] note that the majority of the work focuses on behavioral changes created by persuasive systems, rather than persuading changes in users' attitudes. They also categorize recent papers into one of three types of persuasion context; *the intent, the event,* and *the strategy.*

Explanations in Decision Support. Decision support-related explanations are common in E-Commerce systems where the explanation is intended to help the customer select a specific product (and in the end be happy with the choice). Al-Qaed and Sutcliffe [50] tested user reactions to supporting tools and system advice and found that explanations and suggestions need to be personalized and contextualized in order to influence the user most.

Häubl and Trifts investigate how *interactive decision aids* (i.e. among other things explanations) affect consumer decision making in online shopping environments [51]. They find that there are strongly favorable effects on quality, efficiency, and satisfaction with purchase decisions that are supported by explanations and similar decision aids. Jedetski et al. make similar satisfaction-related observations in tasks performed on websites with decision support systems [52].

3 Attribute Explanations

The core of RS is the items to be recommended; these items can be: person, animal, book, movie, etc. They all have in common that they represent entities which can be described by attributes, e.g. writer of a book or director of a movie. In most cases there is a semantic description available where these attributes are predicate-object relations, e.g. *Coppola is director* of *The Godfather*. The idea is that these attributes serve as explanation for an item when being recommended, like "I think you'll like this movie because of *actor x* and *director y*, even though you don't like *actor z*."

Any attribute explanation for a recommended item should consist of a set of attributes of the recommended item which supports the recommendation and a set of attributes speaking against this item. The information about those attributes should be visualized or highlighted in the user interface, so that the user perceives the pros and cons of the recommended item immediately.

3.1 Items

Semantic knowledge bases like Freebase[1] and DBpedia[2] store structured information about entities. Hence these knowledge sources can be used to retrieve candidate attributes for recommender explanations. To find proper predicates (like for movies *is starring in* or *is director of*) that can serve for recommender

---

[1] http://www.freebase.com
[2] http://dbpedia.org

explanation we suggest to do a survey, where users have to provide this information about item of the same type.

In the sense of semantic representation an item i is a collection of triples, where all subject-predicate-object relations refer to the same subject identifier. All items of the same type share the same semantic predicates and often even the same predicate-object relations.

3.2   Attribute Explanation Model

Let i be a recommended item and A_i a set of attributes of this item. Each $a \in A$ is a predicate-object relation. An attribute explanation for item i consists of two sets of attributes A_i^+ and A_i^-, where A_i^+ includes those a which support the recommendation of i and A_i^- contains those a which speak against the recommendation. No attribute a is part of both A_i^+ and A_i^-, but there might be "neutral" a which are not part of A_i^+ or A_i^-.

3.3   Attribute Ratings as Attribute Explanation Retrieval Approach

As baseline attribute explanation approach we suggest to pass item ratings to the belonging attributes and average this value. The attribute rating r_a is computed with

$$r_a = \frac{\sum_{i \in I_a} \sum_{r \in R_i} r}{\sum_{i \in I_a} |R_i|} \qquad (1)$$

where I_a is the set of items containing attribute a and R_i is a set of ratings for item i.

The retrieved attribute ratings can be seen as a measure of quality which helps to select those attributes which should be used to explain recommendations.

Let m be a model which generates attribute explanations for any item i. The model m needs to retrieve A_i to add selected attributes to the attribute sets A_i^+ and A_i^-. A_i^+ contains all $a \in A_i$ where r_a is known and $r_a > t_1$, where t_1 is a manually selected threshold for retrieving "good" attributes. A_i^- contains all $a \in A_i$ where r_a is known and $r_a < t_2$, where t_2 is the threshold for retrieving "bad" attributes. For instance on a rating scale with 5 options ($\{1, 2, 3, 4, 5\}$) the thresholds can take the values $t_1 = t_2 = 3$.

4   Dataset Retrieval

To create and evaluate attribute explanation retrieval approaches there is a need of collecting ratings for items and belonging attributes. In spite of all of the work conducted in this area, there are no available datasets (to our knowledge) providing the same kind of information as the one presented in this work. We collected user feedback for movies through an online user study, where users were asked to select a movie, rate it and then provide feedback about the role of the attributes which have influenced the rating.

4.1   User Study

The online user study was created to continuously collect data to improve the resulting dataset[3].

To participate, users have to login to the survey, but to respect privacy, users in the final dataset are anonymized.

On the main view, there are three ways of selecting a movie. First, there is the possibility to search for movies. Second, movies can be selected from the members of a Freebase list of rated movies. Third, movies can be selected from a list of the last selected movies of all users. When a movie is selected, the user is directed to the movie page.

The movie page provides additional information about the movie and the possibility to rate it (see Fig. 1). After rating, users are asked to provide the reasons speaking for and against the movie. The help text points out that even a bad movie might have reasons to watch and a good movie might have reasons against watching it. When the user (in Fig. 1) scrolls down, movie attributes from genres, country of origin, music by, story by, produced by, executive produced by, production companies, edited by, rated, adapted from and awards won can be selected as reasons for or against this movie.

For each of these attribute types (e.g. "Performances") the known attributes (e.g. actors) are presented and the user can decide if one specific director, actor, etc. has influenced the given rating. To do so, the user can click on a green or red flag to annotate reasons for or against watching this movie. If an attribute did not influence the rating the user is asked to provide no attribute rating.

Fig. 1. Example movie page with given rating and provided reasons for and against watching it.

---

[3] http://www.dai-labor.de/~scheel/dataset/

4.2 Data Source and Data Structure

The online user study is completely based on Freebase data. Hence all identifiers refer to Freebase entities.

While the user study will go on, there will always be tagged versions which should be used and referenced, so that on the one hand research is done on current data, but on the other hand can be compared to others.

4.3 Data Analysis

We refer to the dataset as RER_{movie} (movie recommender explanation retrieval). The dataset was cleaned. By removing all movie ratings with no rating or reason given. Also all users which provided less than two ratings were removed.

At the time of writing, the dataset contains feedback from 53 users. These users provided 650 movie ratings (consisting of a rating and at least one reason) for 299 different movies. The mean number of movie ratings per user is 12.26 and the mean number of reasons per movie rating is 10.15. The inter-annotator agreement of movie ratings is the mean distance of the ratings of equal movies. It is equal to 1.01 for a maximum rating distance of 4 (the rating interval is between 1 and 5).

The inter-annotator agreement of the reasons can be calculated by looking at the attributes of each movie which have been chosen as reasons by more than one user. The agreement is then calculated by counting how often the reason type (pro or con) matches. Figure 2 shows the agreement per reason type, the overall agreement is 0.85. However, this value hides the fact, that the probability that there are equal reasons is only 0.54. In other words, there is a 54 % chance that a reason which was selected by one user is also a reason for another user, but if both select the same reason, they agree to 85 % that this reason is a reason for or against watching it. Figure 2 shows that the reasons with the most disagreement are rather uncommon movie features like age rating and editor.

In total users provided 6,597 reasons for movies. The most prominent movie attributes speaking for a movie are the genres (35 % of all movie supporting reasons) and the actors (31 %) followed by the director (6 %). The main reasons against movies are actors (26 % of all reasons for not watching movies) and genres (18 %). Hence, in the domain of movie recommendation genres and actors are the main sources for selecting attribute explanations.

The overall percentage of the reasons against movies is low. The reasons can be found in the majority of positively rated movies in the dataset (one star: 27, two stars: 48, three stars: 103, four stars: 217, five stars: 255). Figure 3(a) supports this fact by showing the mean number of reasons per item among the numerical ratings. For a bad movie (one star rating) there are in average 6.9 reasons against and still 3.5 reasons for a movie. For top rated movies (five star rating) there are in average 9.6 reasons for and 1.3 reasons against watching them. It can be seen that positively rated movies have a higher percentage of reason for a movie than against a movie, but in average always at least one reason against a movie can be found. For the worst rating, the number of reasons against

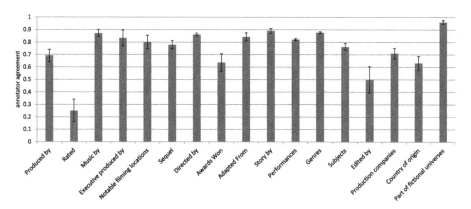

Fig. 2. Annotator agreement on reason types (pro or con) for equal attributes. Annotators agree if they selected the same attribute with the same reason type.

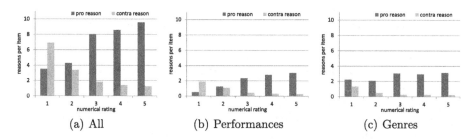

(a) All   (b) Performances   (c) Genres

Fig. 3. Mean number of reasons among the numeric ratings. In Fig. 3(a), each reason was taken, while in Fig. 3(b, c) the reasons are filtered to actors and genres.

movies is higher than the number of reasons for movies. This statement is still valid when only looking at the actors which were chosen as reason in Fig. 3(b), but invalid when looking at the genres in Fig. 3(c). Even if a movie is bad, the user might like the genre.

5 Attribute Explanation Retrieval Performance

Besides the possibility of studying explanation retrieval in general (see Sect. 4.3), the dataset can be used to evaluate approaches for attribute explanation retrieval. We assume that there is a model m returning attribute explanations as described in Sect. 3.2. We first describe how such a model can be evaluated and perform an evaluation on the approach described in Sect. 3.3.

5.1 Evaluation of the Explanation Retrieval

The evaluation on RER_{movie} dataset is done iteratively. For each movie i m provides a set of explanations speaking for the movie and one set for reasons speaking against watching the movie. These sets will be compared with the user given set of reasons for i. Even when there are different sets of reasons available for i, they are evaluated independently. After evaluating all proposed explanation sets, the performance values are averaged to a final performance value.

To compare the user given feedback for each i with the attribute explanations coming from m, the precision and the recall has to be computed first.

$$precision = \frac{TP}{|proposed\ attributes\ from\ m|} \tag{2}$$

$$recall = \frac{TP}{|attributes\ from\ user\ feedback|} \tag{3}$$

TP is equal to proposed attributes which are part of the user feedback and share the same type of reason (pro or con), the precision on the RER_{movie} dataset is defined as the fraction of the attributes in the proposed explanation set that are correctly classified reasons. This means, that a proposed attribute only is counted, when it was selected by the user and when the user and the explanation agree in the fact that it speaks for or against the movie. In general the explanation retrieval approach should return all user given reasons. The recall is needed to punish approaches which only return some attributes to raise the precision value.

A suitable measure for evaluating the performance of explanation retrieval approaches is the F-measure, because it combines precision and recall. For this evaluation, F_1 measure, the harmonic mean of precision and recall, has been applied.

$$F_1 = 2 \cdot \frac{precision \cdot recall}{precision + recall} \tag{4}$$

5.2   Baseline Explanation Retrieval

Given that there is no available dataset including attributes explanation, we couldn't compare our approach to already existing baselines. Then, we applied the baseline attribute explanation retrieval approach from Sect. 3.3 to our RER_{movie} dataset. Additionally, two primitive approaches will be evaluated, which do not judge if attributes speak for or against a movie, but return all attributes as reasons for or against it. These primitive approaches help to reflect the baseline's performance values.

Setting. Section 4.3 results in an observation that there are two types of reasons which have been chosen most: "performances" (actors) and "genres", followed by "directed by". To appreciate this observation, the evaluation is done also on subsets of movie attributes. Besides taking all attributes (ALL) also only genre (GENRES), actor (ACTORS) and director (DIRECTORS) attributes and combinations of these subsets are taken.

The evaluation was done by conducting a five-fold cross validation, where each fold contained a training set consisting of four subsets and a test set consisting of one subset. The baseline recommender candidate used the training set to apply the attribute rating approach, which passed the movie ratings to belonging attributes. If this average value was higher than 3 an attribute was declared as reason for a movie, otherwise as reason against a movie. Note that for instance ALL means all attributes with known attribute rating.

Evaluation. In Fig. 4 six sets of attributes are taken to evaluate selected approaches. While set ALL contains all attributes of a movie, the set GENRES only consists of genre attributes, etc.

Figure 4(a) shows evaluation results for the approach returning all selected attributes as reason for the movies. In contrast to this approach, the approach in Fig. 4(b) tags all selected attributes as reason against a movie. Figure 4(c) shows evaluation results of the approach described in Sect. 3.3.

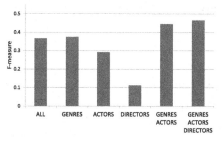

(a) Recommend attributes as reasons for the movies.

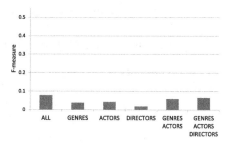

(b) Recommend attributes as reasons against the movies.

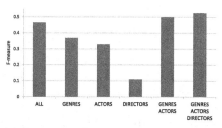

(c) Recommend attributes as reasons for or against the movies according to attribute ratings.

Fig. 4. Proposed baseline explanation retrieval approaches. The task was to generate explanations for each movie in the RER_{movie} dataset. In Fig. 4(a) the returned reasons have been declared as reason for a movie and in Fig. 4(b) against a movie. In Fig. 4(c) attribute ratings (which were passed from movie ratings of the training sets) determine if an attribute speaks for or against a movie.

While the recall of returning all attributes (in average 37.9) in Fig. 4(a) is very high, there is a high number of false positives, leading to a low precision and hence to a low F-measure. The best primitive attribute recommender would be to return all genres of a movie (in average 6.7), claiming they are reasons for watching a movie. This recommender can be improved by adding all actor attributes (in average 16.4) and further improved by adding all director attributes (in average 17.5). According to the F-measure in Fig. 4(b) there is no primitive attribute recommender for recommending attributes speaking against a movie, which is caused by the high percentage of reasons for movies in the dataset.

The best evaluated results can be found in Fig. 4(c) which should be taken as baseline when developing new approaches. When taking all attributes (in average 25.2), this approach reaches the performance of the best primitive approach which only returned all genre, actor and director attributes. When applied to this set of attributes (in average 12.7) the rated attribute approach in Fig. 4(c) performs reaches a F-measure value of 0.53.

The approach in Fig. 4(c) is designed to return reasons for and against movies. The mean percentage of reasons for the movie on the set ALL was 86.2 % and on the set containing all genre, actor and director attributes 89.8 %.

6   Conclusion and Future Work

This work presents a survey of explanations in recommender systems, particularly focused on Hybrid Recommender Systems. On top of the conclusions extracted from this review, we research on improving user's personal perception of recommended items. For a better perception, item attributes have to be presented in such way that the users may recommend or even speak against the item, based on them. These attributes will be presented along with other explanation types, if available.

This work includes a user study in the domain of movies, where users were asked to provide information about self-selected movies. This information includes a numerical rating and a set of attributes speaking for or against the rated movie. Unlike the rest of the works existing in the literature, which infer the attribute rating based on rating given for the items, the one presented here allows that the users directly rate the item attributes. The obtained dataset will allow recommending new items based on the rating given for their attributes. We refer to the resulting dataset as RER_{movie}. It can be downloaded at http://www.dai-labor.de/~scheel/dataset/download. This work includes a user study in the domain of movies, where users were asked to provide information about self-selected movies. This information includes a numerical rating and a set of attributes speaking for or against the rated movie. Unlike the rest of the works existing in the literature, which infer the attribute rating based on rating given for the items, the one presented here allows that the users directly rate the item attributes. The obtained dataset will allow recommending new items based on the rating given for their attributes. We refer to the resulting dataset as RER_{movie}. It can be downloaded at http://www.dai-labor.de/~scheel/dataset/download.

The collected data shows that genres and actors of movies are the most chosen features for reasons speaking for and against watching a movie. The ratio of reasons for and against movies depends on the numerical rating. For a bad movie there are in average 6.9 reasons against and still 3.5 reasons for a movie. For top rated movies there are in average 9.6 reasons for and 1.3 reasons against watching them.

The objective of collecting the data was to create a benchmark dataset for evaluating attribute explanation retrieval. For evaluation, the provided reasons for each movie in the dataset are compared with the proposed explanations coming from the evaluated approaches. This approach, which returns attribute explanations for and against movies, can also be applied to other items and its attributes.

The dataset can be used to create explanation retrieval models. We have shown how to evaluate such models and as an example evaluated an approach which passed ratings for movies to belonging attributes to create a ranked list of attributes which could be used to decide if attributes speak or against watching a movie.

For future work, we will apply learning to rank approaches on the data to receive a better list of ranked attributes or even enable personalized attribute explanation retrieval models. Another direction could be to receive quality values for attributes from external data like the Netflix data.

Although this dataset can be used already, we are still collecting data to be able to create and evaluate personalized explanation retrieval approaches.

References

1. Shani, G., Gunawardana, A.: Evaluating recommendation systems. In: Ricci, F., Rokach, L., Shapira, B., Kantor, P.B. (eds.) Recommender Systems Handbook, pp. 257–297. Springer, New York (2011)
2. McNee, S.M., Riedl, J., Konstan, J.A.: Being accurate is not enough: how accuracy metrics have hurt recommender systems. In: CHI '06 Extended Abstracts on Human factors in Computing Systems, CHI EA '06, pp. 1097–1101. ACM, New York (2006)
3. Vig, J., Sen, S., Riedl, J.: Tagsplanations: explaining recommendations using tags. In: Proceedings of the 14th International Conference on Intelligent User Interfaces, IUI '09, pp. 47–56. ACM, New York (2009)
4. Herlocker, J.L., Konstan, J.A., Riedl, J.: Explaining collaborative filtering recommendations. In: Proceedings of the 2000 ACM Conference on Computer Supported Cooperative Work, CSCW '00, pp. 241–250. ACM, New York (2000)
5. Herlocker, J.L., Konstan, J.A., Terveen, L.G., Riedl, J.T.: Evaluating collaborative filtering recommender systems. ACM Trans. Inf. Syst. 22, 5–53 (2004)
6. Resnick, P., Iacovou, N., Suchak, M., Bergstrom, P., Riedl, J.: Grouplens: an open architecture for collaborative filtering of netnews. In: Proceedings of the 1994 ACM Conference on Computer Supported Cooperative Work, CSCW '94, pp. 175–186. ACM, New York (1994)
7. McSherry, D.: Explanation in recommender systems. Artif. Intell. Rev. 24(2), 179–197 (2005)

8. Ricci, F., Rokach, L., Shapira, B.: Introduction to recommender systems handbook. In: Ricci, F., Rokach, L., Shapira, B., Kantor, P.B. (eds.) Recommender Systems Handbook, pp. 1–35. Springer, New York (2011)

9. Sae-Ueng, S., Pinyapong, S., Ogino, A., Kato, T.: Personalized shopping assistance service at ubiquitous shop space. In: International Conference on Advanced Information Networking and Applications Workshops, pp. 838–843 (2008)

10. Puerta Melguizo, M.C., Boves, L., Deshpande, A., Ramos, O.M.: A proactive recommendation system for writing: helping without disrupting. In: Proceedings of the 14th European Conference on Cognitive Ergonomics: Invent! Explore!, ECCE '07, pp. 89–95. ACM, New York (2007)

11. McSherry, F., Mironov, I.: Differentially private recommender systems: building privacy into the net. In: Proceedings of the 15th ACM SIGKDD International Conference on Knowledge Discovery and Data Mining, pp. 627–636. ACM, New York (2009)

12. Candillier, L., Jack, K., Fessant, F., Meyer, F.: State-of-the-art recommender systems. In: Chevalier, M., Julien, C., Soule-Dupuy, C. (eds.) Collaborative and Social Information Retrieval and Access-Techniques for Improved User Modeling, pp. 1–22. IGI Global, Hershey (2009)

13. Boim, R., Milo, T., Novgorodov, S.: Diversification and refinement in collaborative filtering recommender. In: Proceedings of the 20th ACM International Conference on Information and Knowledge Management, pp. 739–744. ACM, New York (2011)

14. Koren, Y., Bell, R.: Advances in collaborative filtering. In: Ricci, F., Rokach, L., Shapira, B., Kantor, P.B. (eds.) Recommender Systems Handbook, pp. 145–186. Springer, New York (2011)

15. Barman, K., Dabeer, O.: Local popularity based collaborative filters. In: 2010 IEEE International Symposium on Information Theory Proceedings (ISIT), pp. 1668–1672. IEEE (2010)

16. Bellogín, A., Cantador, I., Castells, P.: A study of heterogeneity in recommendations for a social music service. In: Proceedings of the 1st International Workshop on Information Heterogeneity and Fusion in Recommender Systems, pp. 1–8. ACM (2010)

17. Deshpande, M., Karypis, G.: Item-based top-n recommendation algorithms. ACM Trans. Inf. Syst. (TOIS) 22(1), 143–177 (2004)

18. Breese, J.S., Heckerman, D., Kadie, C.: Empirical analysis of predictive algorithms for collaborative filtering. In: Proceedings of the Fourteenth Conference on Uncertainty in Artificial Intelligence, pp. 43–52. Morgan Kaufmann Publishers Inc. (1998)

19. Jannach, D., Zanker, M., Felfernig, A., Friedrich, G.: Recommender Systems: An Introduction. Cambridge University Press, Cambridge (2010)

20. Kelleher, J., Bridge, D.: An accurate and scalable collaborative recommender. Artif. Intell. Rev. 21(3–4), 193–213 (2004)

21. Castellanos, A., Cigarrán, J., García-Serrano, A.: Content-based Recommendation: Experimentation and Evaluation in a Case Study. Conferencia de la Asociación Española para la Inteligencia Artificial (CAEPIA 2013) (2013)

22. Lops, P., de Gemmis, M., Semeraro, G.: Content-based recommender systems: State of the art and trends. In: Ricci, F., Rokach, L., Shapira, B., Kantor, P.B. (eds.) Recommender Systems Handbook, pp. 73–105. Springer, New York (2011)

23. Pazzani, M.J., Billsus, D.: Content-based recommendation systems. In: Brusilovsky, P., Kobsa, A., Nejdl, W. (eds.) Adaptive Web 2007. LNCS, vol. 4321, pp. 325–341. Springer, Heidelberg (2007)

24. Adomavicius, G., Tuzhilin, A.: Toward the next generation of recommender systems: A survey of the state-of-the-art and possible extensions. IEEE Trans. Knowl. Data Eng. **17**(6), 734–749 (2005)

25. Kim, H.-N., Ha, I., Lee, K.-S., Jo, G.-S., El-Saddik, A.: Collaborative user modeling for enhanced content filtering in recommender systems. Decis. Support Syst. **51**(4), 772–781 (2011)

26. Lucas, J.P., Luz, N., Moreno, M.N., Anacleto, R., Figueiredo, A.A., Martins, C.: A hybrid recommendation approach for a tourism system. Expert Syst. Appl. **40**(9), 3532–3550 (2012)

27. Balabanović, M., Shoham, Y.: Fab: content-based, collaborative recommendation. Commun. ACM **40**(3), 66–72 (1997)

28. Pazzani, M.J.: A framework for collaborative, content-based and demographic filtering. Artif. Intell. Rev. **13**(5–6), 393–408 (1999)

29. Vozalis, M., Margaritis, K.G.: Enhancing collaborative filtering with demographic data: The case of item-based filtering. In: 4th International Conference on Intelligent Systems Design and Applications, pp. 361–366 (2004)

30. Jack, K., Duclaye, F.: Improving explicit preference entry by visualising data similarities. In: Intelligent User Interfaces, International Workshop on Recommendation and Collaboration (ReColl), Spain (2008)

31. Berkovsky, S., Kuflik, T., Ricci, F.: Cross-representation mediation of user models. User Model. User-Adapt. Inter. **19**(1–2), 35–63 (2009)

32. Peis, E., del Castillo, J.M., Delgado-López, J.: Semantic recommender systems. Analysis of the state of the topic. Hipertext.net **6**, 1–5 (2008)

33. Ghani, R., Fano, A.: Building recommender systems using a knowledge base of product semantics. In: Proceedings of the Workshop on Recommendation and Personalization in ECommerce at the 2nd International Conference on Adaptive Hypermedia and Adaptive Web based Systems, pp. 27–29 (2002)

34. Cantador, I., Castells, P.: Multilayered semantic social network modeling by ontology-based user profiles clustering: Application to collaborative filtering. In: Staab, S., Svátek, V. (eds.) EKAW 2006. LNCS (LNAI), vol. 4248, pp. 334–349. Springer, Heidelberg (2006)

35. Wang, R.-Q., Kong, F.-S.: Semantic-enhanced personalized recommender system. In: 2007 International Conference on Machine Learning and Cybernetics, vol. 7, pp. 4069–4074. IEEE (2007)

36. Mobasher, B.: Contextual user modeling for recommendation. In: Keynote at the 2nd Workshop on Context-Aware Recommender Systems (2010)

37. Said, A., De Luca, E.W., Albayrak, S.: Inferring contextual user profiles-improving recommender performance. In: Proceedings of the 3rd RecSys Workshop on Context-Aware Recommender Systems (2011)

38. Adomavicius, G., Tuzhilin, A.: Context-aware recommender systems. In: Ricci, F., Rokach, L., Shapira, B., Kantor, P.B. (eds.) Recommender Systems Handbook, pp. 217–253. Springer, New York (2011)

39. Bettini, C., Brdiczka, O., Henricksen, K., Indulska, J., Nicklas, D., Ranganathan, A., Riboni, D.: A survey of context modelling and reasoning techniques. Pervasive Mob. Comput. **6**(2), 161–180 (2010)

40. Sinha, R.R., Swearingen, K.: Comparing recommendations made by online systems and friends. In: DELOS Workshop: Personalisation and Recommender Systems in Digital Libraries'01, pp. -1–1 (2001)

41. Walter, F., Battiston, S., Schweitzer, F.: A model of a trust-based recommendation system on a social network. Auton. Agent. Multi-Agent Syst. **16**(1), 57–74 (2008)

42. Johnson, H., Johnson, P.: Explanation facilities and interactive systems. In: Proceedings of the 1st International Conference on Intelligent User Interfaces, IUI '93, pp. 159–166. ACM, New York (1993)

43. Johnson, H., Johnson, P.: Different explanatory dialogue styles and their effects on knowledge acquisition by novices. In: Proceedings of the Twenty-Fifth Hawaii International Conference on System Sciences, 1992, vol. 3, pp. 47–57 (1992)

44. Tintarev, N., Masthoff, J.: A survey of explanations in recommender systems. In: Proceedings of the 2007 IEEE 23rd International Conference on Data Engineering Workshop, ICDEW '07, pp. 801–810. IEEE Computer Society, Washington, DC (2007)

45. Symeonidis, P., Nanopoulos, A., Manolopoulos, Y.: Moviexplain: a recommender system with explanations. In: Proceedings of the Third ACM Conference on Recommender Systems, RecSys '09, pp. 317–320. ACM, New York (2009)

46. Bilgic, M., Mooney, R.J.: Explaining recommendations: Satisfaction vs. promotion. In: Proceedings of Beyond Personalization 2005: A Workshop on the Next Stage of Recommender Systems Research at the 2005 International Conference on Intelligent User Interfaces, San Diego, CA, January 2005

47. Berkovsky, S., Freyne, J., Oinas-Kukkonen, H.: Influencing individually: Fusing personalization and persuasion. ACM Trans. Interact. Intell. Syst. **2**, 9:1–9:8 (2012)

48. Fogg, B.J.: Persuasive technology: using computers to change what we think and do. Ubiquity **2002** (2002)

49. Torning, K., Oinas-Kukkonen, H.: Persuasive system design: state of the art and future directions. In: Proceedings of the 4th International Conference on Persuasive Technology, Persuasive '09, pp. 30:1–30:8. ACM, New York (2009)

50. Al-Qaed, F., Sutcliffe, A.: Adaptive decision support system (adss) for b2c e-commerce. In: Proceedings of the 8th International Conference on Electronic Commerce: The New e-commerce: Innovations for Conquering Current Barriers, Obstacles and Limitations to Conducting Successful Business on the Internet, ICEC '06, pp. 492–503. ACM, New York (2006)

51. Häubl, G., Trifts, V.: Consumer decision making in online shopping environments: The effects of interactive decision aids. Mark. Sci. **19**, 4–21 (2000)

52. Jedetski, J., Adelman, L., Yeo, C.: How web site decision technology affects consumers. IEEE Internet Comput. **6**, 72–79 (2002)

Features and Classification

Cross-Dataset Learning of Visual Concepts

Christian Hentschel[(✉)], Harald Sack, and Nadine Steinmetz

Hasso Plattner Institute for Software Systems Engineering, Potsdam, Germany
{christian.hentschel,harald.sack,nadine.steinmetz}@hpi.uni-potsdam.de

Abstract. Visual content classification has become a keystone when opening up digital image archives to semantic search. Content-based explicit metadata often is only sparsely available and automated analysis of the depicted content therefore provides an important source of additional information. While visual content classification has proven beneficial, a major concern, however, is the dependency on large scale training data required to train robust classifiers. In this paper, we analyze the use of cross-dataset training samples to increase the classification performance. We investigate the performance of standardized manually annotated training sets as well automatically mined datasets from potentially unreliable web resources such as Flickr and Google Images. Next to brute force learning using this potentially noisy ground truth data we apply semantic post processing for data cleansing and topic disambiguation. We evaluate our results on standardized datasets by comparing our classification performance with proper ground truth-based classification results.

Keywords: Image classification · Cross-dataset learning

1 Introduction

In recent years, automatic classification of visual content such as video and photo data has gained increasing interest from several research communities. The digital era not only made recording and storage cheap and easy but also enabled new distribution channels that made pictorial data available for a larger audience. Authorship of visual content is no longer limited to professionals and Internet community platforms such as *Flickr*[1] are hosting an ever increasing number of private photo collections.

With the growth of visual data came the need to search and retrieve information within these collections. Professional archives and companies such as stock photo agencies have a strong commercial interest in making their content not only accessible but also searchable via Internet. It became apparent that a manual annotation of the depicted content with describing metadata will always be incomplete, subjective and most of all infeasible due to the sheer number of assets even if letting alone these that are added every day.

---

[1] Flickr: http://www.flickr.com/.

© Springer International Publishing Switzerland 2014
A. Nürnberger et al. (Eds.): AMR 2012, LNCS 8382, pp. 87–101, 2014.
DOI: 10.1007/978-3-319-12093-5_4

Research efforts in computer vision target the demands of todays image archives for efficient search and retrieval methods in stored content. Typically, the task is to automatically recognize categories of objects and scenes depicted in photos (i.e. visual concepts) – a task that is a fundamental ability of humans but still an elusive goal when assigned to machines. Various approaches have been presented in literature in recent years. Recognition is usually considered as a classification problem of separating positive from negative examples of a given visual concept. Most approaches rely on supervised machine learning techniques that require a set of manually annotated data for training a model of a specific concept.

However, as manual annotation is labor-intense, most datasets publicly available, despite the undisputed effort that has been accomplished, are rather limited in terms of number of annotated images as well as number of concepts used to describe the depicted content. The breadth of the semantic space covered by the selected visual concepts, however, has important implications for the real-world applicability of automatic visual content classification. Considering the afore-mentioned application scenarios for automatic search and retrieval in large scale photo collections a sufficiently high (preferably unlimited) number of concepts is a minimum requirement. The fewer concepts an automatic classifier is able to recognize, the less useful it is in solving the problems of todays collections as search is limited to the few covered ones.

Hence, a question that arises from these observations is how to significantly increase the number of training data while at the same time limiting the manual effort required to allocate this data. Preferably, in order to meet the goal of supporting arbitrary user queries, the number of covered concepts should be completely independent of manual annotations. In this paper, we present a first step towards this direction by proposing cross-data model training. We first evaluate the straightforward approach of training on additional manually annotated datasets in order to increase the number of covered concepts. The question we target at is whether concepts provided by the various computer vision bench-marking initiatives can be used interchangeably, i.e. whether a specific concept implicitly defined by the training data of one dataset is congruent with the definition of the same concept of another dataset.

Secondly, we investigate the use of training data that requires no manual effort for ground truth generation by using automatically crawled Flickr photos based on matching a specific concept query with user provided tags. We aim at answering the question whether the automatic allocation of labeled data for ad hoc training of a visual concept model can be successful in order to advance towards unlimited user queries. In order to reduce noise within the training data we propose methods for automatic data cleansing based on statistical and semantic analysis of the associated user tags. This goal, however, could only partially be achieved.

This paper is structured as follows: Sect. 2 reviews the relevant literature related to the presented approach. In Sect. 3 the applied bag-of-visual-words method for visual concept recognition is briefly described. The main part of this work is presented in Sect. 4 where the proposed approaches for labeled data acquisition and cleansing are presented. Section 5 provides an evaluation of the proposed approach based on the ImageCLEF 2011 evaluation data. Finally, Sect. 6 concludes the paper and gives a brief outlook to future work.

2 Related Work

The allocation of training data sets for content-based image classification has been the mission for various campaigns, challenges and benchmarking initiatives that aim at providing a comparative evaluation of different approaches. One of the largest efforts in this direction has been the "ImageCLEF Visual Concept Detection and Annotation Task" [13]. The contributed dataset today covers 99 different visual concepts that have been used to manually annotate 18,000 photos, i.e. different concept labels have been assigned to images based on human evaluation whether or not a specific concept is depicted within the image. While the task of annotating 18,000 images is huge the potential benefit for real world scenarios still seems negligible since 99 different concepts are by no means satisfactory to cover a reasonable amount of potential user queries emitted to a photo collection.

Only recently, crowd sourcing strategies such as Amazon Mechanical Turk (MTurk)[2] have provided the possibility to substantially increase the number of sample images, the annotation diversity as well as the number of per-image annotations. The labels provided by the ImageNet project [3] – probably todays largest database of manually annotated photos – are created through a large-scale MTurk process. The ImageNet database is organized according to the WordNet[3] hierarchy: The declared aim of the ImageNet project is to provide on average 1,000 images to illustrate each "synonym set" or "synset" defined in WordNet. In April 2010, ImageNet indexed 14,197,122 images aligned to 21,841 synsets[4].

Next to these clean, manually labeled training data other initiatives focused on the aggregation of potentially weakly labeled datasets compiled by web photo communities such as Flickr or by querying standard web search engines such as Google. Similar to ImageNet the TinyImages [18] collection is arranged around WordNet. By sending all non-abstract WordNet nouns as search queries to image search engines the authors collected 80 million low resolution (32×32 pixels) images each labeled with the respective noun. On average each noun is described by a set of 1,056 images and the average precision within the sets is estimated to be at 10–25 %. The high level of noise makes the dataset less useful for training data acquisition and the lack of additional metadata limits data cleansing to visual-only approaches. Content-based analysis of the images, however, is difficult due to their low resolution.

The Flickr platform provides a public API to query their database in order to retrieve photos that have been tagged by users with a given search term.

---

[2] Amazon Mechanical Turk: https://www.mturk.com/mturk/welcome.

[3] WordNet – A lexical database for English: http://wordnet.princeton.edu/.

[4] ImageNet – Summary and Statistics: http://image-net.org/about-stats.

By means of this API the MIRFLICKR Retrieval Evaluation [11] automatically assembled the MIRFLICKR-1M collection that provides 1 million Flickr images published under the Creative Commons license. Moreover, next to the plain images, the collection also contains the Flickr user tag data if provided by the Flickr users. However, the images are not manually annotated and the tags submitted by Flickr users cannot be considered of similar quality as the ground truth provided by human annotators in the ImageNet initiative. The dataset used in the ImageCLEF benchmark initiative is a manually annotated subset of MIRFLICKR image data.

To the authors best knowledge no effort to analyze the various datasets for congruency in terms of ground truth data has been made yet. The availability of inexpensive web-image data, however, has created considerable interest in the computer vision community to employ this data for training of visual classifiers. In [8] the author uses co-training in order to improve a classifier trained on a small quantity of labeled data. Unlabeled images, which are confidently classified by one classifier are added to the training set of another classifier. Other work aims at prior cleansing of weakly labeled Internet images by means of visual analysis. For example, the authors in [4] train models for parts and spatial configuration of objects without supervision from cluttered web images. The models are later used to re-rank the output of an image search engine. In [9] an iterative framework for visual classification of downloaded web images retrieved through an image search engine is presented.

Text-based outlier removal is performed in [1] where the 100 words surrounding an image link in its associated web page are used for identification of a set of images to be used as visual exemplars for animal classification. Similarly, in [14] and [19] images returned by a web search engine are re-ranked based on the text surrounding the image and metadata features. The top-ranked images are then used as (noisy) training data. With regard to the use of Flickr tags as resource for training data ground truth the authors in [6] use the MIRFLICKR-25000 dataset to train a multiple kernel learning classifier. Tag data as well as visual features are combined and a semi-supervised approach is applied to remove examples that are likely to be incorrectly tagged. Finally, in [17] the authors propose a method to evaluate the effectiveness of a tag in describing the visual content of its annotated images.

3   Content-based Visual Concept Classification

In this section we briefly present the applied Bag-of-(Visual-)Words (BoW) approach for content-based visual concept classification. As the major focus of this paper is not put on improving the various aspects of the BoW method we restrain the presentation to these details required to ensure repeatability of the conducted experiments. For further information we refer to the related work in concept classification (cf., e.g. [2,7,12]).

In our experiments, we extract SIFT (Scale-Invariant-Feature-Transform, [10]) features at a fixed grid of 6×6 pixels on each channel of an image in

RGB color space[5]. By concatenating these features we obtain a 384-dimensional feature vector at each grid point. These features are used to compute a visual vocabulary by running a k-means clustering that provides us with a set of representative visual words (codewords). We compute $k = 4,000$ cluster centers on the RGB SIFT features taken from the training images set. By assigning each of the extracted RGB-SIFT feature of an image to its most similar codeword (or cluster center) using a simple approximate nearest neighbor classifier we compute a normalized histogram of codeword frequencies, i.e. a Bag-of-Words, that is used to describe this image. The combination of SIFT for local image description and the BoW model makes the approach invariant to transformations, changes in lighting and rotation, occlusion, and intra-class variations [2].

Once the image feature vectors have been computed the problem of visual concept recognition can be approached by standard machine learning techniques. Kernel-based Support Vector Machines (SVM) have been widely used in image classification scenarios (cf. [2,15,20]). We use a Gaussian kernel based on the χ^2 distance measure, which has proven to provide good results for histogram comparison[6]. Following Zhang et al. [20] we approximate the kernel parameter γ by the average distance between all training image BoW-histograms. Therefore, the only parameter we optimize in a 4-fold cross-validation is the cost parameter C of the support vector classification. New images can be classified using the aforementioned Bag-of-Words feature vectors and the trained SVM model.

We consider the classification task a one-against-all approach – one SVM per given visual concept is trained to separate the images from this concept from all other given concepts. Hence, the classifier is trained to solve a binary classification problem, i.e., whether or not an image depicts a specific visual concept. This approach provides us with two advantages. First, new concepts can easily be added by simply training a new classifier which is in line with the demand for easy concept extension. Since the features are not adapted to the classification task they can be reused. Second, multiple concepts can be assigned to each image depending on the prediction confidence of each classifier available, again an important property when aiming at preferably unlimited concepts.

4   Cross-dataset Training

Benchmarking datasets for visual content classification consist of a set of labeled training images and a set of evaluation images whose ground truth labels are known only to the authors of the benchmarking initiative. Typically training and evaluation sets are obtained by splitting a larger dataset into two smaller ones (e.g. at a rate of 50 % training and 50 % evaluation data). While this is reasonable in order to provide a certain degree of homogeneity between training and evaluation data, this likewise reduces the already limited and valuable training data.

---

[5] We use the OpenCV 2.4.1 SIFT descriptor implementation: http://opencv.org/.

[6] Our implementation is based on the libsvm-3.1 Library for Support Vector Machines: http://www.csie.ntu.edu.tw/~cjlin/libsvm/.

We consider cross-dataset training as training a model for a given visual concept on a dataset that is completely disjoint in terms of its history of origins from the dataset used to evaluate the classification accuracy of the trained classifier. By following this approach we evaluate to what extent different datasets are congruent in terms of the concept definition implicitly primed by the provided positive and negative samples.

4.1 Evaluation Dataset

For our experiments in cross-dataset learning we decided to use the evaluation set of the "ImageCLEF 2011 Visual Concept Detection and Annotation Task" as ground truth data to estimate the performance of the models trained on different datasets. The dataset has been used in the 2010 and 2011 benchmarking initiative and provides a training dataset of 8,000 photos manually annotated with 99 different visual concepts while the evaluation set comprises 10,000 likewise annotated photos. The ground truth has been publicly released and therefore can be used by researchers to compare their algorithms with others. The choice of this dataset as ground truth for our tests not only provides traceability of our experiments but also comparability to other research results that were published alongside with the benchmarking initiative.

We decided to select two concepts for testing purposes "Bridge" and "Landscape_Nature". The reason being that the average classification performance of the concept "Bridge" was one of the worst among all results submitted by different participants of the benchmark despite a reasonably sized training set (i.e. 105 photos labeled with "Bridge" are available in the training set). The decision for the concept "Landscape_Nature" was based on the observation that most participants in the benchmark performed rather well in classifying photos as members of this concept. A number of 1,362 photos are positively labeled with "Landscape_Nature" in the training set.

4.2 Training Data Acquisition

We use three different datasets for acquisition of positive and negative sample images:

ImageCLEF. As a first step in order to provide a baseline to our experiments we choose the ImageCLEF training set for both concepts to train two models that we evaluate by classifying the photos in the evaluation set. Similar to the evaluators of the ImageCLEF task we compute the interpolated average precision (iAP) as an evaluation measure, i.e. the average precision at 11 recall intervals. We train our models using the BoW method described in Sect. 3. The vocabulary is generated by clustering 800,000 RGB SIFT descriptors randomly sampled from the training set. The classifiers are trained using all photos labeled with the respective concept as positive samples and all other training photos as negatives. The results serve as a baseline (see Sect. 5). Figure 1 plots the distribution of the results of the various participants of the 2011 ImageCLEF challenge as well as the results obtained by our own default BoW approach. We restrained the results to those obtained by visual-only classifiers (i.e. classifiers that operate on the visual image features only and that do not use any additional textual metadata such as Flickr tags, see [13] for more information). The plot shows that our approach resembles the average result of all participants.

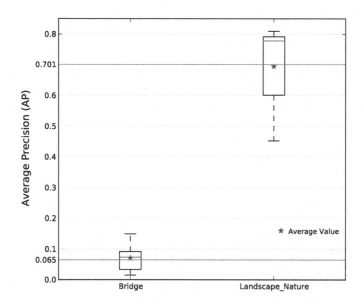

Fig. 1. Results obtained by participants of the 2011 ImageCLEF Visual Concept Detection and Annotation Task (in terms of interpolated average precision). The horizontal lines mark the results obtained by our own default BoW approach (i.e. "Bridge": $AP = 0.065$, "Landscape_Nature": $AP = 0.701$).

ImageNet. As a second training corpus we use sample images from the Image Net project. We map the ImageCLEF visual concept "Bridge" to the ImageNet synset "Bridge, span", which contains at the time of the experiments 1,598 images. The concept "Landscape_Nature" is mapped to the synset "Landscape" comprising 76 images. Negative images are randomly sampled from the validation set of the "ImageNet Large Scale Visual Recognition Challenge 2010", which is available at the ImageNet website after registration. We vary the ratio between positive and negative samples in the training set (see Sect. 5). The BoW vocabulary is generated by clustering $800,000$ RGB SIFT descriptors randomly sampled from all training data. As before, performance is measured using iAP based on the ImageCLEF evaluation set.

MIRFLICKR. Finally, we wanted to know how classification of visual concepts performs when models are trained based on training data crawled from unreliable web resources. We use the MIRFLICKR-1M collection for training data acquisition. Especially, we use only Flickr user tag data as ground truth and thus ignore any manually generated annotations provided by the MIRFLICKR Retrieval Evaluation. All MIRFLICKR-1M tags are preprocessed to be lower case. We select a subset of 100,000 photos to form our training data. Positive sample selection is performed by sub-selecting these photos that are tagged with the term *bridge* as a single word ('bridge') as well as with the term occurring as a substring within a tag ('%bridge%'). Negative sample selection is based on random sub-sampling of all images that were not selected as positive samples. Again, we experiment with different sizes of negative sample sets. Similarly, positive and negative samples are selected for the term *landscape* ('landscape', '%landscape%'). Thus, we obtain two different training sets per concept. The BoW vocabulary generation is based on a random subset of 800,000 RGB SIFT descriptors sampled from all 100,000 training photos. Consistently with the other runs, we measure the performance in terms of iAP computed on the ImageCLEF evaluation set.

4.3   Training Data Cleansing

Naturally, as the Flickr user tag data is not intended to provide a reliable ground truth annotation for classification purposes, the training set annotations derived based on these data must be considered as noisy. In Table 1 a few samples for images taken from the MIRFLICKR-1M dataset are presented that have been tagged using one of the selected visual concepts. Clearly, none of the images actually depicts the concept. A mislabeling or misleading labeling with tags can have various reasons. In [5] the authors analyze different functions tags perform in collaborative tagging systems which holds as well for folksonomies such as Flickr. Among others, tags are used as organizational structure, e.g. the photographer wanted to group all photos he or she has taken when visiting the Golden Gate Bridge. As a matter of fact, the pictures do not necessarily depict the bridge at all. Another reason for a misleading tag can be observed when considering the picture in the third row of Table 1: the photographer might have actually stood on a bridge, so the tag rather describes his viewpoint than the actual concept. Moreover, a photo can be tagged with the term 'bridge' when actually showing only a single pylon or a small part of a bridge such as a rivet. Finally, a tag can be assigned with no visible relation to the depicted content as can be seen in the first and last example image in Table 1.

By means of data cleansing we intend to filter these images that actually show the concept from those that have been mislabeled. We use two different strategies for data cleansing: tag co-occurrence analysis and semantic tag analysis.

Tag Co-occurrence Analysis. In case of the first picture in Table 1 it seems rather unlikely that the tags 'downtown', 'cityscape' and 'skyscrapers' appear frequently together with the tag 'landscape'. We analyze the tag co-occurrences

Table 1. Mislabeled or misleadingly labeled images taken from the MIRFLICKR-1M dataset. Tags that have been used as query tags marked in **<u>bold</u>**.

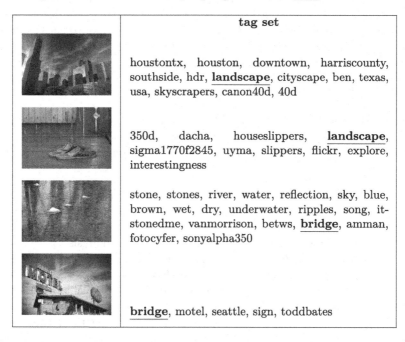

	tag set
	houstontx, houston, downtown, harriscounty, southside, hdr, **landscape**, cityscape, ben, texas, usa, skyscrapers, canon40d, 40d
	350d, dacha, houseslippers, **landscape**, sigma1770f2845, uyma, slippers, flickr, explore, interestingness
	stone, stones, river, water, reflection, sky, blue, brown, wet, dry, underwater, ripples, song, itstonedme, vanmorrison, betws, **bridge**, amman, fotocyfer, sonyalpha350
	bridge, motel, seattle, sign, toddbates

Table 2. Flickr user tags with the most frequent co-occurring tags in the training set.

Query tag	Most frequent co-occurring tags
'bridge'	'river', 'water', 'night', 'sky', ~~'nikon'~~, ~~'hdr'~~, 'city', 'reflection', 'blue', 'clouds'
'landscape'	'nature', 'sky', 'clouds', 'water', 'sunset', ~~'nikon'~~, 'trees', 'paisaje', 'blue', ~~'canon'~~

for each of the two query tags and filter these tags that appear most frequently together with the query tag. These ten most frequent tags are listed in Table 2.

By manual blacklisting, we drop these tags that are technical tags (e.g. the maker of the camera or the tag 'hdr') and further reduce the list to the 5 most frequent remaining tags (['river', 'water', 'night', 'sky', 'city'] and ['nature', 'sky', 'clouds', 'water', 'sunset']). Finally, we sub-select these images from the training set whose tag sets contain at least one of these most frequent non-technical tags co-occurring with the query tag.

Semantic Tag Data Analysis. In this approach, we analyze the semantic relationship between the tags in a tag set of an image and the visual concept that is to be classified. In order to do so, we first manually map the target visual concepts

"Landscape_Nature" and "Bridge" to DBpedia[7] entities (http://dbpedia.org/resource/Landscape and http://dbpedia.org/resource/Bridge). Next, we automatically map every tag from the tag dataset of a Flickr training image to a semantic entity referenced by a DBpedia URI by means of named entity recognition (NER). NER is the process of annotating textual information with semantic entities. It mainly consists of four steps:

- scanning the text (i.e. tag set) for potential named entity terms
- finding entity candidates for each tag
- defining the context of a tag (all other tags from the same tag set as well as the Flickr title of the image, if available, are used as context)
- ranking the entity candidates and determining the relevant candidate in the given context

The last two steps are only necessary in case of ambiguous tags, i.e. where more than one entity candidate was found for a given tag. The ranking algorithm of the entity candidates calculates two scores for every entity candidate of the ambiguous terms: co-occurrence analysis and link graph analysis.

The co-occurrence analysis uses context terms to determine the co-occurrence of an entity candidate and these terms. Therefore, a textual description of the entities is needed. As we use DBpedia entities, the according Wikipedia articles for the entities are used as descriptive texts. The link graph analysis uses the Wikipedia page link graph to find sub cliques representing the given context within this huge graph. This analysis step is based on the assumption that semantically related entities are linked over their Wikipedia articles. The graph analysis algorithm takes into account paths between the entity candidates of the context with a maximum length of 2. Both analysis algorithms calculate a score for every entity candidate and the weighted sum of both scores is used for entity mapping. The entity candidate with the highest score within the given context will be chosen as mapped entity (cf. [16] for further information on entity recognition).

Based on the successfully disambiguated tags we compute the number of tag entities that have a direct link to the entity of the target visual concept within the DBpedia. This score gives us an indicator of how strongly a tag set of a given image is related semantically to the respective concept and thus, how much related the image is. We use several lower bounds of relatedness and select only these images that have at least $\tau \in \{1, 3, 4\}$ tags whose entities exhibit a direct link.

5   Cross-dataset Classification Results

In this section we present and discuss the results that have been obtained using the aforementioned methods for training data acquisition and visual concept classification. Table 3 summarizes the results for the visual concept "Bridge".

---

[7] DBpedia: http://dbpedia.org/About.

Table 3. Results (interpolated average precision, iAP) for classification of the evaluation set using models for the ImageCLEF visual concept "Bridge" trained on different datasets obtained with different sampling strategies.

Training set configuration	iAP	#Pos	#Neg
ImageCLEF (Baseline): 'Bridge'	.065	105	7,895
ImageNet: 'Bridge, span'			
1Pos1Neg	.064	1,598	1,598
1Pos3Neg	.078	1,598	4,794
1Pos5Neg	.078	1,598	7,990
MIRFLICKR: 'bridge'			
1Pos1Neg	.059	874	896
1Pos2Neg	.063	874	1,792
MIRFLICKR: '%bridge%'			
1Pos1Neg	.061	1,289	1,311
MIRFLICKR: 'bridge'+tag-co-occ			
1Pos3Neg, $co-occ \geq 5$	.063	485	1,497
MIRFLICKR: '%bridge%'+NER			
1Pos1Neg, $\tau = 1$	.039	631	643
1Pos1Neg, $\tau = 3$	.048	192	195
1Pos1Neg, $\tau = 4$	.031	93	94

The results obtained by the classifier trained on the ImageNet dataset show an improvement of 20 % in interpolated average precision when compared to the baseline classifier. This can be explained by the significantly increased number of positive training samples: 1,598 images falling into the synset "Bridge, span" are used as positive training samples whereas only 105 positively labeled samples are available in the ImageCLEF training set. In fact, when considering this scale ratio, a much higher improvement would have been expected. We assume, however, that the visual variance in the concept "Bridge" is too large to be successfully trained. Second, we can observe that by increasing the number of negative samples better results were obtained. A saturation seems to be reached when using a set of negative samples that is approx. three times as big as the set of positives. To summarize, ImageNet seems to be congruent with ImageCLEF in terms of the implicit definition of the visual concept 'bridge'. As expected, the results we obtain using the noisy MIRFLICKR training dataset are below the ones obtained using manually labeled training data. The loss in accuracy, however, is rather low: The best result of $iAP = 0.063$ obtained using the uncleansed training data is only 3 % below the result obtained using the baseline classifier. This seems to be a moderate trade-off when comparing the effort needed to provide the different training datasets.

Unfortunately, the proposed methods for training data cleansing do not provide an improvement. While training data refinement using tag co-occurrence analysis ('bridge'+tag-co-occ) provides results similar to the results of the best performing uncleansed sample set, semantic data analysis ('%bridge%'+NER) does not help to improve classification accuracy. Instead, the performance drops even below the noisy classifier.

Named entity recognition is able to map tags such as 'PontNeuf' to the DBpedia 'bridge' entity, which should increase diversity. However, in our current prior data selection strategy, images containing 'PontNeuf' are only considered when the tag 'bridge' is also present, which is not necessarily the case. Thus, in future work, image selection should not be based solely on tags matching the target concept but also on tags matching linked tags provided by NER. Furthermore, since the NER is based on DBpedia and DBpedia itself is based on Wikipedia Infoboxes that are usually not provided for superordinate categories such as 'landscape' and 'bridge' only few direct links between the target concept and the tags can be identified. Entities of frequently co-occurring and tags (and probably frequently co-appearing concepts) such as 'bridge' and 'river' do not exhibit direct DBpedia links and are thus much harder to identify than with simple statistical co-occurrence analysis.

Analysis of the classification results for the concept "Landscape_Nature" (see Table 4) shows a slightly different picture. First of all, the ImageNet-based "Landscape"-classifier does not show superior performance as in the case of "Bridge, span". On the one hand, this can be attributed to the fact that the number of positive samples is significantly lower (i.e. 76). Again, better results were obtained when scaling up the number of negative samples. Furthermore, when looking at example images taken from the ImageNet "Landscape" set (see Fig. 2, top) we notice that compared to the ImageCLEF evaluation data (see Fig. 2, bottom), the visual variance is significantly lower. This example shows

Table 4. Results (interpolated average precision, iAP) for classification of the evaluation set using models for the ImageCLEF visual concept "Landscape_Nature" trained on different datasets obtained with different sampling strategies.

Training set configuration	iAP	#Pos	#Neg
ImageCLEF (Baseline): 'Landscape_Nature'	.701	1,362	6,638
ImageNet: 'Landscape'			
1Pos5Neg	.591	76	380
1Pos20Neg	.607	76	1,520
MIRFLICKR: 'landscape'			
1Pos1Neg	.638	1,704	1,831
1Pos2Neg	.646	1,704	3,662
1Pos5Neg	.647	1,704	9,155
MIRFLICKR: 'landscape'+tag-cooc			
1Pos3Neg, $co-occ \geq 5$	.662	1,059	3,441
MIRFLICKR: '%landscape%'+NER			
1Pos1Neg, $\tau = 1$	.645	1,528	1,641
1Pos2Neg, $\tau = 1$	.655	1,528	3,282
1Pos3Neg, $\tau = 1$	.649	1,528	4,923
1Pos1Neg, $\tau = 3$	.624	310	337

Fig. 2. Examples for images assigned to the synset "Landscape" in ImageNet (top) and to the concept "Landscape_Nature" in ImageCLEF (bottom).

a classical problem in dataset annotations: an annotator typically answers the question of whether or not a concept is *present* in an image (i.e. the Image-CLEF annotations), a user of an image search engine, however, presumably looks for *typical* images visualizing the concept (i.e. the ImageNet approach is to ask whether an image is a typical or wrong example). This discrepancy is only visible when manually analyzing the visual content of the datasets, which makes cross-dataset learning difficult if not infeasible. We state that the Image Net synset "Landscape" is not congruent with the ImageCLEF concept "Landscape_Nature": While the ImageNet set depicts typical landscape scenes, the "Nature" aspect is much stronger within the ImageCLEF data. Future runs therefore should try to enrich the ImageNet data by samples taken from synsets such as "Natural object" or "Geological formation" in order to increase the variance and meet the diversity of the evaluation set.

Similarly to the results obtained for the concept "Bridge", MIRFLICKR-based 'landscape'-classifiers achieved lower accuracy when compared to the baseline. Again, this is an expected outcome due to the assumed noise in the dataset. However, in this case, training data cleansing based on tag co-occurrence analysis ($iAP = 0.662$) outperforms the best classifier obtained without data cleansing ($iAP = 0.647$) by 2 %. Furthermore, differently from the previous results for the concept "Bridge", classifiers based on training data cleansed using semantic tag data analysis outperform those who are based on raw Flickr training data.

The loss in accuracy when comparing the best MIRFLICKR-based 'landscape'-classifier to the baseline classifier is at 6 %. Again, we believe this is an acceptable result, w.r.t. the labeling efforts required.

6   Conclusions

In this paper we have shown that cross-dataset classification of visual concepts is possible and can actually achieve equivalent and even better results than training performed on the same dataset. The performance of the classifier for the "Bridge" concept trained on the manually labeled ImageNet dataset outperformed the classifier trained on the ImageCLEF dataset. Furthermore, the loss in accuracy imposed by the weakly labeled MIRFLICKR training data seems acceptable when considering that no manual effort is required to assemble the data. While weakly labeled datasets such as Flickr do not intend to provide a "clean" ground truth for model training they still represent a very valuable resource for learning useful tag-image relationships simply due to their size.

We furthermore have shown that training data cleansing in fact can help to reduce noise in weakly labeled datasets and thus provides a promising first step towards reliable, inexpensive and unlimited training data. While this is an encouraging result, future research must proof whether our observations hold for a larger classification scenario with larger training data. Therefore, as a next step we intend to train models for all ImageCLEF concepts in order to provide a broader analysis. Besides, our work on training data cleansing was limited to improve reliability of positive samples sets. Likewise, negative sets need closer attention especially when realizing that the probability of users *not* tagging a specific concept is much higher than the probability of tagging it.

Important questions that need to be answered are the aforementioned lack of visual coherence between different datasets and the problem of understanding what a tag actually means with respect to the tagged image. User tag data can have various functions and not necessarily relates to the visual information in an image. The aim of data cleansing is to assemble visually coherent image content, as e.g. full view pictures of bridges rather than images where bridges are only depicted in parts, barely visible due to clutter, low resolution, and strong viewpoint variations or even not depicted at all. By thorough research of the correlation of different user provided tags, especially under consideration of the folksonomy aspects of communities such as Flickr, we intend to improve our data cleansing strategies by semantic analysis.

Acknowledgments. This work was supported in part by means of the German National Library of Science and Technology under the project AV-Portal.

References

1. Berg, T.L., Forsyth, D.A.: Animals on the web. In: Proceedings of the 2006 IEEE Conference on Computer Vision and Pattern Recognition, CVPR '06, vol. 2, pp. 1463–1470. IEEE, Washington, DC (2006)
2. Csurka, G., Dance, C.R., Fan, L., Willamowski, J., Bray, C., Maupertuis, D.: Visual Categorization with Bags of Keypoints. In: Workshop on Statistical Learning in Computer Vision, ECCV, pp. 1–22 (2004)

3. Deng, J., Dong, W., Socher, R., Li, L.J., Li, K., Fei-Fei, L.: ImageNet: a large-scale hierarchical image database. In: CVPR09 (2009)
4. Fergus, R., Perona, P., Zisserman, A.: A visual category filter for google images. In: Pajdla, T., Matas, J.G. (eds.) ECCV 2004. LNCS, vol. 3021, pp. 242–256. Springer, Heidelberg (2004)
5. Golder, S.A., Huberman, B.A.: Usage patterns of collaborative tagging systems. J. Inf. Sci. **32**, 198–208 (2006)
6. Guillaumin, M., Verbeek, J., Schmid, C.: Multimodal semi-supervised learning for image classification. In: IEEE Conference on Computer Vision & Pattern Recognition, CVPR 2010, pp. 902–909. IEEE, San Francisco (2010)
7. Leung, T., Malik, J.: Representing and recognizing the visual appearance of materials using three-dimensional textons. Int. J. Comput. Vis. **43**(1), 29–44 (2001)
8. Levin, A.: Unsupervised improvement of visual detectors using co-training. In: ICCV, pp. 626–633 (2003)
9. Li, L.J., Wang, G., Fei-fei, L.: Optimol: automatic online picture collection via incremental model learning. In: Proceedings of the 2007 IEEE Conference on Computer Vision and Pattern Recognition (CVPR) (2007)
10. Lowe, D.G.: Distinctive image features from scale-invariant keypoints. Int. J. Comput. Vis. **60**(2), 91–110 (2004)
11. Mark J. Huiskes, B.T., Lew, M.S.: New trends and ideas in visual concept detection: the mir flickr retrieval evaluation initiative. In: MIR '10: Proceedings of the 2010 ACM International Conference on Multimedia Information Retrieval, pp. 527–536. ACM, New York (2010)
12. Mbanya, E., Hentschel, C., Gerke, S., Liu, M., Ndjiki-nya, P.: Augmenting bag-of-words - category specific features and concept reasoning. In: CLEF (Notebook Papers/LABs/Workshops) (2010)
13. Nowak, S., Nagel, K., Liebetrau, J.: The clef 2011 photo annotation and concept-based retrieval tasks. In: CLEF (Notebook Papers/Labs/Workshop) (2011)
14. Schroff, F., Criminisi, A., Zisserman, A.: Harvesting image databases from the web. In: Proceedings of the 11th International Conference on Computer Vision (2007)
15. Snoek, C., Worring, M.: Concept-based video retrieval. Found. Trends Inf. Retr. **2**(4), 215–322 (2009)
16. Steinmetz, N., Sack, H.: Named entity recognition for user-generated tags. In: Proceedings of the 8th International Workshop on Text-based Information Retrieval, IEEE CS Press (2011)
17. Sun, A., Bhowmick, S.S.: Image tag clarity: in search of visual-representative tags for social images. In: WSM '09: Proceedings of the First SIGMM Workshop on Social Media, pp. 19–26. ACM, New York (2009)
18. Torralba, A., Fergus, R., Freeman, W.T.: 80 million tiny images: a large data set for nonparametric object and scene recognition. IEEE Trans. Pattern Anal. Mach. Intell. **30**(11), 1958–1970 (2008)
19. Vijayanarasimhan, S., Grauman, K.: Keywords to visual categories: multiple-instance learning for weakly supervised object categorization. In: Proceedings of the IEEE Conference on Computer Vision and Pattern Recognition (CVPR) (2008)
20. Zhang, J., Marszalek, M., Lazebnik, S., Schmid, C.: Local features and kernels for classification of texture and object categories: a comprehensive study. Int. J. Comput. Vis. **73**(2), 213–238 (2006)

Optimized SIFT Feature Matching
for Image Retrieval

Christian Schulze[✉] and Marcus Liwicki

German Research Center for Artificial Intelligence (DFKI),
Trippstadter Str. 122, 67663 Kaiserslautern, Germany
{christian.schulze,marcus.liwicki}@dfki.de

Abstract. Applying SIFT features for retrieval of visual data not only requires proper settings for the descriptor extraction but also needs well selected parameters for comparing these descriptors. Most researchers simply apply the standard values of the parameters without an adequate analysis of the parameters themselves. In this paper, we question the standard parameter settings and investigate the influence of the important comparison parameters. Based on the analysis on diverse data sets using different interest point detectors, we finally present an optimized combination of matching parameters which outperforms the standard values. We observe that two major parameters, i.e., *distmax* and *ratiomax* seem to have similar outcomes on different datasets of diverse nature for the application of scene retrieval. Thus, this paper shows that there is an almost global setting for these two parameters for local feature matching. The outcomes of this work can also apply to other tasks like video analysis and object retrieval.

Keywords: Local features · Matching · Parameter · SIFT · Image retrieval · Video retrieval

1 Introduction

In the last decade, Scale-Invariant Feature Transform (SIFT) gathered more and more attention for image and video recognition and retrieval [17,22]. Several toolkits implementing SIFT are nowadays available [20,21] and SIFT is used for various application domains [6,15,23,24]. Generally, people do not take much care in the selection of the parameters for extracting and comparing local feature descriptors. Often, only the standard values are used. Unfortunately, it might be the case that these standard values do not lead to satisfying results and different parameters could boost the system's performance significantly.

The purpose of this paper is to have better insights into the parameter settings for comparing SIFT descriptors. We will investigate the influence of the two parameters *distmax* and *ratiomax* on datasets of a diverse nature, i.e., images of objects and scenes with significant changes in scale, perspective, and illumination for the application to scene image and object retrieval. In particular, we perform

© Springer International Publishing Switzerland 2014
A. Nürnberger et al. (Eds.): AMR 2012, LNCS 8382, pp. 102–115, 2014.
DOI: 10.1007/978-3-319-12093-5_5

various experiments using the leave-one-out strategy as a grid-search on three datasets employing both SIFT-DoG and multiscale Harris-Laplace detection for selecting SIFT features. Furthermore, we analyze the strengths and weaknesses of local descriptors on different sample images.

The main contribution of this paper is the proposal of a general parameter setting for SIFT comparison parameters in the context of image retrieval. To the best of the authors' knowledge there is no similar work published where the focus is on the parameter setting. In our experiments, we observe that despite the diverse data sets and different detectors, the optimal parameter values for *distmax* and *ratiomax* were always quite similar. This work gives important insights into the comparison of frequently used SIFT features which might be applied by many researchers to boost their retrieval performance.

The rest of this paper is organized as follows. First, Sect. 2 gives an overview over related work. Second, Sect. 3 briefly describes local SIFT feature matching and the most important aspects for this paper. Subsequently, Sect. 4 describes the evaluation procedure and Sect. 5 presents the experimental results. Next, a deeper analysis of the experiments will be presented in Sect. 6. Finally, Sect. 7 concludes the paper and gives an outlook into future work.

2   Related Work

Since the introduction of the SIFT features by Lowe [11] for object recognition, they have been applied to several other vision problems, e.g., content based retrieval [8,13,18], concept detection [14,19], and object tracking [3–5,23]. For the application of SIFT features to these problems a nearest neighbor (NN) matching has to be performed. However, an NN matching does not always result in true positive matches. In fact, depending on the selected parameter values, several false positive matches may appear.

Lowe himself examined the distribution of true and false positive matches in [12], where he draws the conclusion that the ratio between best and second best match should not exceed $ratiomax = 0.8$, due to a strong increase of false positive matches beyond this value. Other researchers have proposed methods for improving the matching of SIFT features by grouping descriptors prior to matching or an increase of the matching speed by preselecting descriptors with an expectation for correctness, based on their local relation to other descriptors [1,2].

Another method of improving the amount of true positive matches is the refinement based on geometric constraints as a post processing step [7,12]. There, the consistency of translation and rotation for the computed matching keypoints is examined to select only plausible matches. This post processing can benefit from an improved initial matching, because a smaller number of matches is then to be examined and a smaller confusion regarding the underlying transformation of the keypoints exists. Nevertheless, a high amount of true positive matches as a precondition has to be achieved by the initial matching step and its settings.

However, most authors applying SIFT features for vision problems do not mention their setup for matching the descriptors, since this is generally not the focus of their work. Still, this is important information for anyone with the aim of setting up such a system.

3 Local Feature Matching

When local features are applied to retrieval or recognition problems, some way of comparing these features is needed. One possibility of achieving this is to compute a distance, i.e., Euclidean distance, between two sets of descriptors. Then the most plausible pairs of descriptors based on their distance value are selected. This selection then represents the set of corresponding descriptors (matches). To achieve good retrieval performance with local features it is therefore important to select matches with a high fraction of true positives.

Lowe proposed in [12] to apply a selection based on a parameter called *ratiomax* to obtain plausible matches. It is applied as a threshold to the ratio of the smallest and second smallest distances between each descriptor of set A and all descriptors of set B. If a predefined value is exceeded, none of the descriptors of set B are considered a match to the particular descriptor of set A. He observed that a *ratio* > 0.8 for the distances leads to an increasing number of false positive matches. This parameter also influences the match result regarding repetitive patterns, i.e., buildings and fences, since it discards matches between descriptors in the presence of further very similar ones.

Another method for selecting good matches is to use the descriptor distances themselves. The *distmax* parameter thresholds these, allowing only descriptors to match if their distance does not exceed it. To apply this parameter, the descriptors should be normalized to unit length before computing their distance. This provides a defined distance value in range of $[0, 1]$ and additionally supports the substitution of the applied distance measure. The main purpose for this parameter is to prevent matching of dissimilar descriptors and thus the occurrence of false positive matches.

These two parameters can also be combined for the selection of matches. For doing so, first the *distmax* parameter is applied followed by the *ratiomax* parameter. Finding a combination for these parameter values that provides maximized retrieval performance is the subject of the experiments presented in Sect. 5.

4 Evaluation Procedure

For evaluating the parameters described in Sect. 3 and selecting their proper combination, we inspect the mean average precision (MAP) of an retrieval system under their influence. The MAP is the mean of the average precision achieved for a number of queries Q and a well known performance measure for systems returning ranked sequences.

$$MAP = \frac{1}{Q} \sum_{q=1}^{Q} AP(q) \qquad (1)$$

The average precision for a single query q is defined as:

$$AP(q) = \frac{\sum_{i=1}^{N} P(i) \times rel(i)}{N_r}, with \ rel(i) = \begin{cases} 1 \text{ if image at } i \text{ is relevant} \\ 0 \text{ else} \end{cases} \quad (2)$$

where N is the number of retrieved images, $P(i)$ is the precision at rank i, $rel(i)$ indicates if the item at rank i is relevant to the query, and N_r is the number of relevant images for the particular query. The precision at rank i is given by nTP, the number of true positives up to i normalized by the rank.

$$P(i) = \frac{nTP}{i} \quad (3)$$

The MAP was computed by performing a complete retrieval within the dataset being evaluated. All images of the dataset were used as query image trying to retrieve all images of the series the query belongs to. For doing so, interest point detection and SIFT descriptor extraction were applied to all images with fixed settings. A nearest neighbor search was then performed to find the matches between query and reference descriptors using the GPU based implementation of siftgpu [21] on the image base. For computational efficiency reasons siftgpu uses the cosine distance to calculate the descriptor distances. Even though this is not the best performing distance measure for the comparison of SIFT descriptors [10], it is still widely used due to its efficient computability and introduces just an offset to the absolute MAP values in these experiments. The list of results is finally ranked according to the number of matches that have been found between the given query image and instances of the dataset. A successive refinement of the matches, as mentioned in Sect. 2, was not done as it would influence the effect of the matching parameters regarding retrieval performance. But the application of further post processing to the set of matches can be expected to benefit from an improved initial selection. To compute the average precision for a single query, the first N_r images of the result were considered exclusively.

Since only two parameters have to be optimized, a grid search seems feasible in this case. Even though the optimization will require more time, a possible presence of local maxima or plateaus might lead to sub optimal results using other procedures like, i.e., evolutionary algorithms or simulated annealing. Applying the grid search ensures a complete view on the investigated parameters without possibly missing the global maximum.

5   Experimental Results

This section presents the results of the grid search experiments for finding an optimal combination for the match parameters *distmax* and *ratiomax*. Table 1 gives an overview on the experiments that are conducted in this section. A deeper analysis of the results will be presented in Sect. 6.

Table 1. Grid search experiments for different data sets and interest point detectors.

Data Set	DoG detection	Harris detection
Holiday	X	X
Inbekio	X	X
Inbekis	–	X

5.1 Holiday Dataset

The Holiday dataset has been assembled and provided by Jegou et al. [9] for the purpose of image search evaluation. It consists of 1491 high resolution images that show fairly different content, ranging from food closeups to landscape and underwater imagery (Fig. 1). The data contains several series of different motives; either objects, scenes, or concepts. The number of pictures for each image series varies between 2 to 10 (most frequently 2 or 3). Originally, Jegou defined only one image of each series as the query image and the other to appear as reference images. In order to get more significant results, however, we will use each of the images of a series as query image in the leave-one-out manner.

Fig. 1. Example images of the Holiday dataset for the variety of visual content.

Despite the variety in visual content the Holiday dataset is specifically designed for the application of local features, such as the SIFT descriptor. It has been observed in previous experiments that adding, e.g., global color information via a weighted sum fusion will degrade the retrieval performance on this dataset. Figure 2 shows some example image series where global image descriptions are likely to fail due to significant changes in scale, perspective, and illumination.

The Holiday dataset, therefore, is well suited for experimenting with different interest point detectors to determine the parameters for nearest neighbor matching. See Fig. 3 for applied interest point detection on Holiday example images. Furthermore the intention of this experiment was to see whether there exists an influence on the performance when using other interest point detectors.

As can be seen in Fig. 4, the performance variation for a given parameter combination for both IP detectors is quite similar. Most important is to notice that the optimal setting for the *distmax* and *ratiomax* parameters for both investigated detectors is nearly at the same position. Exact values for the parameters as well as the achieved performance are presented in Table 2. Another aspect to

Fig. 2. Example images of the Holiday dataset for strong scale change (top), change of perspective (middle) and illumination variance (bottom).

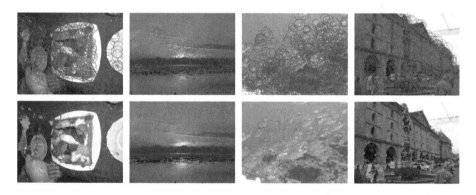

Fig. 3. Interest point detection: DoG (top) and multiscale Harris (bottom) applied to example images of Holiday.

observe is the retrieval performance that can be achieved when only applying the *ratiomax* parameter, which is the case for *distmax* = 1.0.

In fact, a local plateau can be observed in the region of *distmax* = 0.7 and *ratiomax* = 0.7 which appears more prominent for the DoG detector.

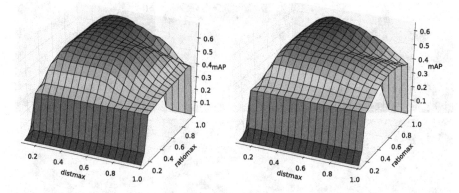

Fig. 4. Performance grid for dataset Holiday, using DoG (left) and multiscale Harris (right) interest point detection with siftgpu for matching

This plateau could possibly pose an issue to greedy search algorithms as mentioned in Sect. 4.

These observations lead to the following new questions:

1. Will the parameter values for the global optimum be the same for both interest point detectors on another dataset?
2. Where is the location of the global optimum on another dataset?
3. Since Holiday images are a mixture of object and scene dedicated images, does the global optimum change for specialized datasets?

To answer these questions, the experiments in the following Sections (5.2, 5.3) have been conducted.

5.2 Inbeki Object Dataset

For the verification of the experimental results performed on the Holiday data, images with very different content and appearance were chosen. The Inbeki Object dataset consists of 725 images in 50 series with a minimal series size of 10. This dataset and the one described in Sect. 5.3 as well as the corresponding ground truth was created by practitioners that utilize image search for their daily work. Due to copyright restrictions these images may not be published and therefore will only be described verbally. For a publicly available dataset please refer to the Holiday dataset of Sect. 5.1.

One of the main differences to the Holiday dataset is the intention under which the Inbeki Object set was assembled. While for the Holiday image series only few samples were selected that pose a challenging retrieval problem regarding scale, perspective and illumination, the Inbeki Object images were collected for arbitrary objects with partially strong changes in perspective. These objects are mostly placed in indoor settings and cover items like wall paintings, pillows, lamps, furniture as well as clothing and jewelery. Some of these objects are poorly textured or even non-rigid which makes this a challenging dataset for interest point detectors. Furthermore, there is a strong variation in image resolution. These differences are born out when, in opposition to the Holiday dataset where adding a global color feature decreases retrieval precision, a performance improvement can be observed for the Inbeki Object dataset when adding color information.

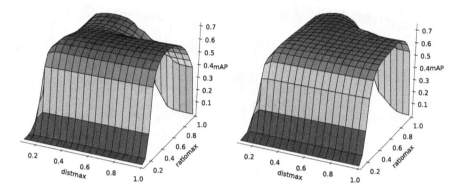

Fig. 5. Performance grid for dataset Inbeki Object, using DoG (left) and multiscale Harris (right) interest point detection with siftgpu for matching

To answer the questions raised in Sect. 5.1, the same retrieval optimization procedure was applied, leading to the results in Fig. 5. As is the case for the results in Fig. 4, a strong similarity between the two plots can be seen here. When compared to the previous plots for the Holiday dataset, these show a different appearance. Still, the optimal matching parameter setting is similar for both detectors and also comparable to the Holiday dataset. This holds especially true for the *distmax* parameter. It can also be seen that for this dataset large regions of similar performance exist which are not strongly affected by either one or both parameters. This seems to correspond to the statement made above regarding the strength of local and global features for this dataset.

5.3   Inbeki Scene Dataset

The Inbeki Scene dataset was examined to confirm the previous results. Here, the grid search using DoG detection was not performed, as the results on the datasets tested before imply that the difference between the detectors can be neglected. Furthermore, applying DoG detection results in a larger number of detected keypoints (see Fig. 3) compared to Harris-Laplace detection, which makes the optimization more time-consuming.

This dataset was composed for the purpose of indoor scene retrieval and consists of 563 images in 36 series where each series contains more than 10 images in high resolution. These images show a strong variation in their foreground (persons) while the background undergoes changes in perspective and visible area. The images of a series mostly show a visual overlap of the background but some are not visually connected to the other images of the series. Due to the content structure of this dataset the application of local features is indicated. As for the Holiday dataset (see Sect. 5.1) the retrieval on this data does not benefit from addition of global features.

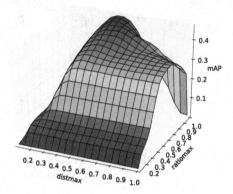

Fig. 6. Performance grid for dataset Inbeki Scene, using multiscale Harris interest point detection and siftgpu for matching

When applying the grid search to this dataset, one can again notice a different appearance of the performance surface compared to the previous ones (Fig. 6). Nevertheless, the global optimum is located at the same position as for the other datasets. It can also be noticed that for the Inbeki Scene data a local maximum exists which is not as distinctive as for the Holiday data.

Table 2. Results for optimized performance (MAP) and the according values of *dist-max (d)* and *ratiomax (r)* as well as the performance achieved by the Lowe proposed ratiomax value for the conducted experiments. Additionally, the loss of performance for a fixed optimized parameter setting is given.

Dataset	Optimized dmax, rmax	Opt. MAP	Loss to opt. d = 0.45, r = 0.95	MAP @ r = 0.8	Loss to opt. @ r = 0.8
Holiday DoG	0.45, 0.9	69.94	0.007	41.22	28.72
Holiday Harris	0.50, 0.85	64.21	0.018	42.62	21.59
Inbeki Object DoG	0.45, 0.5	70.94	0.008	59.62	11.32
Inbeki Object Harris	0.45, 0.95	73.30	–	69.32	3.98
Inbeki Scene Harris	0.50, 0.95	46.49	0.002	32.58	13.91

6   Analysis

As the plots in Figs. 4, 5, and 6 show, a consistent region of optimal performance exists for all tested datasets and interest point detectors. For the Inbeki Object dataset, a large plateau can be observed which most likely originates from a saturation effect due to the difficulties for interest point detectors and local features on this dataset as mentioned in Sect. 5.2. For a better insight on the effect of the parameter settings, we will show some examples of corresponding and non-corresponding images from the Holiday dataset which were matched using the values ($distmax = 1.0$, $ratiomax = 0.8$) and ($distmax = 0.45$, $ratiomax = 0.9$). Note that corresponding images are images which share visual content at least partially, while non-corresponding ones are images which do not have any visual overlap.

Comparing the detected matches of corresponding images in Figs. 7 and 8 only few differences can be observed at first glance. However, taking a closer look reveals that more true positive matches appear for the optimized setup

Fig. 7. Examples of match results of corresponding images for $distmax = 1.0$ and $ratiomax = 0.8$

Fig. 8. Examples of match results of corresponding images for $distmax = 0.45$ and $ratiomax = 0.9$.

for the example on the right side of the figures. This results from the increased value for the *ratiomax* parameter leading to a better matching in the presence of repetitive patterns, like buildings in this example.

The underwater scene instead shows fewer false positive matches, which are due to the application of the *distmax* parameter avoiding matching of descriptors that are not very similar. On the other hand, the two examples on the left show cases were the optimized parameters introduce both, true and false positive matches. However, such cases do not occur very frequently for corresponding image content and thus the overall performance is increased.

Match results for the non-corresponding images are shown in Figs. 9 and 10, where all matches found are considered as false positives. It can be observed that the optimized setting gives fewer matches and therefore the chance for a better ranking of the retrieved images is increased. Here again, the optimized *distmax* parameter has a beneficial influence as it prevents dissimilar matches prior to *ratiomax* evaluation.

Fig. 9. Examples of match results for non-corresponding image content with *distmax* = 1.0 and *ratiomax* = 0.8.

Fig. 10. Examples of match results for non-corresponding image content with *distmax* = 0.45 and *ratiomax* = 0.9.

In addition to the examples shown in Figs. 9 and 10, several cases exist where applying the optimized parameters does not result in any false positive matches for non-corresponding content. Instead, using the standard setting, one to two false positive matches are found.

Of course, there are also cases where the optimized setting will produce match results that appear not as plausible as the ones created with the standard setting, but given the performance increase that can be achieved when averaged over a number of queries, an improved match result is occurring more often.

From the observations made here and in Sect. 5 it can be presumed that the increase in retrieval performance through optimized match parameters does depend on the visual data being compared, but significant improvements are possible. Furthermore, there seems to exist a consistent parameter combination for improved retrieval performance which is independent of the data being analyzed and the applied interest point detector.

Additionally these experiments show that a combination of the DoG and multiscale Harris-Laplace interest point detector can be done without the need to treat the corresponding descriptors differently when selecting matches.

7   Conclusion and Future Work

In this paper we took a closer look into the parameters for matching local features. In particular, the *distmax* and *ratiomax* were investigated on datasets of quite diverse nature. We have found that there seems to be a general global optimum for the parameter values $distmax = 0.45$ and $ratiomax = 0.95$ which was observable on all three datasets and even when using different detectors.

Note that our experiments have only been performed on image data. However, the outcomes could be also applied to video analysis [19] and document retrieval [16]. In fact, we have already received promising results in preliminary experiments on video data using the optimized parameter setting.

In future we plan to further analyze the optimization strategies for the parameter settings of SIFT, because a grid-search is usually very time consuming. In preliminary experiments we observed that greedy methods and genetic algorithms seem to fail and stop early at local maxima. Furthermore, a closer analysis of the influence of the parameters on video data and image sequences will be performed.

Acknowledgment. This research was funded by BMBF grant INBEKI 13N10787.

References

1. Alhwarin, F., Ristić–Durrant, D., Gräser, A.: VF-SIFT: very fast SIFT feature matching. In: Goesele, M., Roth, S., Kuijper, A., Schiele, B., Schindler, K. (eds.) Pattern Recognition. LNCS, vol. 6376, pp. 222–231. Springer, Heidelberg (2010)
2. Alhwarin, F., Wang, C., Ristić-Durrant, D., Gräser, A.: Improved SIFT-features matching for object recognition. In: Proceedings of the 2008 International Conference on Visions of Computer Science: BCS International Academic Conference, VoCS'08, pp. 179–190. British Computer Society, Swinton (2008)

3. Brox, T., Rosenhahn, B., Gall, J., Cremers, D.: Combined region- and motion-based 3d tracking of rigid and articulated objects. IEEE Trans. Pattern Anal. Mach. Intell. **32**(3), 402–415 (2010)
4. Eckmann, M.: Sifting for better features to track: exploiting time and space. Lehigh University (2007)
5. Fazli, S., Pour, H.M., Bouzari, H.: Particle filter based object tracking with SIFT and color feature. In: Proceedings of the 2009 Second International Conference on Machine Vision, ICMV '09, pp. 89–93. IEEE Computer Society, Washington, DC (2009)
6. Garz, A., Sablatnig, R., Diem, M.: Layout analysis for historical manuscripts using SIFT features. In: Proceedings of the 2011 International Conference on Document Analysis and Recognition, ICDAR '11, pp. 508–512. IEEE Computer Society, Washington, DC (2011)
7. Islam, S.M.S., Davies, R.: Refining local 3d feature matching through geometric consistency for robust biometric recognition. In: 2009 Digital Image Computing: Techniques and Applications, pp. 513–518 (2009)
8. Jain, A.K., Lee, J.E., Jin, R., Gregg, N.: Content-based image retrieval: an application to tattoo images. In: 2009 16th IEEE International Conference on Image Processing (ICIP), pp. 2745–2748. IEEE (2009)
9. Jegou, H., Douze, M., Schmid, C.: Hamming embedding and weak geometric consistency for large scale image search. In: Forsyth, D., Torr, P., Zisserman, A. (eds.) ECCV 2008, Part I. LNCS, vol. 5302, pp. 304–317. Springer, Heidelberg (2008)
10. Ling, H., Okada, K.: EMD-L_1: an efficient and robust algorithm for comparing histogram-based descriptors. In: Leonardis, A., Bischof, H., Pinz, A. (eds.) ECCV 2006. LNCS, vol. 3953, pp. 330–343. Springer, Heidelberg (2006)
11. Lowe, D.G.: Object recognition from local scale-invariant features. In: Proceedings of the International Conference on Computer Vision, ICCV '99, vol. 2, pp. 1150–1157. IEEE Computer Society, Washington, DC (1999)
12. Lowe, D.G.: Distinctive image features from scale-invariant keypoints. Int. J. Comput. Vision **60**(2), 91–110 (2004)
13. Mikolajczyk, K., Schmid, C.: A performance evaluation of local descriptors. IEEE Trans. Pattern Anal. Mach. Intell. **27**(10), 1615–1630 (2005)
14. Mühling, M., Ewerth, R., Freisleben, B.: On the spatial extents of SIFT descriptors for visual concept detection. In: Crowley, J.L., Draper, B.A., Thonnat, M. (eds.) ICVS 2011. LNCS, vol. 6962, pp. 71–80. Springer, Heidelberg (2011)
15. Sivic, J., Zisserman, A.: Video Google: a text retrieval approach to object matching in videos. In: Proceedings Ninth IEEE International Conference on Computer Vision (ICCV), vol. 2, pp. 1470–1477 (2003)
16. Smith, D., Harvey, R.: Document retrieval using image features. In: Proceedings of the 2010 ACM Symposium on Applied Computing, SAC '10, pp. 47–51. ACM, New York (2010)
17. Snoek, C.G.M., Worring, M.: Concept-based video retrieval. Found. Trends Inf. Retr. **2**(4), 215–322 (2009)
18. Ulges, A., Schulze, C.: Scene-based image retrieval by transitive matching. In: Proceedings of the 1st ACM International Conference on Multimedia Retrieval, ICMR '11, pp. 47:1–47:8. ACM, New York (2011)
19. Ulges, A., Schulze, C., Koch, M., Breuel, T.M.: Learning automatic concept detectors from online video. Comput. Vis. Image Underst. **114**(4), 429–438 (2010)
20. Vedaldi, A., Fulkerson, B.: VLFeat: An open and portable library of computer vision algorithms (2008). http://www.vlfeat.org/

21. Wu, C.: SiftGPU: a GPU implementation of scale invariant feature transform (SIFT) (2007). http://cs.unc.edu/~ccwu/siftgpu
22. Zheng, Q.-F., Wang, W.-Q., Gao, W.: Effective and efficient object-based image retrieval using visual phrases. In: Proceedings of the 14th Annual ACM International Conference on Multimedia, MULTIMEDIA '06, pp. 77–80. ACM, New York (2006)
23. Zhou, H., Yuan, Y., Shi, C.: Object tracking using SIFT features and mean shift. Comput. Vis. Image Underst. **113**(3), 345–352 (2009)
24. Zhou, X., Zhuang, X., Yan, S., Chang, S.-F., Hasegawa-Johnson, M., Huang, T.S.: SIFT-bag kernel for video event analysis. In: Proceedings of the 16th ACM International Conference on Multimedia, MM '08, pp. 229–238. ACM, New York (2008)

Representativeness and Diversity in Photos via Crowd-Sourced Media Analysis

Anca-Livia Radu[1,2,3]([☒]), Julian Stöttinger[1], Bogdan Ionescu[2],
María Menéndez[1], and Fausto Giunchiglia[1]

[1] DISI, University of Trento, 38123 Povo, Trento, Italy
{radu,julian,menendez,fausto}@disi.unitn.it
[2] LAPI, University "Politehnica" of Bucharest, 061071 Bucharest, Romania
bionescu@alpha.imag.pub.ro
[3] Military Technical Academy, Bucharest, Romania

Abstract. In this paper we address the problem of user-adapted image retrieval. First, we provide a survey of the performance of the existing social media retrieval platforms and highlight their limitations. In this context, we propose a hybrid, two step, machine and human automated media analysis approach. It aims to improve retrieval relevance by selecting a small number of representative and diverse images from a noisy set of candidate images (e.g. the case of Internet media). In the machine analysis step, to ensure representativeness, images are re-ranked according to the similarity to the "most common" image in the set. Further, to ensure also the diversity of the results, images are clustered and the best ranked images among the most representative in each cluster are retained. The human analysis step aims to bridge further inherent descriptor semantic gap. The retained images are further refined via crowd-sourcing which adapts the results to human. The method was validated in the context of the retrieval of images with monuments using a data set of more than 25.000 images retrieved from various social image search platforms.

1 Introduction

The people's desire to try to preserve important moments in their lives has led to a fast and continuous growing of online personal digital image collections and to an intrinsic desire for the automatic indexing and searching of these media assets, the so called image retrieval. At the same time, the high number of public image search engines providing an image retrieval system cannot keep up with the tremendous number of available online images. They have certain limitations since most of them rely on keywords-based and/or GPS-based search. Keywords-based image search is inspired by text search techniques, relying on the images' surrounding text (e.g. comments, titles, tags or other description of the images) [1]. But the surrounding text is not particularly accurate, mostly because people usually tag all pictures from a collection with a particular word. On the other hand, GPS-based image search may also lead to bad results, since

A. Nürnberger et al. (Eds.): AMR 2012, LNCS 8382, pp. 116–130, 2014.
DOI: 10.1007/978-3-319-12093-5_6

geo-tagging is not always accurate and typically doesn't refer to the position of the query object, but to the position of the photographer. Thus, most of the time, when we type a keyword or a set of GPS coordinates into an image search engine, images are not perfectly returned in a descending order of their representativeness or, even worse, some of them are not related to the subject at all.

Research efforts have been made towards developing re-ranking techniques in order to solve the above-mentioned search limitations. Most of the existing re-ranking methods only try to refine the retrieved images. Though, users are interested in taking possession of not only accurately representative images, but also diverse images that can depict the query object in a comprehensive and complete mode, covering different aspects of the query. In this respect, we propose the following approach: given a query term and a set of GPS-coordinates (latitudes and longitudes) we aim to select a small set of most representative and diverse images that image search engines can provide us. The procedure consists in an automatic media image analysis that uses only visual information in images. In order to finally maximize results' refinement, a crowd-sourcing process is performed, since perfectly translating the query-text or query-coordinates to a semantic meaning is yet unreachable. Our method assumes that among the downloaded images, a big amount depict the subject in a clear way. Another assumption that we make is that a big part of the best images retrieved from the image search engines are among the first returned.

The remainder of the article is organized as follows: Sect. 2 discusses related work and situates our approach accordingly. Section 3 analyses the limitations of the existing image search engines. Section 4 presents the proposed approach for selecting a representative and diverse set of images. Experimental validation is presented in Sect. 5, while Sect. 6 concludes the paper.

2 Related Work

There are a number of specific fields like re-ranking, relevance feedback and automatic geo-tagging that are related to our work. The papers concentrating on *re-ranking* are the closest to our work. For instance, the approach in [2] builds clusters of images and then ranks them according to ratio of inter-cluster distance to intra-cluster distance and according to cluster connectivity. Inside each cluster, images are ranked according to the similarity with images from the same cluster and dissimilarity to random images outside the cluster. A similar approach [3] considers to be representative images that are most similar with images in the same cluster, but different to images in an external class build by using, e.g. a keyword-based search for word "test" on *Flickr*. In [4], the proposed method measures the "visual consistency" among the images and re-rank them on the basis of this consistency. In other words, a probabilistic model is learned in an unsupervised manner and given the learnt model, the likelihood ratio is computed for each image. This ratio will be used to rank all the images in the data set.

Relevance feedback (RF) is another tangent domain to our work to which a high interest has been given in recent years. A typical RF scenario can be formulated as follows: for a certain retrieval query, an user provides feedback by marking the results as relevant or non-relevant. Then, using this information, the system re-computes a better representation of the information needed. One of the earliest RF algorithms attempts to update the query features by adjusting the position of the original query in the feature space according to the positive and negative examples and their associated importance factors [5]. More recently, machine learning techniques have been introduced to RF, e.g. Support Vector Machines, classification trees or boosting techniques. In this case the RF problem can be formulated as a two-class classification of the negative and positive samples.

The availability of GPS-enabled cameras triggered the interest in validating *automatic geo-tagging techniques*, i.e. providing automatic GPS localization of recorded media using text and visual features (see MediaEval benchmarking - Placing task [6]). In [7], a combination of textual and visual features are employed for trying to decide which of ten landmarks in a certain city is the subject of an image. In order to do that, for each of the ten landmarks, a classifier is build with positive examples images of a given landmark and negative examples images from other landmarks. A similar approach is presented in [8] in which a prediction of geographic location using a nearest-neighbour classification visual features-based is employed. The method is limited to a sub-set of images tagged with at least one area name.

The approach proposed in this paper is at the intersection of these three fields, but goes beyond the state of the art along these dimensions:

- **diversity:** adding the diversity requirement to the existing re-ranking methods;
- **enabling better crowd-sourcing:** providing a quality starting point to a future extended crowd-sourcing study;
- **representativeness:** providing users not only with correspondence between images and locations, but also with a set of representative and diverse images for each location in order to obtain a clear and complete understanding.

The novelty of our method with respect to the three adjacent fields is also enhanced by its cultural aspect with a practical use: selecting a small number of representative and diverse images for a high number of Italian monuments may prove of high interest in tourist world. In the same time, Sect. 3 is the first published study about possible retrieved mistakes in search engines which emphasizes the drawbacks of the existing textual and location image search platforms.

3   Image Representativeness

In this section we present a detailed study of the actual performance of the existing social image retrieval platforms. For exemplification, we have selected three of the most famous: *Picasa, Flickr* and *Panoramio*. As case study we use the application domain of this work - the search of Italian monuments. We search for images of monuments using both keywords (e.g. name of the monument) and GPS tags (retrieved from *Wikipedia*). For practical reasons we limit to retaining only the first 100 retrieved images for each of the three image search engines. At a simple inspection of the retrieved images, we can sustain that accurate pictures of an object and/or place can be made both during day and during night and also can use different viewpoints, scaling and orientations, as long as they depict the subject in a clear and distinctive way. Regardless the accurate text tags and GPS localization, the search engines tend to fail due to the following situations:

| (a) | (b) | (c) | (d) | (e) | (f) |

| (g) | (h) | (i) | (j) | (k) | (l) |

Fig. 1. Exemplification of possible occurring mistakes (letters correspond to the cases presented in the text). Image sources: *Picasa, Flickr* and *Panoramio*

(a) **People in focus:** Pictures with people visiting the inside or the outside of the monument, but with no monument in it (Fig. 1a). This mistake is caused by the inaccurate images' surrounding text made by people who tag an entire album photo with a particular word. For this reason, three different images, one containing a representative view of the monument, the second one a person totally occluding the monument or standing close of it and a third one a person inside of the monument will probably receive same tags.

(b) **People in front:** Pictures with people in front of the monument (Fig. 1b). People appear in images depicting monuments because of the inaccurate images' surrounding text and also because it is difficult to capture monuments with no person on the image. Some monuments are very famous, others are less famous, but they all attract visitors.

(c) **Unfocused images:** Pictures with a far sight of the monument (Fig. 1c). When the photographer is taking a distant shot or a high altitude shot of a monument, he can add or modify the GPS coordinates and most of the times he will geo-tag the image with the GPS coordinates of the monument, even though the image depicts only a far sight of the monument. Another reason for this mistake is that images are often tagged with the GPS coordinates or the name of the most famous place/object that appears in the image, even when the image is not focused on it.

(d) **Reproductive art objects:** Pictures with drawings, posters, paintings or sculptures of the monument (Fig. 1d). Most likely, images that contain reproductive objects of a monument will be assigned the name of the monument among other tags, causing a confusion and thus this type of mistake to occur.

(e) **Surrounding or inside map:** Pictures with inside maps of the monument or maps with the region where the monument is located (Fig. 1e). Images containing maps of a place are usually tagged with the name of that place and thus retrieved when searching a monument by its name.

(f) **Advertising:** Books, postcards, articles, tickets or other similar writings or objects related to the monument (Fig. 1f). Inevitably, most of the images with writings about a monument will be accompanied by surrounding text containing words related to the monument.

(g) **Inside pictures:** Pictures of monuments' interior (Fig. 1g). The first cause for this mistake involves people usually tagging all images taken during a visit to a monument with one word, regardless the images' content. The second reason refers to the fact that both the photographer and the inside of the monument being photographed have the same location and will be correctly geo-tagged with the GPS coordinates of the monument, but incorrectly retrieved when searching for images with the monument.

(h) **Detailed images:** Pictures with only small parts of the monument (e.g. statues, different objects from inside or outside of the monument, etc.; see Fig. 1h). The cause of this type of mistake is the same as for the previous case.

(i) **Accommodation and restaurants:** Pictures with hotels and restaurants to go to during the monument visit (Fig. 1i). This mistake appears because images with hotel rooms or restaurants in the vicinity of monuments are tagged with names of monuments since the owners make use of their names to attract customers, because it can happen for some hotels or restaurants to be close enough to the monument to fit into the search radius or because the atmosphere of people's vacation is highly influenced by both hotels and restaurants and people upload images containing them but add tags containing the name of the monuments.

(j) **Other places:** Pictures with different places close to a certain monument but with no monument in it (Fig. 1j). The two main reasons of this mistake are related to the inaccurate tagging of places located or not in the vicinity of the monument caused, firstly, by people who just simply make use of the monument's name to describe their photos and, secondly, by photographers placed on top of the monument while shooting down or away.

(k) **Other monuments:** Other monuments located in the same region or not, some of them belonging to the same category of monuments (Fig. 1k). The reasons for this type of mistake can be the coincidence of names between two different monuments or the use of the name of the monument of interest for describing, comparing other monuments.

(l) **Meaningless objects:** Pictures with objects that have absolutely nothing to do with the monument (Fig. 1l). This is the most general error that can occur when searching for images with monuments and it is based on all the causes previously described: inaccurate text-based image search because of the poor correlation that exists between surrounding text and the visual image content and also inaccurate image search by GPS coordinates since manually or automatically geo-tagging do not consider the content of the image.

4   The Proposed Approach

To address these issues and to qualitatively refine the results, we use the following approach:

- extract from the Internet the target photos for a certain query. We aim to select a set of candidate photos;
- perform a fully automated image analysis whose goal is to select a subset of most significant and in the same time diverse images;
- perform the final refinement on the remaining set of photos via crowd-sourcing.

Each step is presented in the sequel.

4.1   Selecting the Monuments' Photos

Following the study in Sect. 3, we further try to define and classify monuments from most prominent to least prominent and then to establish a connection between these attributes and each type of mistakes listed in Sect. 3. We assume that the more popular the monument is, the more hits will be returned by Google. These results are however only quantitative. To have also an idea of their relevance, we assess also the number of correct images returned by specialized search engines (e.g. Picasa, Flickr, Panoramio). Figure 2 shows this information. Concerning the prominence, we found that the least three prominent monuments in the list are *"Victor Emanuel II"*, *"Aselmeyer Castle"*, *"Papal Archbasilica of St. John Lateran"* with 2.270, 6.970, 16.000 number of returned images (mainly due to the restricted local history they are connected to), while the most common three monuments are *"Two Towers"*, *"New Gate"*, *"Juliet's House"* with 361.000.000, 279.000.000, 98.800.000 number of returned images. The most handy assumption that someone can do is that most prominent monuments will bring almost only representative images when searching them by keywords. Before trying to adopt or reject such an assumption, lets try first to make a simple analysis of the monuments' names. It easily leads us to considering three categories of names for prominent corresponding monuments on *Google*:

- names that contain common words which coincide with aspects/objects from real life. In this case, the returned images when searching monuments by keywords will present many shots of monuments' homonyms and less shots of the real monuments. The most handy examples are *"Two Towers"* (many pictures with shots from *"The Lord of the Rings: The Two Tower"* movie or pictures with other two towers) and *"Saint Mary of the Flower"* (pictures with statues of Saint Mary or different flowers);

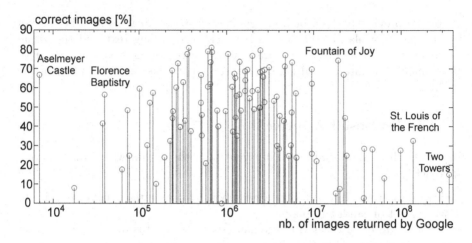

Fig. 2. Monuments' prominence: percentage of correct images returned by social image sharing platforms vs. number of images returned by Google (on a log scale; each vertical line corresponds to a monument).

- names that are not sufficiently precise so that the returned images to make reference to a single monument, but to worldwide monuments. The simplest examples are *"Cathedral Square"* (pictures with different cathedrals' squares) and *"Fountain of Joy"* (pictures with several fountains or with people having fun). For these two examples, there are, definitely, many cathedrals which have frontal squares, or many fountains that gather people and joy. These cases of ambiguity lead to inevitable random results among the returned images;

- names that mostly contain simple distinctive nouns that are known worldwide. As example we can mention *"Pantheon"* and *"Verona Arena"*. In both cases, when searching images by keywords, we obtain a relatively high number of representative images. It is obvious that the number of images returned for prominent monuments is larger than the less prominent monuments, but the aspect that interests us in the first instance is the quality of images. A conclusion on this aspect is given by the above classification. For the first two categories, the quality of the returned images is not sufficiently high. On the other hand, for prominent monuments in the third category, the returned images are not just many but also of high quality. Differently from the prominent monuments, some names lead to classify the corresponding monuments in low prominent for two main reasons:

- names contain complex words combination and the probability of having images tagged with all these words is small. The most handy examples are *"Basilica of Our Lady Help of Christians"* and *"Santa Maria della Spina"*;

- names contain simple, but distinctive nouns that are not very known for people. We can mention as examples *"Aselmeyer Castle"*, *"Basilica di San Zeno"* and *"Garisenda Tower"*.

In conclusion, the fact that a monument is more prominent does not guarantee the relevance of the results (see *Two Towers* in Fig. 2). More accurate are the results for the least prominent ones since the reduced number of uploaded images tend to be more representative.

4.2 Automated Image Analysis

The proposed method tries to select from a given set S of N noisily ranked images returned by social image search engines (search performed by keywords and GPS-coordinates) the best representative images that will present the query subject in a diverse manner. The following mechanism is employed:

- **Step 1**: The first step consists of determining, for each of the N images in set S, a description of the underlying visual content. Considering the application of our approach (retrieval of monuments pictures), we use color and feature descriptors for representing images by 92 dimension feature arrays as a combination between Colour Naming Histogram [9] (11 components) and Histogram of Gradients (81 components). Then, to assess image similarity, we compute the Euclidean distance between their corresponding feature arrays. Further, we construct a Synthetic Representative Image Feature (SRI) by taking the average of all distances.
- **Step 2**: Furthermore, for each image, I_i, the average of the Euclidean distances to the rest of *(N-1)* images in the set is computed, resulting a global array of N average values. The value of SRI is subtracted from the new array which is further sorted in ascending order and the position of each value in the sorted array will be the new rank to the corresponding image.
- **Step 3**: Considering the second assumption that our algorithm makes regarding most of the best downloaded images being among the first returned, the final ranking of images will consider both the initial ranks given by the image search engines and the new ranks computed at Step 2. Thus, the average between the two ranks of each image is computed, resulting another array of N average values. The new array is sorted in ascending order and images will be arranged according to their final position in the sorted array.
- **Step 4**: All re-ranked images are clustered in M clusters using *k-means* method. The value for M has been experimentally chosen to 15 in order to get the best results.
- **Step 5**: For the set of images inside of each cluster $C_j, j = 1, ..., M$, steps 1, 2 and 3 are reiterated and the SRI_j value is computed for each cluster and images are re-ranked according to their similarity with it. In this way, first ranked image in each of the M clusters is considered to be the representative image for its cluster. Totally, there will be M such images $(RI_j, j = \{1, 2, ..., M\})$.
- **Step 6**: From all the M representative images for the clusters, a small set of best ranked images (ranking according to the final rank computed in Step 3) will be chosen as the representative and diverse images for the set of N images.

4.3   Crowd-Sourced Image Analysis

A pilot crowd-sourcing study was performed in order to qualitatively quantify
and also perform a final refinement on the results obtained in the automated
image analysis process. The study aimed at assessing the level of representativeness and diversity of the selected set of images. The selected set of 701 pictures
related to 107 Italian monuments were annotated and clustered by twenty-one
participants (15 masculine, mean age = 31.6 years) of a local Italian University.
Participants were rewarded with a coupon.

The study consisted of two tasks. For the first one, participants were asked
to annotate with "1" all pictures which showed, partially or entirely, the outside
part of the monument. Pictures containing people were accepted if the outside
of monument, or part of it, was clearly depicted. Participants were asked to
annotate with "0" all pictures which did not show, partially or entirely, the
outside part of the monument. Alternatively, participants could indicate they
could not decide whether the picture contained the outside part of the monument. Comments could be added to all the annotations. For each monument, an
example image was provided for reference. The example image was chosen by
the researchers and represented a prototypical image of the monument. For the
second task, participants were asked to cluster images annotated as representative. Pictures belonging to the same cluster should depict the monument from
the same perspective and share light conditions. At the end of the study, users
were asked to fill in a short questionnaire where issues related to task design
were addressed.

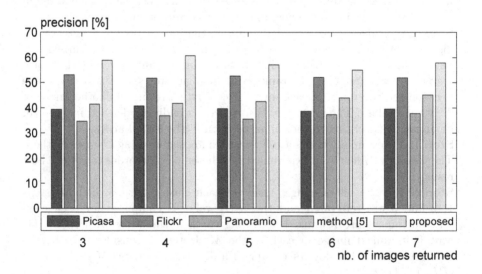

Fig. 3. Performance comparison.

5   Experimental Results

As previously mentioned, to validate our approach presented in Sect. 4, we use a particular application domain that is the search for pictures with Italian monuments. We use a data set of over 25.000 images retrieved from *Picasa, Flickr* and *Panoramio* using both keywords and GPS based search. We use 107 Italian monument locations, from the least known to the grand public to the most famous ones, and a data set of over 25.000 images retrieved from Picasa, Flickr and Panoramio using both keywords and GPS based search. For each of the monuments we attempted to retain the first 100 retrieved images per search engine (thus around 300 per monument - for some of the locations this number may be inferior depending on the availability of images). These images are used as input images for the proposed approach. The assessment of its performance is presented in the sequel.

5.1   Representativeness Results

The first validation experiment consists on assessing the representativeness of the images from the precision point of view:

$$precision = \frac{tp}{tp + fp} \tag{1}$$

where tp is the number of true positives and fp the false positives. A ground truth was determined by manually labelling all pictures in the data set. This task was carried out by an expert with extensive knowledge of these particular monument characteristics and localization. Figure 3 compares the results obtained with the proposed approach against the initial retrieval given by the three image search engines and the approach proposed in [2]. The results are analysed independently for each of the 107 monument queries, but we report the global average precision. For experimentation, we vary the number of representative and diverse images returned by our algorithm from 3 to 7. The best precision is obtained for 4 returned images - 60.7 % - while the lowest precision is around 55 % (in the case of 6 images). There is a slight tendency that the precision may decrease by the number of returned images. Globally, there is an obvious improvement over the approach in [2], ranging from 11 % to 19 %; and also compared to the initial retrieval, e.g. more than 16.3 % for *Picasa*, 2.9 % for *Flickr* and 17.6 % for *Panoramio*, respectively.

   In-line with the statements in Sect. 3 we assess the precision also according to each category of prominence (for exemplification we use the case of 7 returned images). We obtain very interesting results as for the least prominent monuments the precision is up to 70.8 %, for average prominent monuments is around 63.3 % while for most prominent ones is only 25 % (the monuments were divided in these categories based on the number of images returned by Google search - thresholds were set empirically, see Fig. 2). The accuracy of retrieval in the case of prominent monuments is significantly lower than for the rest because of their names containing common words, easily mistakable with everyday aspects or objects (see Sect. 3).

In Fig. 4 we depict an example of good retrieval (for *"Palazzo Pubblico"* monument and 7 returned images) where images are all relevant snapshots of the target monument. On the other hand, Fig. 5 presents a typical case when the method tends to be less efficient. This is typically the case when among the initial retrieved images very few are representative and the representative ones have no high ranks.

5.2   Diversity Results

The second step of the evaluation highlights the precision and the level of completeness in monuments' view for the diversity part obtained when using a ground-truth built employing a number of 21 people to manually label all pictures in the set. The level of completeness was computed as:

$$completness = \frac{nc}{tp} \tag{2}$$

where nc are the total number of clusters that can be obtained from the true positive images.

Thus, the precision obtained when using the crowd-sourcing study is 48.14 % and the level of completeness is 88.53 %. The level of accuracy is lower than the accuracy obtained using the ground truth built by the expert in Sect. 5.1. Instead, the diversity in monuments' images is sustained by the high level of completeness on which all participants agreed. The inter-rate agreement using Kappa statistics was calculated for each pair of participants.

In general, the data indicated a low level of agreement among annotators. The average accuracy was 0.47, with a maximum of 0.78 and minimum of 0.183.

Fig. 4. Results - *Palazzo Pubblico*. Image sources: reference *Wikipedia*; others *Picasa*, *Flickr* and *Panoramio*

Fig. 5. Results - *Egg Castle*. Image sources: reference *Wikipedia*; others *Picasa*, *Flickr* and *Panoramio*

In our understanding, the low level of accuracy and inter-rater agreement is related to participants' different levels of familiarity with the monuments and to the task's design. The results of the pilot study suggest that participants' level of familiarity with the monument influences their answers. The description of the task required to annotate with "1" those pictures which showed, partially or entirely, the outside part of the example picture of the monument. However, many participants annotated with "1" images which represented the monument from a different point of view, not visible in the example image. A few participants reported to have used external services, such as *Wikipedia* and *Google*.

These results reflect the need for better defining the scope of related pictures. Providing an example image can be too restrictive, since it only depicts one point of view. An alternative approach could consist in providing a pointer to a source of information, such as the *Wikipedia*'s page of the monument. Collecting users' familiarity with the monument could contribute to the assignment of different levels of trust to the annotations. Although most of the participants reported that the task description was clear (mean = 6.22 in a 7-point Likert scale), some annotations did not follow the instructions. Theoretically, all pictures representing the inside of the monument should be annotated with "0". However, many participants annotated images clearly depicting the inside of a monument with "2" (i.e., they could not decide whether the picture contained the outside part of the monument). Furthermore, some annotations were inconsistent within participants (images depicting the inside of the monument were annotated with "0" and "2" by the same participant).

In Fig. 6 we depict an example of results obtained using the whole machine and human chain. To have a reference of the improvement, we illustrate the results in the case of the queries presented in Figs. 4 (showing a relevant re-ranking) and 5 (showing a case when the media analysis tends to fail due to the limited number of relevant pictures returned by the search engine). One may observe that the refinement of Fig. 4 is accomplished by keeping 6 out of 7 pictures as being representative and in the same time diverse for the query

(a)

(b)

Fig. 6. Final results as machine - human media analysis chain output (a) *Palazzo Pubblico*; (b) *Egg Castle*

subject (Fig. 6a). The outlier picture was mostly annotated by people as also being representative, but showing high similarity to another selected picture, thus not contributing in diversity requirement. In this case, the human analysis process tends to increase the diversity among the already representative pictures. On the other hand, the less relevant results from Fig. 5 are improved by selecting 2 out of 7 pictures that present the subject in a representative and diverse manner (Fig. 6b). In this case, diversity already exists among images and the human analysis process tends to increase images' relevance.

The machine and human automated media analysis as a whole provides a better mean for adapting the results to the human requirements. In this chain, the automated media part plays a critical role as a pre-filtering step that diminish the time, pay and cognitive load and implicitly people's work in the crowd sourcing part. This makes profitable to have the crowd part as an automated human computation step in the whole chain, although expensive (both in terms of time and costs) to run directly on the initial results returned by the search engines.

6   Conclusions

In this article we introduced a method for the selection of a small number of representative and diverse images from a set of contaminated images automatically retrieved from several image search engines. It aims to select only images containing correct and complete view of the query subject. In order to maximize the quality of our method' results from both representativeness and diversity point of view, we have adapted our method to human constraints by means of crowdsourcing. To this end, we first highlighted the improvements brought by our method to the initial retrieval in comparison to the method in [2] by employing a ground-truth built by an expert and, second, the high level of diversity in images by employing a ground-truth obtained using a crowd-source study (human participants were asked to position the retrieved images in the same cluster if they share the same perspective and light condition). Future work will mainly consists on adapting the approach to the large scale media analysis constraints.

Acknowledgments. This research is partially supported by the CUbRIK project, an IP funded within the FP7/2007–2013 under grant agreement n287704 and by the Romanian Sectoral Operational Programme Human Resources Development 2007–2013 through the Financial Agreement POSDRU/89/1.5/S/62557.

References

1. Bartolini, I., Ciaccia, P.: Multi-dimensional keyword-based image annotation and search. In: ACM International Workshop on Keyword Search on Structured Data, New York, USA, pp. 1–6 (2010)
2. Kennedy, L.S., Naaman, M.: Generating diverse and representative image search results for landmarks. In: International Conference on World Wide Web, New York, NY, USA, pp. 297–306 (2008)
3. Popescu, A., Moëllic, P.A., Kanellos, I., Landais, R.: Lightweight web image reranking. In: ACM International Conference on Multimedia, New York, NY, USA, pp. 657–660 (2009)
4. Fergus, R., Perona, P., Zisserman, A.: A visual category filter for Google images. In: Pajdla, T., Matas, J.G. (eds.) ECCV 2004. LNCS, vol. 3021, pp. 242–256. Springer, Heidelberg (2004)
5. Nguyen, N.V., Ogier, J.M., Tabbone, S., Boucher, A.: Text retrieval relevance feedback techniques for bag of words model in CBIR. In: International Conference on Machine Learning and Pattern Recognition (2009)
6. Larson, M., Rae, A., Demarty, C.H., Kofler, C., Metze, F., Troncy, R., Mezaris, V., Jones, G.J. In: MediaEval 2011 Workshop at Interspeech 2011, vol. 807, CEUR-WS.org, 1–2 September 2011
7. Crandall, D.J., Backstrom, L., Huttenlocher, D., Kleinberg, J.: Mapping the world's photos. In: International Conference on World Wide Web, New York, NY, USA, pp. 761–770 (2009)
8. Hays, J., Efros, A.A.: Im2gps: estimating geographic information from a single image. In: IEEE International Conference on Computer Vision and Pattern Recognition (2008)
9. Van de Weijer, J., Schmid, C., Verbeek, J., Larlus, D.: Learning color names for real-world applications. IEEE Trans. Image Process. **18**(7), 1512–1523 (2009)

Location Context

Exploring Geospatial Music Listening Patterns in Microblog Data

David Hauger and Markus Schedl$^{(\boxtimes)}$

Department of Computational Perception, Johannes Kepler University, Linz, Austria
{david.hauger,markus.schedl}@jku.at

Abstract. Microblogs are a steadily growing, valuable, albeit noisy, source of information on interests, preferences, and activities. As music plays an important role in many human lives we aim to leverage microblogs for music listening-related information. Based on this information we present approaches to estimate artist similarity, popularity, and local trends, as well as approaches to cluster artists with respect to additional tag information. Furthermore, we elaborate a novel geo-aware interaction approach that integrates these diverse pieces of information mined from music-related tweets. Including geospatial information at the level of tweets, we also present a web-based user interface to browse the "world of music" as seen by the "Twittersphere".

Keywords: Microblogs · Geospatial music taste · Music listening patterns

1 Introduction

Due to their continuously growing importance and usage, social media provide a valuable source of user-generated and user-related information. Especially microblogs – due to their nature of being less conversational and providing means to share activities, opinions, experience, and information [20] – are well-suited for discovering breaking news and for user-centric information retrieval [24,31,33].

Since its advent in 2006, `Twitter`'s [8] popularity has been continuously growing, resulting in being today's most popular microblogging service. According to `Twitter`'s last official announcements in March 2011 they claimed to have more than 200 million registered users creating a billion posts per week [1]. Given this remarkable user base, it is no surprise that `Twitter` has already been used for various information retrieval and datamining tasks, including analyzing the spread of diseases [22], detecting earthquakes [25] and hot topics [29], recommendation of information sources [9] and ranking tweets according to the relevance of the user [14,32]. There have also been attempts to identify spam users based on the temporal entropy of tweets containing URLs [30].

One of the many types of information posted via tweets, i.e. messages on the `Twitter` platform limited to 140 chararacters, is information on the music

© Springer International Publishing Switzerland 2014
A. Nürnberger et al. (Eds.): AMR 2012, LNCS 8382, pp. 133–146, 2014.
DOI: 10.1007/978-3-319-12093-5_7

a user is currently listening to. This information may be provided either manually (e.g. included in personal comments) or automatically by plugins for music players or music portals [7]. This research aims at identifying geospatial music listening patterns of the music-tweeting community (although these users are not necessarily representative for the total population).

Section 2 presents related work on microblog mining and geospatial visualization of musical information. In Sect. 3 we present a novel approach to mine `Twitter` posts for music listening-related information. Additionally, we suggest the use of genre-based clustering and propose a method to co-occurrence-based similarity estimation to organize and visualize the extracted information. Section 4 illustrates how geospatial music listening data may be supportive for various tasks. We present a number of use cases and the user interface of a visualization framework to interactively browse and dynamically explore the world of tweeted music listening events.

2   Related Work

The work at hand, as far as we are aware of, is the first to provide a framework to explore the `Twitter` "world of music" and to visualize geospatial music listening patterns in an interactively explorable environment.

Related work may be categorized into work related to mining microblog data and the geospatial visualization of musical information.

2.1   Mining Microblog Data

Hardly any research has been conducted on the intersection between microblog mining and music information retrieval (MIR). Among the few works, Schedl et al. [28] analyze artist popularity on the country level, using term frequencies of `Twitter` posts as one source of information. Zangerle et al. [34] compute inverse document frequency on a fulltext index to map tweets to artists and tracks. The authors propose a co-occurrence-based approach to construct a song recommender system. Schedl and Hauger [27] use microblog data from all cities with more than 500,000 inhabitants in order to calculate deviations of musical taste from the mainstream on country and city level.

General work on microblog mining includes the following: Java et al. [16] analyze microblogs from `Twitter`, `Jaiku`, and `Pownce` in order to study network properties and friendship relations as well as intentions of using those systems. Moreover, they report on geographical distributions of `Twitter` users and the growth of the network. Furthermore, Java et al. aim to identify trends and communities based on keywords. Kwak et al. [17] extend Java et al.'s approach to trend detection by gathering tweets mentioning `Google`'s most frequently used search terms and analyzing the re-tweeting behavior. The authors particularly stress the recentness as one of the major advantages of this source of information.

There is a wide field of different applications that exploit information shared via microblogs. Exploiting geospatial data, De Longueville et al. [13] used data from `Twitter` for forest fire detection. Lee et al. [18] mined `Twitter` for information on earthquakes and plotted them on a world map. As most of those tweets had no information on geo-coordinates attached, they used city names to define positions of tweets. As mentioned in their paper, geo-coordinates are hardly available as they require GPS-enabled devices – which is one of the reasons why they have not been exploited earlier. Bollen et al. [11] mined `Twitter` for emotion-related terms in order to calculate the "public mood", which was then linked to the emergence of stock markets trying to predict future trends.

2.2   Geographic Visualization of Musical Information

Most visualization approaches for musical information are based on various types of content- or context-based features (or similarity measures). These features are mapped to visual aspects such as position, color, distance, or font size. Geographic information is usually not taken into account. However, Raimond et al. [23] combine information from different sources to retrieve geospatial information on artists in order to be able to locate them on a map. Similarly, Govaerts and Duval [15] aim to detect artist origin and plot the results on a map. Another possibility to link music to geographical information is presented by Byklum [12], who searches lyrics for geographical content like names of cities or countries.

A different approach for combining music and geospatial information is presented by Park et al. [21]. They started from geospatial positions and tried to generate music matching the selected environment, based on ambient noise, surroundings, traffic, etc.

As far as we know, geospatial information has not yet been scientifically used to visualize listening patterns, which is most probably due to the fact that this is a relatively new type of information available.

3   Methodology

3.1   Data Acquisition and Processing

For the work reported in this paper we used the `Twitter` Streaming API to retrieve tweets with geospatial coordinates available (preliminary analysis showed that this applies to less than 3 % of the tweets). Between September 2011 and August 2012 we crawled `Twitter` for potentially music-related hashtags, e.g. `#nowplaying`, `#np`, `#itunes`, `#musicmonday` and `#thisismyjam`. The most frequently used music-related hashtag `#nowplaying` and its abbreviation `#np` have already been proven successful to determine music listening-related tweets [26]. During these nine months we retrieved 2,337,489 tweets including both one of the hashtags mentioned above and geospatial information.

However, microblog data is not standardized, neither in terms of the content nor concerning the usage of hashtags. For instance, `#nowplaying` is also used to refer to activities other than music listening (among others, sports events, movies, or games), to a much smaller extent though. Moreover, some tweets are music-related, but contain no information that could be used for our purposes (e.g. "#nowplaying my favorite songs again and again...").

Having obtained the tweets, our goal was to parse and analyze the content to extract artist information. Dictionary-based text matching algorithms and word stemming [10] are not suited to process this type of data, as artist names may match common speech terms. This results in "I", "You", "Me", and "Love" as the most popular, often erroneously detected, artists in our tests, using a list of artists from `freebase` [6]. Artist names that are part of other artist names also pose a serious problem.

In order to overcome these difficulties, we elaborated an alternative approach. Preliminary observations revealed that music-related tweets often contain patterns, such as:

- *song title* by *artist name* [on *some platform*]
- *artist name*: "*song title*"
- *song title* #*artist name*
- *song title* − *artist name*
- *artist name* − *song title*

Therefore, we decided to adopt a multi-level, pattern-based approach, matching only potential artist names against the artist dictionary. Starting with the specific patterns listed above and continuing the search with more general ones (e.g. any term separated by special characters) in case the mentioned ones could not be applied, we were able to eliminate erroneous detections of common speech terms and account for the problems with artist names occurring as substrings in other artist names.

However, relying exclusively on artist information and ignoring song titles still left us with some remaining ambiguity. For instance, the tweet "#np Lena − Satellite" matches the patterns "*artist name* − *song title*" and "*song title* − *artist name*", with both "Lena" and "Satellite" being valid potential artist names [27].

Consequently, we decided to add track information. For the approach described in this paper we used the `musicbrainz` database [3] as knowledge base for artist names and related song titles.

Applying the approach just described, we were able to map 697,614 of the retrieved tweets (29.8 %) to 97,515 unique tracks by 20,567 unique artists ("Drake" being the most popular one with 12,998 tweets).

In the following, we present different approaches to facilitate exploration of music collections, which we implemented in the proposed UI.

3.2   Genre-Based Clustering

Aiming to visualize geospatial music listening activities, we had to come up with a meaningful color-mapping. The first approach presented in this paper organizes tweets in a number of clusters, where a cluster may represent, e.g., genre, mood, country, or language and each cluster is assigned a specific color. As genre classification is the most traditional way of organizing music, our default clustering is based on genres.

Earlier work made use of `allmusic`'s [5] 18 major genres to categorize music [27]. Since `allmusic` over-emphasizes the "Pop and Rock" genre (with more than 60 % of the artists being assigned to it), using these genre labels would result in one big heterogeneous cluster encompassing many different styles, which might be not very helpful to the users.

Therefore, we decided to employ tag-based clustering. For each artist we gathered the available tags from `last.fm` [2]. In order to group artists by genre we filtered the tags using a list of 1,944 known genres from `freebase` [6]. Applying non-negative matrix factorization (NNMF) [19], we split the artists (and genre tags) into k clusters, k ranging from 10 to 20 in our experiments, which seemed a reasonable range. The top-20 genre tags for 10 clusters are listed in Table 1. A higher number of clusters increases their homogeneity, but results in a higher number of necessary colors, increasing visual clutter. To chose a tradeoff between granularity and diversity of colors we allow users to set the number of clusters manually.

3.3 Similarity Estimation

In addition to the approach for static color mapping using a clustering algorithm described above, we also implemented a dynamic visualization approach. One possibility to explore music collections is to find songs by artists similar to a seed artist. Therefore, we calculate similar artists to display and apply a color mapping expressing the similarity scores with respect to a selected seed artist.

To calculate the similarity between two artists i and j we used the co-occurrence-based similarity function

$$sim(i,j) = \frac{cooc_{i,j}}{\sqrt{occ_i \cdot occ_j}}$$

with occ_i being the total number of occurrences of artist i, and $cooc_{i,j}$ being the number of co-occurrences of artists i and j. The co-occurences of i and j are defined as the number of users twittering about songs by artist i as well as about songs by artist j. This similarity function has already been proven successful [27].

3.4 Visualization and User Interaction

To visualize the geographic distribution of tweets the coordinates can be mapped to a 2D representation of a world map (cf. Fig. 1). In Sects. 3.2 and 3.3 two different approaches for similarity-based color-mapping have been proposed. Interaction possibilities can be categorized as follows:

- Interactions with the visualized tweets: Each tweet is represented by a small square on the map. Hovering it with the mouse opens an information window (see Fig. 1) presenting information on artist and song title. Further, it is possible to apply a variety of filters (e.g. date, genre, artist name, track title)

Table 1. Groups of genre tags using NNMF for 10 clusters.

Cluster	Assigned genre tags (top 20)
1	Electronic, House, Electronica, Dance, Techno, Electro, Trance, Downtempo, Synthpop, Minimal techno, Progressive House, Deep house, Tech house, Drum and bass, Breakbeat, Electropop, Dub, Dubstep, Electro house, Electroclash
2	Rock, Alternative, Alternative rock, Hard rock, Progressive rock, Classic rock, Heavy metal, Psychedelic rock, Grunge, Garage rock, Christian, Alternative metal, Progressive, Stoner rock, Nu metal, Christian rock, Post-grunge, Rock and roll, Southern rock, Modern rock
3	Indie, Indie rock, Indie pop, Post-punk, Lo-fi, Emo, Britpop, Dream pop, Math rock, Power pop, Indietronica, Indiepop, Noise pop, Chamber Pop, Piano rock, Twee pop, Dance-punk, Neo-Psychedelia, Hamburger Schule, Jangle pop
4	Experimental, Ambient, Noise, Psychedelic, Dark ambient, Drone, IDM, Industrial, Post-rock, Avant-garde, Instrumental, Glitch, New Age, Noise rock, Contemporary classical, Breakcore, Space rock, Electroacoustic, Darkwave, Krautrock
5	Hip-Hop, Rap, hip hop, Underground hip hop, Underground, Gangsta rap, Reggae, Dirty South, Turntablism, Southern rap, Grime, Dancehall, G-funk, Horrorcore, Crunk, Ragga, Reggaeton, Memphis rap, Chicano rap, Experimental hip hop
6	punk, Punk rock, Pop punk, Ska, Street punk, Ska punk, Garage punk, Garage, Anarcho-punk, Skate punk, Folk punk, Streetpunk, Psychobilly, Skacore, Horror punk, Riot Grrrl, Melodic, Celtic punk, Deathrock, Christian punk
7	Folk, Singer-songwriter, Acoustic, Celtic, Folk rock, Country, Americana, World, Irish, Indie folk, Traditional, Bluegrass, Neofolk, Medieval, Ethnic, Freak folk, New Weird America, Trad, Folk metal, Acoustic rock
8	Pop, Rnb, Pop rock, New Wave, J-pop, Disco, Eurodance, Soft rock, Turkish, Anime, Latin pop, K-pop, Europop, Ballad, Russian pop, Turkish pop, C-pop, Asian, Gospel, Teen pop
9	Jazz, Funk, Soul, Fusion, Blues, Lounge, Piano, Acid jazz, Free jazz, Swing, Smooth jazz, Nu jazz, Jazz fusion, Soundtrack, Contemporary Jazz, Easy listening, Vocal jazz, Bossa nova, Classical, Big band
10	Hardcore, Metalcore, Metal, Hardcore punk, Death metal, Post-hardcore, Thrash metal, Screamo, Gabber, Black metal, Grindcore, Melodic hardcore, Straight edge, Deathcore, Melodic death metal, Progressive metal, Hardcore techno, Mathcore, Thrashcore, Power metal

- Interactions with the underlying map: This includes basic navigation and zooming as well as opportunities to geographic filtering.
- Interactions for statistical purposes: To facilitate analysis we offer tools to calculate play counts on different levels (song, artist, and genre) along with the mentioned filters and temporal aggregation.

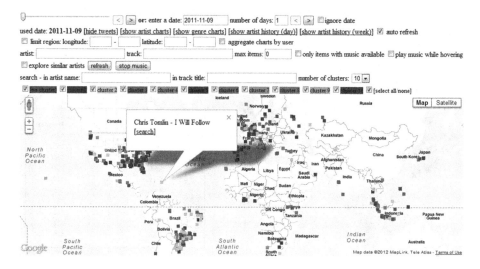

Fig. 1. Visualization of all tweets with the mouse hovering one tweet in Puerto Rico. Map image provided by `Google Maps` [4], ©Google 2012.

3.5   Implementation

To retrieve the tweets provided by `Twitter`'s Streaming API we use the command line tool `curl` requesting all tweets with geospatial information. After filtering these tweets for music related hashtags the textual information of the tweets is analyzed according to the patterns mentioned in Sect. 3.1. Potential artist names and according song titles are automatically matched against our `musicbrainz` server, i.e. a `Postgres` server hourly updating to provide a current copy of the `musicbrainz` database. If a match for a certain tweet is found, this information is written to a `MySQL` database. Old tweets that could not be mapped to known songs are regularly checked again, as they may refer to songs contained in a later update of the `musicbrainz` database.

For visualization and the interactive user interface as described in Sect. 3.4, we decided to create an overlay visualization for `Google Maps`, using the navigation functionality provided by the `Google Maps` API [4]. For the client-side part we relied on web technologies including HTML5 and AJaX.

4   Exemplary Use Cases

In order to illustrate how users might want to explore the "world of music" using geospatial information and the concepts described above, the current section presents a number of use cases as well as approaches to achieve these goals.

4.1   Acoustically Exploring the World of Music

As the most natural way of exploring music is listening, the proposed system aims to provide short mp3-snippets for the tracks referred to by the tweets.

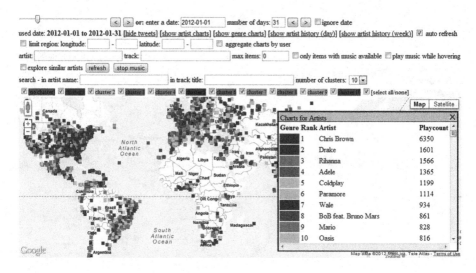

Fig. 2. Visualization of all music tweets and play counts aggregated on artist level for January 2012. Map image provided by `Google Maps` [4], ©Google 2012.

To this end, we matched the tweets to a collection of 2.3 million tracks, resulting in available snippets for 12,070 of the 60,651 identified tracks. To facilitate aural exploration, a "play" button is displayed in the respective information windows. Additionally, the user interface offers a mode in which snippets are automatically played when hovering the corresponding item. Furthermore, it is possible to set a filter to omit tweets without a snippet attached.

4.2 Detecting Globally Popular Artists

As shown in Fig. 2 it is possible to display the play counts for all artists. Alternatively, to reduce the effect of very active `Twitter` users promoting their favorite artists, the charts may be aggregated on user level, i.e. the charts refer to users twittering about these artists instead of particular play counts.

To explore temporal dynamics, charts may be generated for customizable time windows, which enables, for instance, daily or weekly charts. Moreover, the filters mentioned in Sect. 3.4 can be applied.

4.3 Detecting Local Trends

In addition to global popularity estimations, analysis may be restricted to tweets of a certain geographic area. The current version of the system allows to set a rectangular bounding box as shown in Fig. 3, which allows to calculate local charts. This may help to identify local trends as well as popular local artists.

When visually exploring the map, some trends might be surmised. For instance, the first overview already gives the impression that the cluster consisting of Hip-Hop, Rap, etc. is relatively wide-spread in the United States (cluster

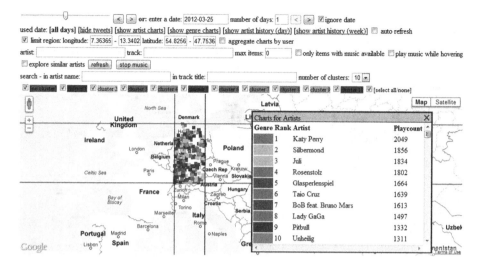

Fig. 3. Visualization of music tweets and play counts for a geographic region (roughly corresponding to Germany). Map image provided by `Google Maps` [4], ©Google 2012.

5 in our examples), whereas South America shows a strong preference for the Rock cluster (cluster 2 in our examples). As the user interface of our framework allows to (de-)select single clusters, we can compare those two clusters directly to each other as shown in Fig. 4. Here we can see an arbitrarily selected period of three days where we can observe the previously mentioned pattern. Selecting these two areas and comparing their genre charts to each other (see Figs. 4(b) and (c)) reveals that for the given period of time, cluster 2 is indeed three times as popular as cluster 5 in South America, but cluster 5 is 5.6 times as popular as cluster 2 in the United States of America. This pattern remains consistent for the whole period of observation. Further investigation reveals that also France shows a relatively high occurrence of Hip-Hop/Rap, whereas Spain and Italy (like South America) have a much stronger Rock cluster.

4.4   Exploring an Artist's Popularity

Having compared artist or genre distributions, one might be interested in detailed information on a specific artist. In addition to filtering tweets, it is possible to display the play counts for the different tracks by an artist. Figure 5 displays the play counts for songs by Madonna and shows how popularity changes with new releases. In this case, the release of the album "MDNA" (a popular track of which is "Girl Gone Wild") in March 2012 and the pre-release in February 2012 can be seen well in the resulting charts. Optionally, these charts can be restricted to evaluate only tweets from within a geographic region.

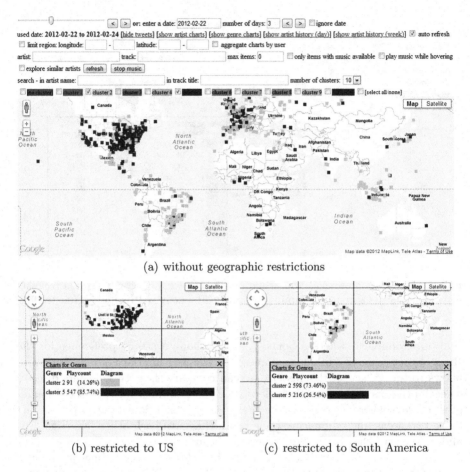

(a) without geographic restrictions

(b) restricted to US   (c) restricted to South America

Fig. 4. Visualization of music tweets of two genre clusters for a period of three days. Map image provided by `Google Maps` [4], ©Google 2012.

4.5 Retrieving Similar Artists

Another means of music exploration is by similar artists. The proposed system offers a "similar artist mode", where users can enter a seed artist (tweets of this seed are displayed in black on the world map). According to the co-occurrence-based similarity function $sim(i, j) = \frac{cooc_{i,j}}{\sqrt{occ_i \cdot occ_j}}$ (see Sect. 3.3), the 50 most similar artists are calculated. The similarity scores are mapped to the range $[0, 255]$. The resulting values are subsequently mapped to the RGB color space using the red channel only.

Figure 6 shows similar artists as well as a popularity chart among these similar artists. Clicking on artist names results in a new query using this artist as the new seed artist. This offers a multimodal data view, combining popularity and similarity information.

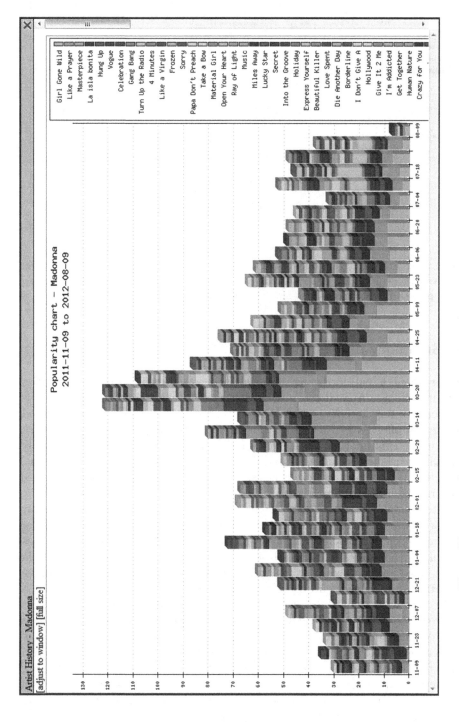

Fig. 5. Play counts for a single artist ("Madonna") on track level, aggregated by week

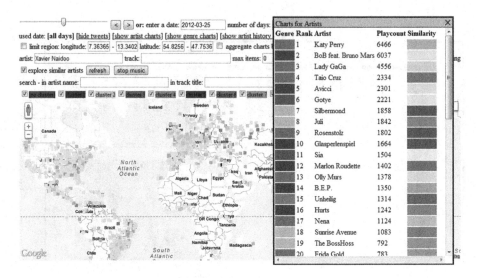

Fig. 6. Map and play counts for a seed artist ("Xavier Naidoo", black) and the 50 most similar artists, ranging from red (most similar) to white (least similar). Map image provided by `Google Maps` [4], ©Google 2012.

5   Summary and Future Work

We proposed a pattern-based approach to extract music listening activities from microblogs. Applying this approach to a data set covering nine months of microblogging activity gathered via `Twitter`'s Streaming API, we indexed tweets that offer geospatial information. In addition, we presented a framework to visualize this information and elaborated a user interface for interactively exploring world-wide music listening histories and detecting listening patterns. Using tag information we implemented genre-based clustering and used these clusters as source of information for the graphical representation. Alternatively, to detect similar artists, we implemented a visualization of the artists most similar to a selected seed artist using co-occurrence-based similarity measures. This approach could additionally be used or extended by various other types of similarity measures (e.g., based on term weight or on features obtained via signal-based audio processing), and might serve as an alternative way of proposing artists and/or tracks in dynamic playlist generation.

In order to be able to test users' hypotheses on observable listening patterns we provide possibilities to filter the data set by geographic coordinates. As a possible extension we could use information on the geographic boundaries of political regions to perform evaluations on country level. As already mentioned, genre tags are only one of many ways of clustering music – so we are exploring a variety of different clustering features and algorithms. Furthermore, we could make use of URLs or other links contained in the tweets. Via real-time processing of tweets, we could relate this information to album releases and concert tours, and further analyze temporal dynamics of artist popularity. As part of future work we will also look into building personalized music retrieval models, for which geolocalized information on music consumption might serve to incorporate cultural specifics in listening activity.

Acknowledgments. This research is supported by the Austrian Science Funds (FWF): P22856-N23 and Z159.

References

1. http://blog.twitter.com/2011/03/numbers.html. Accessed August 2012
2. http://last.fm. Accessed August 2012
3. http://musicbrainz.org. Accessed August 2012
4. https://developers.google.com/maps/. Accessed August 2012
5. http://www.allmusic.com. Accessed August 2012
6. http://www.freebase.com. Accessed August 2012
7. http://www.spotify.com. Accessed August 2012
8. http://www.twitter.com. Accessed August 2012
9. Armentano, M., Godoy, D., Amandi, A.: Recommending information sources to information seekers in twitter. In: International Workshop on Social Web Mining, Co-located with IJCAI 2011 (2011)
10. Baeza-Yates, R., Ribeiro-Neto, B.: Modern Information Retrieval. Addison Wesley, New York (1999)
11. Bollen, J., Mao, H., Zeng, X.: Twitter mood predicts the stock market. J. Comput. Sci. **2**(1), 1–8 (2011)
12. Byklum, D.: Geography and music: making the connection. J. Geogr. **93**(6), 274–278 (1994)
13. De Longueville, B., Smith, R.S., Luraschi, G.: "OMG, from here, I can see the flames!": a use case of mining location based social networks to acquire spatio-temporal data on forest fires. In: Proceedings of the 2009 International Workshop on Location Based Social Networks, LBSN '09, pp. 73–80. ACM, New York (2009)
14. Duan, Y., Jiang, L., Qin, T., Zhou, M., Shum, H.-Y.: An empirical study on learning to rank of tweets. In: Huang, C.-R., Jurafsky, D. (eds.) Proceedings of the 23rd International Conference on Computational Linguistics (COLING 2010), pp. 295–303. Tsinghua University Press, August 2010
15. Govaerts, S., Duval, E.: A web-based approach to determine the origin of an artist. In: Hirata, K., Tzanetakis, G., Yoshii, K. (eds.) Proceedings of the 10th International Society for Music Information Retrieval Conference (ISMIR 2010), pp. 261–266. International Society for Music Information Retrieval (2009)
16. Java, A., Song, X., Finin, T., Tseng, B.: Why we twitter: understanding microblogging usage and communities. In: Proceedings of WebKDD and SNA-KDD, San Jose, CA, USA, August 2007
17. Kwak, H., Lee, C., Park, H., Moon, S.: What is Twitter, a social network or a news media? In: Proceedings of the 19th International Conference on World Wide Web, WWW '10, pp. 591–600. ACM, New York (2010)
18. Lee, C.-H., Yang, H.-C., Chien, T.-F., Wen, W.-S.: A novel approach for event detection by mining spatio-temporal information on microblogs. In: International Conference on Advances in Social Networks Analysis and Mining (ASONAM 2011), pp. 254–259, July 2011

19. Lee, D.D., Seung, H.S.: Learning the parts of objects by non-negative matrix factorization. Nature **401**(6755), 788–791 (1999)
20. Oulasvirta, A., Lehtonen, E., Kurvinen, E., Raento, M.: Making the ordinary visible in microblogs. Pers. Ubiquit. Comput. **14**(3), 237–249 (2010)
21. Park, S., Kim, S., Lee, S., Yeo, W.S.: Online map interface for creative and interactive MusicMaking. In: Proceedings of the 2010 Conference on New Interfaces for Musical Expression (NIME 2010), Sydney, Australia, pp. 331–334 (2010)
22. Paul, M.J., Dredze, M.: You are what you tweet: analyzing twitter for public health. Artif. Intell. **38**, 265–272 (2011)
23. Raimond, Y., Sutton, C., Sandler, M.: Automatic interlinking of music datasets on the semantic web. In: Linked Data on the Web (LDOW2008) (2008)
24. Romero, D.M., Meeder, B., Kleinberg, J.: Differences in the mechanics of information diffusion across topics: idioms, political hashtags, and complex contagion on twitter. In: Proceedings of the 20th International Conference on World Wide Web (WWW 2011), pp. 695–704. ACM, New York (2011)
25. Sakaki, T., Okazaki, M., Matsuo, Y.: Earthquake shakes twitter users: real-time event detection by social sensors. In: Proceedings of the 19th International Conference on World Wide Web (WWW 2010), May 2010
26. Schedl, M.: Analyzing the potential of microblogs for spatio-temporal popularity estimation of music artists. In: Proceedings of IJCAI: International Workshop on Social Web Mining, Barcelona, Spain, July 2011
27. Schedl, M., Hauger, D.: Mining microblogs to infer music artist similarity and cultural listening patterns. In: Proceedings of the 21st International World Wide Web Conference (WWW 2012): 4th International Workshop on Advances in Music Information Research: "The Web of Music" (AdMIRe 2012), Lyon, France (2012)
28. Schedl, M., Pohle, T., Koenigstein, N., Knees, P.: What's Hot? estimating country-specific artist popularity. In: Proceedings of the 11th Internat, Society for Music Information Retrieval Conference (ISMIR 2010), Utrecht, Netherlands, August 2010
29. Sharifi, B., Hutton, M.-A., Kalita, J.: Summarizing microblogs automatically. In: Proceedings of NAACL HLT, June 2010
30. Steeg, G.V.: Information theoretic tools for social media. In: Making Sense of Microposts (#MSM2012), p. 1 (2012)
31. Teevan, J., Ramage, D., Morris, M.R.: #TwitterSearch: a comparison of microblog search and web search. In: Proceedings of the 4th ACM International Conference on Web Search and Data Mining (WSDM'11), Hong Kong, China, February 2011
32. Weng, J., Lim, E.-P., Jiang, J., He, Q.: TwitterRank: finding topic-sensitive influential twitterers. In: Proceedings of the Third ACM International Conference on Web Search and Data Mining, WSDM '10, pp. 261–270. ACM, New York (2010)
33. Wu, S., Hofman, J.M., Mason, W.A., Watts, D.J.: Who says what to whom on twitter. In: Proceedings of the 20th International Conference on World Wide Web (WWW 2011), pp. 705–714. ACM, New York (2011)
34. Zangerle, E., Gassler, W., Specht, G.: Exploiting twitter's collective knowledge for music recommendations. In: Making Sense of Microposts (#MSM2012), pp. 14–17 (2012)

A Study into Annotation Ranking Metrics in Community Contributed Image Corpora

Mark Hughes[1], Gareth J.F. Jones[2], and Noel E. O'Connor[1($\boxtimes$)]

[1] CLARITY: Centre for Sensor Web Technologies, Dublin City University,
Dublin 9, Ireland
mhughes@computing.dcu.ie, noconnor@eeng.dcu.ie
[2] Centre for Next Generation Localisation, Dublin City University, Dublin 9, Ireland
gjones@computing.dcu.ie

Abstract. Community contributed datasets are becoming increasing common in automated image annotation systems. One important issue with community image data is that there is no guarantee that the associated metadata is relevant. A method is required that can accurately rank the semantic relevance of community annotations. This should enable the extracting of relevant subsets from potentially noisy collections of these annotations. Having relevant, non-heterogeneous tags assigned to images should improve community image retrieval systems, such as Flickr, which are based on text retrieval methods. In the literature, the current state of the art approach to ranking the semantic relevance of Flickr tags is based on the widely used tf-idf metric. In the case of datasets containing landmark images, however, this metric is inefficient and can be improved upon. In this paper, we present a landmark recognition framework, that provides end-to-end automated recognition and annotation. In our study into automated annotation, we evaluate 5 alternate approaches to tf-idf to rank tag relevance in community contributed landmark image corpora. We carry out a thorough evaluation of each of these ranking metrics and results of this evaluation demonstrate that four of these proposed techniques outperform the current commonly-used tf-idf approach for this task. Our best performing evaluated approach achieves a significant F-Measure increase of .19 over tf-idf.

Keywords: Image annotation · Landmark recognition · Tag relevance

1 Introduction

Web sites that store and organise personal image collections online such as Flickr[1] have very large volumes of personal images in their databases. Flickr currently has over five billion personal photos stored online with an average of 3–5 million images being uploaded daily. Unfortunately, the proliferation of shared photographs has outpaced the technology for searching and browsing

---

[1] Flickr: www.flickr.com

© Springer International Publishing Switzerland 2014
A. Nürnberger et al. (Eds.): AMR 2012, LNCS 8382, pp. 147–162, 2014.
DOI: 10.1007/978-3-319-12093-5_8

such collections. With this very large and growing body of information, there is a clear requirement for efficient techniques to structure it and present it to users.

Many consumers, tourists in particular, capture large numbers of images in destinations that they visit, and upon return, share these images online with friends and family. One popular genre of images that are being uploaded to online image repositories, and the genre that we focus on in this paper, are photographs containing famous landmarks from around the world. Due to drawbacks in image classification technology, in most cases it is not possible to automatically classify high level semantic information from these images (such as to label them with the name and location of a landmark) based on image content alone.

Due to technological constraints, for high-level semantic image retrieval queries, retrieval systems are forced to rely on text based retrieval methods based on captions created by users, with little or no formal rules on objectivity or detail. This can lead to retrieval errors (an example of which can be seen in Fig. 1) due to homogeneous and subjective captions, and in some cases no caption provided at all. Homogeneous captions result in poor reliability of individual items in search, and subjective labels are unlikely to be useful for users other than the captioner.

The average consumer, taking a picture with their digital camera or smartphone generally does not pay much attention to how images are stored, organised and retrieved. They simply want a fast and reliable automated technology that allows them to photograph an image and at a later stage retrieve, view and share that image. They don't wish to spend large amounts of time, in what they regard as the monotonous task of providing textual descriptions for images before uploading them to a web site of their choice. Therefore, an automated approach to this task is desirable. In this paper, we present an automated solution to this problem.

Community datasets are undoubtedly a useful resource for image matching processes as many of them contain manually created metadata describing the content and context of each image. There are however, many problems associated with their use. The main issue with these datasets is the unreliability of the relevance and accuracy of their metadata. The aim of this paper is to explore methods to retrieve subsets of semantically relevant tags with which to annotate a test image, from noisy collections of community data, retrieved through the use of an image recognition framework.

The paper is split into two main sections, the first of which is a description of an image recognition framework that we implemented to gather a set of candidate image annotations from a community dataset which can be used to annotate a test image. The second section describes a number of approaches that we evaluate with the aim of selecting a subset of the candidate annotations containing those most semantically relevant to a test image. The paper concludes with a thorough evaluation of each of these approaches.

Fig. 1. An example of the problems associated with homogeneous tags in text based image retrieval systems. Pictured is the top three ranked results (ranked from left to right) returned from Flickr (24-Jan-2012) when searching for images of the famous landmark, the statue of liberty using the query text: statue of liberty.

2   Background

In the field of Computer Vision, the detection and description of salient image regions is now relatively mature and several algorithms exist that can detect salient regions and create a highly discriminative feature descriptor to describe the region. These feature descriptors can then be applied to find corresponding regions in multiple visually similar images with a high level of accuracy and some invariance to rotation, affine and lighting differences.

Comparing large amounts of these descriptors using a brute force approach is processor intensive and several alternatives have been suggested in the literature to allow for fast comparisons of large numbers of images. Lowe [10] applied an approximate nearest neighbour algorithm based on a kd-tree data structure called best bin first. In similar work, Nister [8] suggested the use of a hierarchical k-means structure, which we adopt in this work as part of an image matching framework to return visually similar images based on a query image. We carry out an evaluation of this approach using our image corpora in Sect. 3.

In a related field, work has been carried out analysing how to best extract representative textual tags from clusters of images in community contributed datasets. Kennedy et al. [1] explored different methods to structure Flickr data, and to extract meaningful patterns from this data. Specifically, they were interested in selecting metadata from image collections that might best describe a geographical region. In similar work [2], they focused these techniques on extracting textual descriptions of geographical features, specifically landmarks, from large collections of Flickr metadata. Tags are clustered based on location, and using a tf-idf approach tags are selected so as to correlate with nearby landmarks.

Ahern et al. [3] employ a tf-idf approach on sets of Flickr tags to create a visualisation of representative tags overlaid on a geographical map. They call this system the 'World Explorer', and it allows users to view unstructured textual tags in a geographically structured manner.

Sigurbjornsson and Van Zwol [6] developed a technique to augment a user defined list of tags with an additional set of related tags extracted from large collections of Flickr imagery. They adopt a co-occurrence methodology retrieving tags that regularly co-occur with each user defined tag within the dataset.

Xirong et al. [4] combine visual information with a tf-idf scoring metric to estimate tag relevance within a dataset of Flickr images. For each test image, they carry out a visual search procedure to find its nearest neighbours visually within the dataset. They show that by calculating co-occurrences of tags within visually similar images, it is possible to estimate relevant tags for a query image over using text based methods alone with a higher probability.

Most of the approaches to date have focused on variations of text-retrieval based models using a tf-idf scoring approach to choose relevant representative tags from a cluster of metadata [5]. We have found that tf-idf is not an optimal ranking metric when dealing with a corpus of landmark images due to high repetition of well-known landmark terms, therefore, the aim is to improve upon tf-idf, by analysing alternative statistical methods to select semantically relevant sets of tags for a test image.

3   Landmark Identification

In order to carry out an investigation into automated annotation, firstly a method to extract a set of candidate tags for a test image must be applied. In this paper we implement an image recognition framework using a collection of geotagged images from Flickr as a training corpus. This framework will analyse a test image and retrieve relevant images from the corpus based on visual similarity. The short textual annotations that accompany these retrieved images (denoted as tags) are then considered to be a list of candidate tags with which we aim to select semantically relevant subsets to use as annotations that describe our test image.

For the purposes of this work, we use training and test corpora consisting of images containing commonly photographed landmark images. The focus is on landmarks due to the significant contribution that they make to a large scale public photo repository such as Flickr (eg. Flickr search for Eiffel Tower returns over 450,000 images, Flickr search for Empire State returns over 370,000 images (June 2011)). Landmarks also tend to have a unique visual appearance that leads to high discrimination values between different landmarks.

3.1   Image Corpora

It was desired to create a dataset of geo-tagged imagery that covered an entire metropolitan region of a large city. The city of Paris was chosen, mainly because in certain regions within the city there is a high distribution of landmarks. Additionally, the Parisian region is one of the most densely populated regions that is represented on Flickr with regards to geo-tagged photographs (490,000 in Paris region as of June 2011).

Our training corpus of geo-tagged images was harvested using the publicly available Flickr API. When using the Flickr API, users can provide a text query which is used by the Flickr system to return images relevant to that query. To return possible landmark images, the Flickr system was queried with a list of generic words that might indicate a landmark is present in an image, such as landmark, church, bridge, building, facade etc.

To filter out non-landmark images from the corpus, an approach based on the use of stop words was adopted. A list of tags was created, each of which occurs frequently in sets of images that do not contain landmarks and do not tend to occur in sets of images that do contain landmarks. This list of tags is labelled as stop words and images containing one of these stop word tags were then filtered out from the data set.

To build a list of stop words, an image set collected from Flickr consisting of 1000 images was manually inspected and classified as containing a large landmark. This set was labelled as S_1. A further set of 1000 images that did not contain a large landmark, but rather depicted an event or different types of objects, people, and animals was also collected and denoted S_2.

For the set S_1 a list of all associated tags was extracted and denoted as T_1. A second list of tags T_2 was created containing all the tags associated with images in S_2. All tags contained in $T_2 \backslash T_1$ were considered possible candidate tags, however the presence of a tag in $T_2 \backslash T_1$ alone is not enough to indicate that the tag would suggest a non-landmark image. It was decided therefore, to selected the tags that occurred the highest number of times in T_2 but not T_1. The final set of stop words was selected based on the tag frequency of each possible candidate tag from $T_2 \backslash T_1$. The frequency was calculated using the following formula:

$$tf_i = \frac{t_i}{|T_2 \backslash T_1|}$$

where t_i is the number of occurrences of the tag i in the list $T_2 \backslash T_1$. If the term frequency was above a threshold of .005 (roughly translating to a frequency of 10), the tag was marked as a candidate tag.

Any image within our corpus containing one of these candidate tags was filtered out. In total, we downloaded just under 200,000 geo-tagged images from Flickr in the Paris region and over 100,000 were filtered out using this approach, leaving a final training corpus consisting of 90,968 images. From informal empirical inspection this tag filtering approach seems to work quite well, with the majority of images in our dataset depicting a place or landmark.

3.2   Spatial Filtering

Spatial information provides a useful method for filtering the search space for image retrieval processes, particularly when searching for images containing landmarks as they have a fixed location. The first stage of our image matching framework is to filter out non-candidate image matches based on spatial information. While all of the images within our corpus have associated geotags, there is no guarantee as to their accuracy. It is therefore not known in advance what level of spatial filtering will be optimal to ensure the best balance between precision and recall. In this section we carry out some experimentation to ascertain the optimal spatial filtering parameter for our image matching process.

There has been some work in recent years evaluating geotag accuracies such as work carried out by Girardin [11] and Hollenstein [12]. Both of these evaluations however were based on statistical information without any manual inspection and therefore can only be considered as estimates. It remains difficult to garner a precise measurement of the level of accuracy of each individual geotag, since only the person who captured the image can be sure to a high degree of accuracy where they were located at the time of image capture. Based on this premise, it is assumed that some element of manual inspection and a fundamental level of local geographical knowledge is required to measure precisely the level of accuracy contained in this metadata. Inspired by the work of Hollenstein, a detailed manual analysis was carried out on a subset of the images contained within the corpus to provide a reliable and accurate measurement of geo-tag precision. A subset comprising of 673 images of 4 landmarks were selected to be analysed. Based on local knowledge of the region, each of these images was estimated to have been photographed within very close proximity (approximately 100 m) to four different landmarks in Paris (Paris Opera House, Arc De Triomphe, Louvre Pyramid and Pont Neuf Bridge).

The geographical centre point of each of these landmarks was noted, and a bounding box with side lengths of 200 m was created surrounding the centre. The geo-tags of each of these 673 images were examined, and for each one the distance between the geo-tag value and the associated bounding box was calculated to measure the accuracy of each geo-tag. To calculate distances between a geo-tag and a geographical bounding box the Haversine formula is used.

The results of this analysis are quite interesting (presented in Table 1), in that they indicate that the geo-tags within this dataset are generally accurate to within a relatively small radius. These results show that the majority of geo-tags that were examined are accurate to within 200 m from our bounding boxes (over 80 %). This is not as accurate as a modern, high end GPS receiver (generally accurate to within 10 m, depending on the strength of the connection and line of sight), but should be accurate enough to allow for efficient filtering of unwanted images in our image search framework.

We also carried out a set of queries determining the average percentage of the image corpus that remains after spatial filtering at each distance threshold. We find that the search space starts to grow rapidly once the radius has a value of 500 m or more. Based on the results in Table 1, therefore, we choose a value of 250 m to use as our spatial radius in our image classification process (Sect. 3.3), which represents the best balance between precision and recall.

3.3   Image Based Landmark Classification

The first aim of the work described in this paper is based on the automated recognition of landmark images. To achieve this goal, an approach based on the nearest neighbor matching of SURF [7] interest point features is utilised. Brute force matching of interest points is notoriously time consuming and possibly intractable when searching large image corpora, therefore, in this work, an alternative approach is utilized based on approximate nearest neighbor search.

Table 1. Results describing the number of correct geo-tags for each spatial radius, along with the percentage of correct geo-tags from the subset of those examined

Distance	No. of images	% Images	% Search Space
50 m	372	55.2 %	1.2 %
100 m	506	75.1 %	2.1 %
200 m	545	80.9 %	3.7 %
250 m	552	82 %	4.6 %
500 m	578	85.8 %	8.8 %
1000 m	599	89 %	17.6 %
2000 m	625	92.8 %	32.5 %

We use an approach proposed by Nister [8] called a hierarchical vocabulary tree to index large numbers of images features and allow for fast nearest neighbour search. A hierarchical vocabulary tree is a tree structure that is built upon a large visual word vocabulary [9]. In this work, we use a vocabulary size of 250,000. The hierarchical vocabulary tree is a form of a hierarchical k-means algorithm, where the inputs consist of visual words, and the clusters centres outputted from each k-means invocation, are used as the pivots of the tree structure.

The algorithm quantises the vocabulary into k smaller subsets at each level using the k-means clustering algorithm on each partition independently. Each quantisation takes place recursively on smaller subsets of data. Instead of the k parameter determining the final number of leaf nodes, k determines the branch factor of the structure.

To classify a test image, firstly, its spatial data is analysed and only images that are located within a geographical radius of 250 m are retrieved from the image corpus, followed by a hierarchical tree based SURF feature matchibng process. Each child node within the tree represents a vocabulary feature, and is given an identification number. SURF features are extracted from the test image and propagated down the tree structure, each feature being assigned an ID based on its path down the tree. This list of IDs is then compared against the list of IDs associated with all retrieved corpus images and identical IDs correspond as a match. Corpus images are then ranked based on the number of correspondences, and if that number is above a threshold the corpus image is considered a match.

A re-ranking procedure was then carried out using brute force SURF feature matching based on the distance ratio measure using a ratio value as .7. This was carried out as a confirmation stage to eliminate false positives. As point to point matching is an expensive process, the SURF re-ranking process was carried out only on the top images returned from the vocabulary tree that had a number of tree-based correspondences above a threshold k where we evaluated 4 values for k (5, 15, 25 and 35).

Fig. 2. Landmark recognition example: Rose Window in Notre Dame Cathedral

Fig. 3. The top 4 results retrieved for this image using just the vocabulary tree without SURF re-ranking. While all top ranking images are located within the same structure, they do not contain the same part of the structure present in the query image

Fig. 4. The top 4 results retrieved for this image using the vocabulary tree followed by SURF re-ranking. As can be seen in the Figure, the SURF re-ranking significantly improves the accuracy of the object matching retrieving images containing the specific window in the query image

3.4   Landmark Image Recognition Evaluation

To test the effectiveness of the landmark recognition system, a test set was created using the Flickr API. This test collection consisted of 1000 images containing landmarks as the main object within the image, photographed within the Parisian region. The precision of the system was evaluated using 4 separate metrics: Precision, precision calculated over the top 3 images (Precision(3)), top 5 images(Precision(5)) and top 10 images (Precision(10)). We test the precision of the system using two approaches, with and without a SURF re-ranking process. We also analyse a number of values for threshold t which determines the number of tree correspondences is required for a corpus image to be considered a match (t=5, 15, 25 and 35) (Tables 2 and 3).

From the results in Table 4, it is evident that the object matching approach performs with a very high level of precision. The spatial filtering process ensures that the search space is significantly reduced without which the precision would be expected to fall. As expected, the SURF re-ranking process adds a significant improvement over using tree corresponding matches alone. An example of this can be seen in Figs. 2, 3 and 4.

Table 2. Classification results: hierarchical vocabulary tree. The threshold represents the number of tree correspondences required in order for a corpus image to be considered a match

Threshold t	5	15	25	35
Precision(Overall)	0.319	0.514	0.612	0.697
Precision(3)	0.695	0.778	0.822	0.851
Precision(5)	0.595	0.711	0.774	0.814
Precision(10)	0.493	0.647	.0.720	0.776

Table 3. Classification results: hierarchical vocabulary tree with SURF correspondence re-ranking

Threshold t	5	15	25	35
Precision(Overall)	0.875	0.936	0.925	0.937
Precision(3)	0.961	0.989	0.988	0.996
Precision(5)	0.934	0.978	0.981	0.984
Precision(10)	0.905	0.955	.0.968	0.970

4   Tag Selection Schemes

For each test image that we process, the image recognition process will return a large set of tags that are associated with each matched image from our corpus. We call this set of tags a result set. There is a significant challenge in retrieving semantically relevant annotations from these result sets, as Flickr tags are notoriously noisy and much of the data is heterogeneous and semantically non relevant. An example of this can be seen in Fig. 5. Homogeneous captions are observed to be a common occurrence where a user uploading a large number of images will use the same caption to describe the whole set, which creates obvious problems. The goal is to create a method that will optimally select semantically relevant tags for an image that replicate those that might be selected by a human annotator. For any query image, there could be a large number of retrieved images returned from the image corpus by our image recognition framework. Each of these retrieved images will have its own set of textual tags, with no guarantee that any tag is relevant. It remains a challenge, when captioning query images, to select a relevant tag or set of tags that might best semantically describe the landmark depicted within the query image.

4.1   TF-IDF Tag Selection

Following previous work [1,3–5], a method based on the term frequency - inverse document frequency' (tf-idf) approach is implemented and used as a baseline to evaluate all other proposed tag selection approaches. One important measurement in determining the importance of a candidate tag is its level of 'uniqueness' or 'specificity' across the entire corpus. This method assigns a higher score to tags that have a high term frequency within a result set, and a lower frequency across the entire corpus. Additionally, it will assign a low score to any tag that occurs regularly across the corpus.

The tf-idf metric is a combination of the term frequency metric defined in Sect. 4.3, and a metric called the inverse document frequency (idf). The document frequency of a tag t is defined as the number of images within the corpus that contain t. To scale the weight of the document frequency, an inverse document frequency of a tag is defined as

$$idf_t = \log \frac{N}{df_i}$$

where df_i is the document frequency of the tag i and N is the total number of images within the corpus. The tf-idf metric is then formulated as:

$$tf - idf_t = tf_t \times idf_t$$

Each tag within an image result set was assigned a tf-idf score using this metric, and tags were ranked in a descending order with the top k tags selected as the most representative or relevant.

4.2   Proposed Approaches

In this paper we implement and evaluate a number of alternate tag selection schemes to the commonly adopted tf-idf approach. From analysing the structure of the data, five different types of selection schemes were identified.

– Frequency Based
– Image Similarity Ranking
– Tag Ranking Based
– Geographical Based
– Co-Occurrence Based

Tag Selection Based on Term Frequency. The first approach evaluated was based on selecting the tag with the highest term frequency score within

2008	travel	louvre	paris621pams
2009	capital	buttress	arrondisement
toureiffel	city	church	tourist
gothic	military	rosewindow	vacation
otw	gioconda	arcdutriumph	winter
rosetón	religion	window	francia
basillica	stained	places	iglesia
street	seine	overseas	window
abroad	europeanunion	sainte	cité
saintechapelle	gotik	pedroyana	vidriera
french	abbey	temple	gargoyle
architecture	daarklands	soe	dragondaggerphoto

Fig. 5. A randomly selected subset of candidate tags retrieved by our image recognition framework for the test image shown. As can be seen from the figure, a large percentage of the tags are heterogeneous and of little semantic value to the test image.

a result set. Term frequency (TF) is calculated by the number of times a tag appears within a result set, divided by the total number of images within the result set. Tags were ranked based on descending term frequency scores, which essentially corresponds to the terms with majority representation within a result set at the top of the ranking.

The tf score was calculated by using the following formula:

$$tf_i = \frac{t_i}{q}$$

where t_i is the number of times a tag i appears within a result set, and q represents the total number of images within the result set.

Tag Selection Based on Image Similarity Rankings. Each result set is ranked based on visual similarity to the query image, with the highest ranking images having the highest number of SURF correspondences. It would seem logical to analyse whether this visual relationship with an image corresponds to contextual similarity within the associated tags. The higher the rank of an image, the more likely it is that the image is a correct match. An incorrectly matched image is more likely to contain irrelevant tags, therefore it seems plausible that the higher ranked images have a higher probability of containing relevant tags. To evaluate this hypothesis, a tag selection scheme based on the ranked position of each matched image was carried out. The higher the rank of an image, the larger the weight associated with its corresponding tags. Two weighting schemes were implemented, both of which were based on a mixture of tag frequency within a result set and image ranking.

The first scheme places a large importance on a small number of high ranked images, while the weight associated with images lower down the ranking system is decremented significantly, to such an extent that the lowest ranked images are effectively irrelevant. The score assigned to each tag t was calculated as follows:

$$score(t) = tf_i \times \sum_i^n w_i \quad \text{where} \quad w_i = \frac{1}{r}$$

where r is the rank of the image i in each result set.

The second ranking based scheme provided a more balanced weight across all ranked images. The weight associated with lower ranked images is decremented more slowly. This scheme can be formulated as:

$$Score(t) = tf_i \times \sum_i^n w_i \quad \text{where} \quad w_i = 1 - \frac{r}{q}$$

where r is the rank of the image i, and q is the total number of images within the ranked result set.

Tag Selection Based on Ranked Term Frequency. Using the Flickr interface, when users are prompted to create tags to describe the content of an image, it could be assumed they will enter the tags that they deem most relevant to the image in descending order. This order is preserved within the data, and therefore could be considered as a ranked list. It is possible that these tags could be heterogeneous making the ranking redundant for a single image.

It is logical to assume, however, that if there is a high level of correlation between high ranking tags over a result set of images, that these correlated tags could be deemed most relevant semantically. An evaluation was carried out across all top ranking tags within each result set. Similarly to the ranked image approach evaluated, two different ranking schemes were utilised. The first ranking scheme places a large weight on tags that were ranked near the top of the lists. Tags that are ranked at the lower ends of the list are assigned a weight so low that they are effectively disregarded. This ranking scheme can be formulated as:

$$Score(t_j) = \sum_i^n w_j \quad \text{where} \quad w_j = 1 - \frac{1}{r}$$

where n is the total number of images within a result set in which the tag t_j appears and r is the rank of the tag t_j in image i.

The second ranking approach placed a more balanced weight distribution across all tag ranking positions. The variation in weights between top ranking and lower ranking tags is smaller than in the first ranking metric. This second approach is formally defined as:

$$Score(t_j) = \sum_i^n w_j \quad \text{where} \quad w_j = 1 - \frac{r}{q}$$

where n is the total number of images within a result set that the tag t_j appears in, r is the rank of the tag t_j in an image i, and q is the total number of tags retrieved for image i.

Tags were then ranked in a descending order based on $Score(t_j)$. The top ranked k tags were then chosen as the most representative tags for the retrieved image result set.

Tag Selection Based on Geographical Distribution. Combining the geographical and textual based metadata that accompanies each image within the training corpus, should improve tag selection precision, as not only does a geo-tag have a semantic relationship with an image, it also has a semantic relationship with the associated textual metadata.

By calculating the spatial distribution of a tag throughout the whole corpus, it is hypothesised that it is possible to predict a relevant tag with a higher probability. A tag with a geographical distribution based over a small geographical area is more likely to describe a landmark within that area, rather than a tag with a citywide geographical distribution.

To indicate the geographically diverse distribution of each tag, a metric calculating the standard deviation was utilised. It is formally calculated using the following formula:

$$dev_i = \sqrt{\frac{1}{N} \sum_{i=0}^{N} (x_i - \bar{x})^2}$$

where x_i is the geographical location for an ith instance of a tag and $\bar{x}$ is the mean geographical location of the tag. All standard deviation values are normalised in the range 0–1.

The actual score calculated for each tag is a combination of the tag frequency within the image result set and the geographical variation of the tag. This can be formally defined as:

$$score_i = tf_i \times (1 - dev_i)$$

It was found from experimentation that using a weighted value for tf_i performed better. Based on this, two weights were evaluated:

$$score_i = w(tf_i) \times (1 - dev_i)$$

where w is equal to 2 and 4.

Tag Pair Co-Occurrences. The final metric that we evaluate is based on the co-occurrence of pairs of tags across each result set. It is believed that it is more likely that semantically relevant tags (e.g. eiffel and tower) would be more likely to appear in pairs in multiple sets of tags than generic tags (e.g. holidays and 2007). We calculate a metric that ranks pairs of tags by the number of times they co-occur.

To calculate the co-occurrence metric, we create a co-occurrence matrix M of size $N \times N$ where N is the number of unique tags returned from within a result set of images, a set denoted as T. The value of position ij in M is number of times tags i and j (where $i, j \in T$) appear together in each retrieved image. The top k pairs of co-occurring tags were then extracted from M (where different values for k were evaluated: 1,2,3,4 and 5) and these tags were then re-ranked based on their tag frequency within the retrieved result set of images, where a higher frequency represents a higher ranking.

5   Tag Selection Evaluation

A subset of 100 images was randomly selected from our corpus to be used as a test set to analyse retrieved tags from our image matching framework. To evaluate our proposed approaches, a benchmark selection of tags representing the ranked lists of images returned for each image in this subset was created. Tags associated with each image out of each of these ranked results were analysed manually. This benchmark consisted of a total of 602 retrieved images with an average of just under 6 tags per test image, resulting in a total of 3444 tags. Each tag was deemed semantically relevant or irrelevant to the original test image using the following protocol:

- **Relevant** A classification of relevant is given to a tag that contains a high level semantic description. It must contain the name of the main landmark or surrounding geographical area (localised, not on a city wide scale), such as 'Notre Dame Cathedral' or 'Place de la Concorde'. A tag was also deemed relevant if it contains a mid-level semantic description of the content within an image. For example, if a tag describes the type of landmark or location depicted, it is deemed relevant. E.g. 'Cathedral', 'Facade' or 'Fountain'.
- **Non-Relevant** A classification of non relevant is given to a tag that contains temporal information or a low-level semantic description of an image. Examples of a low-level semantic tag might be 'outdoor', 'sky', 'night', 'river', 'park', 'vacation', 'honeymoon' or 'trip'. The tags 'Paris' and 'France' would also be deemed non-relevant, as the entire corpus is located within these locations.

Each approach evaluated analysed different numbers (k) of top ranked tags outputted by our proposed approaches, where $k = [1, 2, 3, 4, 5]$. It is believed that for this task a balanced performance between precision and recall is desired. Based on this, the F-Measure metric is seen as the most important in the evaluation stage. The F-Measure is the weighted harmonic mean of precision and recall. A table displaying the overall F-Measure scores for each evaluated approach is displayed in Table 4.

From the results in Table 4, it is evident that the tag selection scheme based on tag pair co-occurrences performed with the most desirable level of precision and recall out of all evaluated approaches. When utilising a value of 3 for k, where k is the number of selected tags to annotate a test image, there is a recall score of 94 % an average precision score of 66 % which indicates that for

Table 4. The F-Measure results outputted by each of our evaluated approaches

Ranking Approach	$k = 1$	$k = 2$	$k = 3$	$k = 4$	$k = 5$
TF*IDF	.43	.51	.57	.58	.57
Term Frequency	.51	.58	.63	.65	.65
Image Ranking $(\frac{1}{r})$	.35	.47	.55	.56	.56
Image Ranking $(1 - \frac{r}{q})$	.34	.43	.51	.50	.47
Tag Ranking $(\frac{1}{r})$	.22	.43	.50	.52	.56
Tag Ranking $(1 - \frac{r}{q})$	.58	.70	.70	.69	.67
Geographical $(w = 2)$	.60	.73	.75	.72	.68
Geographical $(w = 4)$	.57	.72	.73	.72	.53
Co-Occurrence	**.75**	**.76**	**.76**	**.73**	**.71**

every test image, there is on average 2 semantically relevant tags assigned to it. The selection schemes based on geographical distributions also performed well, outperforming the traditional tf-idf approach, as did the tag ranking based metric which would support the hypothesis that users would enter tags that they discern to be more semantically relevant before heterogeneous tags.

All but one of our proposed alternative approaches to the tf-idf method outperform it in this task (Term Frequency, Tag Ranking, Geographical distribution and Tag pair co-occurrences). Interestingly, the addition of the idf metric to create tf-idf, actually hinders performance over using the tf measure alone. It is thought that the tf-idf metric performs poorly in this dataset due to the high distribution of landmarks. A commonly photographed landmark such as 'The Eiffel Tower' will have a high distribution within the dataset and therefore will have a low idf score which will bias the metric against commonly occurring but semantically relevant tags.

Of all the proposed approaches, the metric based on image similarity rankings performed the worst. It is believed that this is due to the high precision demonstrated by the image matching framework in Sect. 3. The images returned by the system tend to be very accurate which in turn would make the metric redundant.

6   Conclusions

In this paper, several methods were proposed to garner semantic knowledge about a test image from a collection of noisy community metadata. The majority of textual tags associated with this community data is heterogeneous, subjective, and bears minimal semantic relevance from an information retrieval perspective to the content of an image.

Due to the poor performance of the tf-idf ranking metric in this task, the aim of this paper was to propose alternative approaches to tf-idf for rank the relevance of Flickr tags within a visually similar result set. The results of this evaluation were extremely positive, it can be seen all but one of our proposed approaches (Term Frequency, Tag Ranking, Geographical distribution and Tag pair co-occurrences) outperform the state of the art tf-idf method that has widely been used for similar purposes [2–4], in this task.

Additionally, as part of this work, we also carried out a detailed manual analysis of the accuracy of geo-tags within our dataset and demonstrated that in this domain, they are accurate to within 200 m over 80 % of the time. Based on this information, we could build an image recognition framework that achieved precision scores of over .9.

References

1. Kennedy, L., Naaman, M., Ahern, S., Nair, R., Rattenbury, T.: How flickr helps us make sense of the world: context and content in community-contributed media collections. In: MULTIMEDIA '07: Proceedings of the 15th international conference on Multimedia, pp. 631–640 (2007)
2. Kennedy, L., Naaman, M.: Generating diverse and representative image search results for landmarks. In: WWW '08: Proceeding of the 17th international conference on World Wide Web, pp. 297–306 (2008)
3. Ahern, S., Naaman, M., Nair, R., Yang, J.: World explorer: visualizing aggregate data from unstructured text in geo-referenced collections. In: Proceedings of the Seventh ACM/IEEE-CS Joint Conference on Digital Libraries, pp. 1–10 (2007)
4. Xirong, L., Snoek, C., Worring, M.: Annotating images by harnessing worldwide user-tagged photos. In: Proceedings of the 2009 IEEE International Conference on Acoustics, Speech and Signal Processing, pp. 3717–3720 (2009)
5. Mahapatra, A., Wan, X., Tian, Y., Srivastava, J.: Augmenting image processing with social tag mining for landmark recognition. In: Lee, K.-T., Tsai, W.-H., Liao, H.-Y.M., Chen, T., Hsieh, J.-W., Tseng, C.-C. (eds.) MMM 2011 Part I. LNCS, vol. 6523, pp. 273–283. Springer, Heidelberg (2011)
6. Sigurbornsson, B., Van Zwol, R.: Flickr tag recommendation based on collective knowledge. In: WWW '08: Proceeding of the 17th International Conference on World Wide Web, pp. 327–336 (2008)
7. Bay, H., Tuytelaars, T., Van Gool, L.: SURF: speeded up robust features. In: Leonardis, A., Bischof, H., Pinz, A. (eds.) ECCV 2006, Part I. LNCS, vol. 3951, pp. 404–417. Springer, Heidelberg (2006)
8. Nister, D., Stewenius, H.: Scalable recognition with a vocabulary tree. In: IEEE Computer Society Conference on Computer Vision and Pattern Recognition, pp. 2161–2168 (2006)
9. Sivic, J., Zisserman, A.: DVideo Google: a text retrieval approach to object matching in videos. In: Ninth IEEE International Conference on Computer Vision 2003, Proceedings, pp. 1470–1477 (2003)
10. Lowe, D.: Distinctive image features from scale-invariant keypoints. Int. J. Comput. Vis. **60**, 91–110 (2004)
11. Girardin, F., Blat, J.: Place this photo on a map: a study of explicit disclosure of location information. In: UbiComp (2007)
12. Hollenstein, L.: Capturing vernacular geography from georeferenced tags. Masters thesis, University of Zurich (2008)

Language and Semantics

Semantic Indexing for Efficient Retrieval of Multimedia Data

Xiaoqi Cao$^{(\boxtimes)}$ and Matthias Klusch

German Research Center for Artificial Intelligence,
Stuhlsatzenhausweg 3, Campus D3.2, Saarbrücken 66123, Germany
{xiaoqi.cao,klusch}@dfki.de

Abstract. We present a novel approach, called SemI, to semantic indexing of annotated multimedia objects for their efficient retrieval. The generation of multimedia indices with SemI relies on the semantic annotation of these objects with references to concepts formally defined in standard OWL2 and semantic services described in OWL-S. For scoring the annotated multimedia data in these indices an appropriate semantic similarity measure makes use of approximated logical concept abduction in order to alleviate strict logical false negatives. Efficient query answering over SemI indices is performed with the use of Fagin's threshold algorithm. The results of our comparative experimental evaluation reveals that SemI-enabled multimedia retrieval can significantly outperform representative approaches of LSA- and RDF-based semantic retrieval in this domain in terms of precision at recall, averaged precision and discounted cumulative gain.

Keywords: Semantic-based indexing · Semantic-based multimedia retrieval

1 Introduction

A large number of approaches to mere syntactic and logic-based feature extraction and annotation of multimedia such as images, audio files, 3D models, videos and other kinds of multimedia objects (MO) is available [2,6,12,14,15]. Main challenges of creating indices for an efficient retrieval of such annotated MOs are to decide what kinds of object features shall be indexed and how to best index them for their efficient retrieval from these indices. In this respect, the integration of multimedia with semantic technologies [10] appears promising but still in its infancy.

Regarding the first challenge, most of the previous efforts consider the semantic indexing of the content, formation, event, and meta-property annotations of MOs. Only few works like [13] notice also the functional and behavioral information of MO but do not index them for an efficient MO retrieval. For example, an image of a firework in a public event can be labeled with a formally defined concept *fireworks* and also be annotated with a semantic service (SS) named

© Springer International Publishing Switzerland 2014
A. Nürnberger et al. (Eds.): AMR 2012, LNCS 8382, pp. 165–180, 2014.
DOI: 10.1007/978-3-319-12093-5_9

celebration which is grounded in a respective 3D animation script for the image display. However, to the best of our knowledge, no approach to semantic multimedia retrieval utilizes both the conceptual and service-based functional and behavioral information about MOs.

To address the second challenge, there exist quite a few scoring approaches which rely on non-logic-based semantic tagging and clustering of multimedia metadata properties such as [9,11,15,16,18]. However, these methods have the risk of mis-categorizing or mis-ranking multimedia items due to syntactic mis-matching and ambiguity of keywords. For example, an image of a red coloured MacBook computer tagged with the label *Red Apple* can be mis-categorized into the class *Apple* of images about apples (fruit) with a high score based on mere string matching. RDF[1]-based approaches to semantic multimedia retrieval like [1,8,17] and logic-based indexing [19] solve this scoring problem by logical reasoning on annotations with formally defined concepts in a given formal ontology. However, these more advanced approaches are subject to misclassifications caused by strict rather than approximated logical reasoning or SPARQL[2] pattern matching.

To cope with the both challenges, we propose an approach of semantic indexing, called SemI, for an efficient retrieval of semantically annotated multimedia content. The generation of multimedia indices with SemI relies on the semantic annotation of content with references to logic-based concepts in standard OWL2 and semantic services. For scoring the annotated multimedia data in these indices we devised a semantic similarity measure which makes use of approximated logical concept abduction in order to alleviate strict logical false negatives. Efficient query answering over SemI indices is performed with the use of Fagin's threshold algorithm. The results of our comparative experimental evaluation reveal that SemI-enabled multimedia retrieval can significantly outperform representative approaches of LSA- and RDF-based semantic retrieval in terms of precision at recall, averaged precision and discounted cumulative gain.

We introduce, in Sect. 2, the background and preliminaries in terms of basic notions and definitions required to understand the SemI approach which is described in detail in Sect. 3. The main results of our comparative experimental evaluation of multimedia retrieval performance over SemI are presented in Sect. 4. We discuss related work and conclude the paper in Sects. 5 and 6.

2   Background and Preliminaries

In this section, we briefly introduce the basic notions and definitions which are required to understand the SemI approach presented in the next section.

Semantic annotation of multimedia. Semantic annotation of MOs is to support human- and machine-understandable description of their semantics. The comparison problem of MOs in diverse formats therefore has chance to

---

[1] http://www.w3.org/RDF/
[2] http://www.w3.org/TR/rdf-sparql-query/

Toledo Car :
MOC: τ (Toledo) := Vehicle^P ⊓ Car^P ⊓
=4 hasWeels.Wheel^P ⊓ ∀canCarry.Passenger^P
⊓ ¬ ∀ownership.Private^P;
SS: { transport [
In: (Passenger psg, Location lc), Out: (),
Prec: (availableFor(psg) ∧ reachable(lc)),
Eff: (at(psg, lc))] }
Data: <Scene>...
<Shape>...</Shape> ... </Scene>

Fig. 1. Semantic annotation of a 3D model *Toledo Car*.

be relieved. The semantics of the logic-based semantic annotation of objects are defined in a logic-based ontology language like standard OWL2 or RDFS that are common sense in the semantic web. We call such concepts C, which are used to describe the semantics of a multimedia object, as multimedia object concept (MOC). The complete logical expression of a MOC in the given ontology O is denoted as $\tau(C, O)$ ($\tau(C)$ in short) and contains only logical operators (conjunction ⊓, negation ¬) and quantifiers (universal ∀, exists ∃) over a set of primitive terms ($\cdot^P$). A primitive term can be either a primitive concept indicating the basic type of C, a primitive role term which defines the relation of C with another concept, or a role cardinality restriction with primitive concepts C^P. For example, the logical expression of the MOC *Toledo* of a 3D model *Toledo Car* is shown in Fig. 1 where $Vehicle^P$ and Car^P are primitive concepts; $\forall canCarry.Passenger^P$ and $\neg\forall onwership.Private^P$ are primitive roles; $= 4hasWheels.Wheel^P$ is a role cardinality restriction.

In addition to the static representation of object semantics with formally defined ontological concepts, we also support the functional description of object behaviours with appropriate semantic services. These services are supposed to be grounded in executable service programs such as 3D animation scripts for respective functionalities like the opening or closing of a door. For this purpose, we assume that these semantic services are described in OWL-S[3] with profiles including input/output (In/Out) parameters, preconditions and effects (Prec/Eff). The type of any In/Out parameter is a concept defined in OWL2 while the precondition and effect are specified in conjunctive normal form of 2-valued predicates over variables and individuals. For example, the functionality of the car model *Toledo* in Fig. 1 is captured by an appropriate semantic service named **transport**. The service accepts variables *psg*, *lc* of concepts *Passenger* and *Location* as input, and its execution has the effect described by the predicate *at(psg, lc)*, means the *Passenger psg* is at *Location lc*. Given a request *rs* for services and an offered service *os*, the process of semantic service matchmaking [7] is to compute the matching degrees between both service descriptions by means of determining the logical subsumption relations between their input and output concepts as well as their logical plug-in relation between their

---

[3] http://www.w3.org/Submission/OWL-S/

preconditions and effects. Our approach to off-line semantic indexing of MOs which are annotated with semantic concepts and services make particular use of methods for computing approximated logical concept subsumption relations and service matchmaking.

Logical abduction-based concept subsumption. A concept C (logically) subsumes a concept D (denoted as $D \sqsubseteq C$) if all individuals of D are of the type expressed by C. If the MOC C of a requested MO is subsumed by the MOC D of a MO x, this request is regarded to be fully satisfied by x in terms of the conceptual description. For example, in Fig. 1, the concept $Toledo$ is subsumed by the concept $\tau(Car) := Vehicle^P \sqcap Car^P$. However, strict logical subsumption may lead to false negatives caused by only minor conflicts in the logical definitions which the user might tolerate: the concept $FamilyCar$, $\tau(FamilyCar) := Vehicle^P \sqcap Car^P \sqcap \leq 4hasWheels.Wheel^P \sqcap \forall canCarry.$ $Goods^P \sqcap \forall canCarry.Passenger^P \sqcap \forall ownership.Private^P \sqcap$ $\forall hasNickname.Name^P$, does not strictly subsume $Toledo$ since the latter contains an incompatible part $G = \neg \forall onwership.Private^P$ and a missed part $M = \forall canCarry.Goods^P \sqcap \forall hasNickname.Name^P$ despite its compatible part $K = Vehicle^P \sqcap Car^P \sqcap = 4hasWheels.Wheel^P \sqcap \forall canCarry.Passenger^P$ w.r.t. $FamilyCar$. Logical concept abduction [4] addresses this problem by means of rewriting the incompatible part G with the compatible and missed terms such that the rewritten - hence approximated - concept can be strictly subsumed. For example, the incompatible part of $Toledo$ w.r.t. $FamilyCar$ can be rewritten as $G_R = \forall onwership.Private^P$ such that the approximated concept $Toledo' = (G_R \sqcap M \sqcap K) \sqsubseteq FamilyCar$. Based on the rewritten terms discovered during concept abduction, the approximated score $s_{ab,\sqsubseteq}(\cdot,\cdot)$ of concept abduction-based subsumption can be derived. We discuss this in Sect. 3.

Preliminaries. In the following we provide the definitions of basic notions and assumptions as follows:

Definition 1: *Semantically annotated multimedia object.*
A semantically annotated multimedia object x is defined by the tuple: $x = [id,$ $sd, \tau(C, O_p), SS, da]$ where id denotes the UUID of the object x provided by publisher p; sd the syntactic description of x; $\tau(C, O_p)$ the complete logical expression of the MOC C in the MO ontology O_p in OWL2-DL[4]. O_p is maintained by the publisher p of x, and SS denotes the set of semantic services (cf. Definition 2) in OWL-S which are used to describe the functionality of x; da refers to the data of media object x like a MPEG-4 file or the set of triangle data for rendering as described by a X3D scene graph object. ■

Definition 2: *Semantic service of an annotated multimedia object.*
A semantic service ss of a semantically annotated MO x is defined by the service tuple: $ss = [URI, sd, In, Out, Prec, Eff]$ where URI denotes the URI of service description file in OWL-S; sd the syntactic service description; In the set of input parameters; Out the set of output parameters; $Prec$ the logic expression of the

---

[4] http://www.w3.org/TR/owl-features/

precondition of ss; Eff the logical expression of the effect of ss. All concepts in In and Out (predicates used in $Prec$ and Eff) are defined in a service annotation ontology O_{sp} (predicate set A_{sp}) of the MO publisher. Let $ss.In$, $ss.Out$, $ss.Prec$ and $ss.Eff$ denote $ss[i]$, $ss[o]$, $ss[p]$ and $ss[e]$, respectively. ∎

Definition 3: *Query for annotated multimedia objects.*
A query q for semantically relevant MOs is defined by the query tuple: $q = [sd, \tau(C, O_{req}), SS, N]$ where sd denotes the syntactic (textual) description of the desired MOs; $\tau(C, O_{req})$ is the complete logical expression of the query concept C in the ontology O_{req} of the requester req; SS the set of semantic services (functionalities) that the desired MOs should have; N the number of most relevant multimedia objects requested. ∎

Let $\mathcal{X}$ denote the set of semantically annotated MOs which are maintained in some MO repository MOR and available for indexing. Strictly speaking, a publisher of each MO $x \in \mathcal{X}$ can have its own ontology O_p, O_{sp} and A_{sp} for defining the multimedia object concepts as well as the parameter concepts and predicates. For the sake of simplicity, we assume that the MOR relies on integrated ontologies O, O_s and predicate set A_s, respectively.

3   Semantic Indexing and Retrieval

In this section, we first provide a brief overview of SemI approach which is followed by more detailed presentation.

3.1   Overview

Semantic indexing and retrieval of annotated multimedia objects is done in two subsequent steps: (a) Off-line construction of two indices, one for concepts and one for services, and (b) on-line query processing over these indices.

Off-line construction of indices. Each MO x will be separately ranked into multimedia object concept index I_{MOC} and semantic service I_{SS} index.

- The index I_{MOC} is formed by a set of ranked lists. Each list $R(C')$ ($C' \in O$) corresponds to a concept $C' \in O$ (cf. Definition 1) and contains a list of pairs $[x.id, d(x, C')]$ ($d(x, C') \in [0,1]$) for all $x \in \mathcal{X}$ in the descending order of $d(x, C')$. The scoring value $d(x, C')$ bases on the (abduction-based) approximated concept subsumption relation $s_{ab,\sqsubseteq}(C, C')$ which will be introduced in detail later on.
- The index I_{SS} consists of two subindices: the In/Out concept subindex I_{IO} and the Prec/Eff subindex I_{PE} which base on the sets of input/output parameter concepts in O_s, respectively, precondition/effect predicates defined in A_s. The In/Out subindex contains a set of ranked lists for concepts $C_s \in O_s$: $R(C_s)[i] \in I_{IO}$ and $R(C_s)[o] \in I_{IO}$ which are additionally labeled with suffix $[i]$, respectively, $[o]$. Media object x is ranked in $R(C_s)[i]$ ($R(C_s)[o]$) if a service input (output) parameter C'_s of $ss \in x.SS$ is sufficiently semantically similar

with $C_s \in O_s$: $s_{ab,\sqsubseteq}(C'_s, C_s) \geq \theta$ $(\theta \in \mathbb{R}, \theta \in)$ [0,1]. The score $d_s(x, C_s)[i]$ $(d(x, C_s)[o])$ of x in $R(C_s)[i]$ $(R(C_s)[o])$ is estimated based on the approximated subsumption similarity $s_{ab,\sqsubseteq}(C'_s, C_s)$ and the frequency of C'_s in the services SS of x;

Likewise, the Prec/Eff subindex also comprises a set of ranked lists: two lists $R(\alpha)[p] \in I_{PE}$ and $R(\alpha)[e] \in I_{PE}$ labeled with $[p]$, respectively $[e]$ correspond to one predicate $\alpha \in A_s$. Media object x is ranked in $R(\alpha)[p]$ $(R(\alpha)[e])$ if the non-negative form of α appears in the precondition (effect) of a service $ss \in x.SS$. The score $d_a(x, \alpha)[p]$ $(d_a(x, \alpha)[e])$ of x in $R(\alpha)[p]$ $(R(\alpha)[e])$ is the plausibility value of α in the precondition (effect) of a service $ss \in x.SS$.

Query processing over indices. Based on the indices constructed off-line, a query q for desired MO can be processed efficiently. To do this, two sub-queries in terms of MOC and semantic service are processed in parallel. These sub-processes yield two ranked lists $\mathcal{R}_{MOC}(q)$ and $\mathcal{R}_{SS}(q)$ of MOs. The final ranking of relevant MOs is generated by executing TA process [5] on the $\mathcal{R}_{MOC}(q)$ and $\mathcal{R}_{SS}(q)$. We detail this process in Sect. 3.4.

3.2   Approximated Concept Subsumption

During off-line index construction, the multimedia object concept (or the service parameter concept) C of a MO x is compared with each defined concepts in $C' \in O$ (O_s) via abduction based approximated concept subsumption. That is to compute a score value $s_{ab,\sqsubseteq}(C, C') \in [0,1]$ of C being subsumed by C' under concept abduction of C w.r.t. C'. This makes sense in particular when measuring to what extent a request C can be satisfied by an offer C' or a MO concept C can be categorized into the type C'. For this purpose, the first step is to find, by concept contraction [4], the incompatible G, missed M and compatible K parts of C $(C = G \sqcap K)$ w.r.t. C'. Subsequently, G, M and K are used by concept abduction that computes an approximated concept C_{app} fully subsumed by C'. Denote C^P a primitive term in C, which conflicts with a primitive term $\bar{C}^P$ (named as counter-part) in C'; $|C|$ the number of conjunctive primitive terms in C; $PC(C)$ the set of primitive concepts of C; $PR(C)$ the set of primitive roles of C; $PRE(C)$ the set of role cardinality restrictions of C. The score value $s_{ab,\sqsubseteq}(C, C')$ can be calculated as follows:

$$s_{ab,\sqsubseteq}(C, C') = \frac{|K|}{|C'|} \cdot (1 - s_{acf}(C, C')),$$

$$s_{acf}(C, C') = \frac{1}{|C|} \sum_{C^P \text{ in } G \text{ or } M} \left(s_{cf}(C^P, C') \cdot w(C^P, C') \right),$$

$$s_{cf}(C^P, C') = 1, \quad \text{if} C^P \text{ appears in } M \text{ or } PC(C) \cup PR(C), \tag{1}$$

$$s_{cf}(C^P, C') = \frac{rg(C^P) \backslash rg(\bar{C}^P)}{rg(C^P)}, \text{if } C^P \in PRE(C).$$

where $\frac{|K|}{|C'|}$ is the fraction of compatible part of C w.r.t. C'; s_{acf} the averaged strength of conflicts between C and C'; $s_{cf}(C^P, C')$ the strength of atomic conflict about C^P in C w.r.t. C', which can be further computed as follows: When C^P is in M, it will surely appear in the approximated concept C_{app} of C w.r.t. C'; while in case that C^P is a primitive concept or role in G, any conflict on C^P raises entire rewriting of C^P during concept abduction; if C^P is a role cardinality restriction in G, the conflict strength is the fraction of uncovered range of C^P w.r.t. its counter-part $\bar{C}^P$ in C'. The function $rg(C^P)$ computes the restricted numeric range of $C^P \in PRE(C)$. Each atomic conflict strength $s_{cf}(C^P, C')$ is weighted by $w(C^P, C')$ ($\sum_{C^P \text{ in } G \text{ or } M} w(C^P, C') = 1, w(C^P, C') > 0$) which is determined by estimating the importance of the conflict on C^P w.r.t. C': $w(C^P, C') = \frac{impt(\bar{C}^P, C')}{\sum_{C'^P \text{ in } G \text{ or } M} impt(\bar{C}^P, C')}$. Denote C'_{lp} the direct parent concept of C' in O (O_s); C'' the mutated concept of C', which is computed by replacing $\bar{C}^P$ with C^P if C^P is in G, or removing C^P from C' if C^P is in M. The binary function $impt(\bar{C}^P, C') \in \{a, b\}$ ($0 < a < b \leq 1$) determines the importance of $\bar{C}^P$ in terms of keeping the hierarchy of C' in O (O_s). It returns b if $C'' \sqsubseteq C'_{lp}$ is false; a otherwise. In other words, if the replacement of $\bar{C}^P$ (in C') with C^P or the removal of C^P (from C') makes C'' to be no longer a subclass of C'_{lp}, the conflict on C^P between C and C' then has greater negative impact of C being subsumed by C'.

3.3   Semantic Indexing

The first step of SemI approach is the off-line index creation for MOs with respect to their semantic annotation with concepts and services.

MOC index construction. I_{MOC} comprises a set of ranked lists. A list $R(C') \in I_{MOC}$ about a concept $C' \in O$ (cf. Definition 1) contains a list of pairs $\{(x.id, d(x, C'))\}$ in the descending order of the similarity score $d(x, C')$ between the MO concept C of x and C'. This value is computed by abduction based approximated subsumption above that checks to what extent that x can be categorized to C': $d(x, C') = s_{ab, \sqsubseteq}(x.C, C')$. To index a MO x in I_{MOC} is to compute the relevancy score $d(x, C')$ between x and all $C' \in O$ and insert the pair $\{x.id, d(x, C')\}$ into the corresponding list $R(C')$.

Semantic service index construction. Each semantic service $ss \in x.SS$ provided by a MO x can have input/output parameter sets, precondition and effect. According to SemI, x will be indexed in IO and PE subindices, respectively.

- *IO subindex construction*: It is possible that a concept $C_s \in O_s$ is semantically similar with an input or output parameter C'_s of $ss \in x.SS$. To distinguish the parameter directions, we create two ranked lists $R(C_s)[i]$ and $R(C_s)[o]$ for one concept $C_s \in O_s$. $R(C_s)[i]$ ($R(C_s)[o]$) is labeled with suffix $[i]$ ($[o]$) indicating that the ranked MO have a service input (output) parameter concept C'_s that is sufficiently semantically similar with concept $C_s \in O_s$: $s_{ab, \sqsubseteq}(C'_s, C_s) \geq \theta$ ($\theta \in [0,1], \theta \in \mathbb{R}$). Please note that the suffix only declares the parameter direction, rather than affects the concept definition. $R(C_s)[i]$ ($R(C_s)[o]$) is a list of pairs $\{(x.id, d_s(x, C_s)[i])\}$ $(\{(x.id, d_s(x, C_s)[o])\})$ in the descending order of $d_s(x, C_s)[i]$ $(d_s(x, C_s)[o])$. Each entry associates x with a relevancy score $d_s(x, C_s)[i]$ $(d_s(x, C_s)[o])$ between x and $C_s \in O_s$. We show the computation of $d_s(x, C_s)[l]$ $(l \in \{i, o\})$ as follows:

$$
\begin{aligned}
d_s(x, C_s)[l] &= max_{C'_s \in ss[l]} d_c(C'_s, C_s)[l], \\
&\quad ss \in x.SS \\
d_c(C'_s, C_s)[l] &= fr(C'_s)[l] \cdot s_{ab,\sqsubseteq}(C'_s, C_s), \\
fr(C'_s)[l] &= \frac{|x.SS_{C'_s}[l]|}{|x.SS|} \cdot max_{ss \in x.SS_{C'_s}} \frac{n(C'_s, ss[l])}{|ss|}, \\
\text{subject to:} \quad & s_{ab,\sqsubseteq}(C'_s, C_s) \geq \theta, \theta \in [0, 1], \theta \in \mathbb{R}.
\end{aligned}
\tag{2}
$$

where $x.SS_{C'_s}[l] \subseteq x.SS$ is the subset of services that use C'_s as input ($l = i$) or output ($l = o$) parameter; $n(C'_s, ss[l])$ the appearance number of C'_s in the input ($l = i$) or output ($l = o$) parameter set of ss; $|ss|$ the total number of parameters of ss. $d_c(C'_s, C_s)[l]$ means concept-level relevancy between C'_s and $C_s \in O_s$. It is further estimated by the appearance frequency ($fr(C'_s)[l]$) of C'_s in $x.SS$ and the semantic similarity $s_{ab,\sqsubseteq}(C'_s, C_s)$.

When constructing the service index, the relevancy score $d_s(x.C_s)[l]$ will be computed for each MO x, by considering each parameter concept C'_s of each $ss \in x.SS$ with any concept $C_s \in O_s$. As a result, x can be indexed by multiple ranked lists based on the parameter direction and $s_{ab,\sqsubseteq}(C'_s, C_s)$.

- *PE subindex construction*: Each defined predicate $\alpha \in A_s$ corresponds to two lists $R(\alpha)[p]$ and $R(\alpha)[e]$ in I_{PE}. A MO x is indexed in $R(\alpha)[p]$ $(R(\alpha)[e])$ if the non-negative form of α appears in $ss.Prec$ $(ss.Eff)$ of any service $ss \in x.SS$. Each $R(\alpha)[p]$ $(R(\alpha)[e])$ consists of a list of pairs $\{(x.id, d_a(x, \alpha[p])\}$ $(\{(x.id, d_a(x, \alpha)[e])\})$ in the descending order of $d_a(x, \alpha)[p]$ $(d_a(x, \alpha)[e])$. The value of $d_a(x, \alpha)[p]$ $(d_a(x, \alpha)[e])$ is the plausibility $pl(\alpha, x)[p]$ $(pl(\alpha, x)[e])$ of the predicate α over the preconditions (effects) of all services of x. Let $l' \in \{p, e\}$; $A_s(x)[l']$ the set of non-negative predicates that appear in the preconditions or effects of the services provided by x; and $\mathcal{H} = 2^{A_s(x)[l']}$ the power set of $A_s(x)[l']$:

$$
\begin{aligned}
pl(\alpha, x)[l'] &= 1 - Bel_{A_s(x)[l'] \setminus \alpha}(x), \; Bel_H(x)[l'] = \sum_{h \subseteq H} v(h), \\
v_H(x)[l'] &= \frac{n_H(x)[l']}{n_{\mathcal{H}}(x)[l']}, \text{ s.t.: } v(\emptyset) = 0, \sum_{H \subseteq \mathcal{H}} v(H) = 1, \\
n_{\mathcal{H}}(x)[l'] &= \sum_{H \subseteq \mathcal{H}} n_H(x)[l'], \; n_H(x)[l'] = \sum_{\alpha \in H} n_\alpha(x)[l'], \\
n_\alpha(x)[l'] &= \sum_{ss \in x.SS} P_\alpha(x.ss[l']|\alpha).
\end{aligned}
\tag{3}
$$

where $P_\alpha(x.ss[l']|\alpha)$ is the probability of $x.ss[l']$ being true given that α is true. This value can be computed via the truth table of the Prec/Eff formula $x.ss[l']$. Intuitively, if the truth of a predicate $\alpha[l']$ has larger probability of making $x.ss[l']$ true, then object x is more suitable for being indexed (with greater score value) in $R(\alpha)[l']$. This increases the chance of x to be selected as relevant for a given query q if $\alpha[l']$ appears in the service Prec/Eff of the requested objects.

3.4   Semantic Retrieval

Once semantic indices are created, any on-line query q for relevant multimedia objects can be answered efficiently. Key idea is to process in parallel two sub-queries of q over the indices I_{MOC} and I_{SS}. The results are two ranked lists $\mathcal{R}_{MOC}(q)$ and $\mathcal{R}_{SS}(q)$ of MOs which are semantically relevant for q in terms of query (requested MO) concept and semantic services, respectively. The final ranking is computed by applying Fagin's algorithm TA [5]. We detail the sub-query processing in each aspect and the final aggregation as follows:

MO concept sub-query processing. If the logical expression $\tau(C, O_{req})$ of the requested MO concept C of a query q is not empty, the sub-query for the MO concept will be processed. For this purpose, the process classifies $\tau(C, O_{req})$ as a concept $C' \in O$ into the integrated ontology O and then returns the corresponding rank list as $\mathcal{R}_{MOC}(q)$ of candidate objects that are (partially) relevant to q regarding the requested MO concept C.

Semantic service sub-query processing. If $q.SS$ of a query q is not empty, the sub-query for the semantic service aspect will be processed: Firstly, for each $ss \in q.SS$, a rank list $R(ss)$ of MOs that are relevant to ss is computed. For this purpose, the indices I_{IO} and I_{PE} are searched in parallel and the resulted ranked lists $R(ss)[io]$ and $R(ss)[pe]$ are then merged, which yields to $R(ss)$. Finally, all lists $R(ss)$ of all $ss \in q.SS$ are further merged, which leads to the ranked list $\mathcal{R}_{SS}(q)$ of MOs that partially match q in terms of the requested semantic services.

- *Searching index I_{IO} for each $ss \in q.SS$:* This subprocess first retrieves in parallel a set of ranked lists $\{R(C'_s)[l]\}$ ($l \in \{i, o\}$). Each list corresponds to a distinct parameter concept $C'_s[l]$ in $ss[l]$. For this purpose, the logical expression of each distinct concept C'_s in $ss[i]$ ($ss[o]$) is classified to a concept $C_s \in O_s$ and its corresponding ranked list with suffix $[i]$ ($[o]$) is retrieved. Subsequently, TA [5] is performed on $\{R(C'_s)[l]\}$ to compose a ranked list $R(ss)[io]$ of MOs that are relevant to q in respect of the IO parameters of the requested service ss.

   Let m the cardinality of $\{R(C'_s)[l]\}$ for ss. TA performs a sorted scan over all lists $\{R(C'_s)[l]\}$ from top to bottom. The i-th scan fetches the score values at the i-th positions of all lists in $\{R(C'_s)[l]\}$. Besides, it employs a m-ary function t for computing the aggregated relevancy score and threshold. The general form of t is given in Fagin's work [5] which leaves space to applications for further customization. In our context, we define t as the weighted average of the score vector s that is fetched from input lists per scan. The weight v_j of the j-th list in $\{R(C'_s)[l]\}$ refers to the appearance count of its corresponding concept C'_s in either $ss[i]$ or $ss[o]$:

$$t(s) = \frac{\sum_{j=1}^{m} v_j \cdot s_j}{\sum_{j=1}^{m} v_j} \qquad (4)$$

Each scan may find a new object x_n that does not exist in the current $R(ss)[io]$. To insert x_n into $R(ss)[io]$, it is necessary to compute the aggregated relevancy score $s(x_n, ss)[io]$ of x_n to q w.r.t. the In/Our parameters of $ss \in q.SS$: From each ranked list in $\{R(C'_s)[l]\}$, TA collects (possibly by random access) the so far missed $d_s(x_n.id, C'_s)$ of x_n; and further applies the t function on all $d_s(x_n.id, C'_s)$ in order to compute $s(x_n, ss)[io]$. TA maintains a threshold value T for determining its termination, which is updated with the t function value over the latest scanned values after each scan. TA terminates, if $T \leq s(x, ss)[io]$ for all the ranked objects x in $R(ss)[io]$.

- *Searching I_{PE} for each $ss \in q.SS$*: The searching of I_{PE} for ss results in two sets of ranked lists $\{R(\alpha)[l']\}$ ($l' \in \{p, e\}$) for every non-negative predicate α in $ss[l']$. In addition, it merges the ranked lists in each set into a list $R(ss)[l']$ of MOs that are relevant to ss in terms of $ss[l']$. For this purpose, multiple pairs of the same object x in different lists are merged. The score value $s(x, ss[l'])$ of each ranked object x in $R(ss)[l']$ is computed by applying Gödel minimum t-norm and maximum t-conorm functions according to the conjunctive, respectively disjunctive relations between the predicates in $ss[l']$:

$$\begin{aligned} s(x, ss[l']) &= min_{cla \in ss[l']}(s(x, cla[l'])), \\ s(x, cla[l']) &= max_{\alpha \in cla}(d_a(x, \alpha)[l']). \end{aligned} \tag{5}$$

where $cla[l']$ denotes a clause of disjunctive predicates. Finally, the search process merges $R(ss)[p]$ and $R(ss)[e]$ in order to compute $R(ss)[pe]$ of MOs that are relevant to ss in terms of precondition and effect. Finally, the completion of the parallel computation above triggers the merging of $R(ss)[io]$ and $R(ss)[pe]$, by which the ranked list $R(ss)$ of MOs relevant to q in terms of $ss \in q.SS$ is computed. The relevancy score $s(x, ss)$ of x in $R(ss)$ is the convex combination of the corresponding scores in $R(ss)[io]$ and $R(ss)[pe]$:

$$s(x, ss) = \phi \cdot s(x, ss[io]) + \psi \cdot s(x, ss[pe]), \text{ s.t.: } \phi + \psi = 1. \tag{6}$$

For this, real positive values ϕ and ψ are used for weighting IO and PE aspects respectively. They can vary in specific systems with different concerns.

- *Merging $R(ss)$ for all $ss \in q.SS$*: The subprocess on I_{SS} merges the resulted ranked lists $R(ss)$ of all $ss \in q.SS$. The entries in different lists are merged if they have the same object id. The relevancy score $s(x, q.SS)$ for x with respect to $q.SS$ is the average of the scores $s(x, ss)$ of x in $R(ss)$ for each service ss:

$$s(x, q.SS) = \frac{1}{|q.SS|} \sum_{ss \in q.SS} s(x, ss). \tag{7}$$

Finally, the merged list are resorted in descending order of $s(x, q.SS)$ yielding the ranked list $\mathcal{R}_{SS}(q)$ of objects partially relevant to q with respect to $q.SS$.

Aggregation of rankings. The final ranking is then executed by applying the TA algorithm to $\mathcal{R}_{MOC}(q)$ and $\mathcal{R}_{SS}(q)$. TA terminates if the threshold is not larger than the least score of the N-th (cf. Def.3) entry in the total ranking, or the lists above are all scanned over. During the TA and merging processes, if the score of a MO x is missing in some rank list, the lowest score in that list is used.

Example. We illustrate the principled working of SemI approach by a example:

Semantic indexing: If considering to semantically indexing the 3D model *Toledo Car* in Sect. 2 (abbr. x). the conceptual description and semantic services of x will be indexed separately in I_{MOC}, respectively I_{SS}.

- The former process computes the similarity score $s_{ab,\sqsubseteq}(Toledo, C')$ between the MOC *Toledo* of x and each concept C' in the MOC ontology O. Assume *FamilyCar* (abbr. FC) in Sect.2 is defined in O, which does not strictly subsume *Toledo*. In addition, the direct parent concept *PrivateCar* $\in O$ (abbr. PC) of FC is defined as: $\tau(PC) = Vehicle^P \sqcap Car^P \sqcap \forall ownership.Private^P \sqcap \forall hasNickname.Name^P$. The rank score $d(x, R(FC)) = s_{ab,\sqsubseteq}(Toledo, FC)$ of x w.r.t. the ranking $R(FC)$ can be computed by Formula (1) as follows: Based on the incompatible part $G = \neg \forall onwership.Private^P$ and the missed part $M = \forall canCarry.Goods^P \sqcap \forall hasNickname.Name^P$ of *Toledo* w.r.t. FC, we compute $s_{cf}(\neg \forall onwership.Private^P, FC) = 1$, $s_{cf}(\forall canCarry.Goods^P, FC) = 1$ and $s_{cf}(\forall hasNickname.Name^P, FC) = 1$. For their corresponding weights, the indexing process computes the mutated concept FC' of FC based on the conflict of *Toledo* w.r.t. FC. Assume the range of the importance function is $\{0.1, 0.9\}$. If replacing $\forall onwership.Private^P$ in $\tau(FC)$ with $\neg \forall onwership.Private^P$, the mutated concept FC' is no long subsumed by PC. It follows that $impt(\forall onwership. Private^P, FC') = 0.9$, which further determines: $w(\forall onwership. Private^P, FC) = w(\forall hasNickname.Name^P, FC) = \frac{0.9}{0.9+0.1+0.9} = 0.47$, $w(\forall canCarry.Goods^P, FC) = 0.06$; Subsequently, the averaged conflict strength is $s_{acf} = \frac{1}{|Toledo|} \cdot \sum_{C^P \text{ in } G \sqcap M} (s_{cf}(C^P, FC) \cdot w(C^P, FC)) = \frac{1}{5}(1 \cdot 0.9 + 1 \cdot 0.1 + 1 \cdot 0.9) = 0.38$. Finally, $s_{ab,\sqsubseteq}(Toledo, FC) = \frac{|K(Toledo,FC)|}{|FC|} \cdot (1 - s_{acf}(Toledo, FC)) = \frac{3}{7} \cdot (1 - 0.38) = 0.26$. A pair $(x, 0.26)$ is inserted into the ranked list $R(FC) \in I_{MOC}$.

- Semantic service indexing process ranks x into I_{IO} and I_{PE}. Assume $\theta = 0.25$ and $s_{ab,\sqsubseteq}(x, People) = 0.5$ where the concept *People* is defined in the service ontology O_s, which is the approximated subsumption score of *Passenger* w.r.t. *People*. To determine the appearance frequency (cf.Formula (2)) of the *Passenger* w.r.t. $x.SS$: $fr(Passenger)[i] = \frac{|1|}{|1|} \cdot max_{ss \in x.SS_{Passenger}[i]} \{\frac{1}{2}\}$ $= 0.5$. The concept-level relevancy between *Passenger* and *People* $\in O_s$ is $d_c(Passenger, People)[i] = fr(Passenger)[i] \cdot s_{ab,\sqsubseteq}(x, People) = 0.5 \cdot 0.5 = 0.25$. The concept level relevancy $d_c(Location, People)[i]$ is ignored as the approximated score $s_{ab,\sqsubseteq}(Location, People) = 0.1$ is smaller than θ. Finally, $d_s(x, People)[i] = d_c(Passenger, People)[i] = 0.25$. Assume that $d_s(x, Place)[i] = 0.63$. The pair $(x, 0.25)$ $((x, 0.63))$ is inserted into the ranking $R(People)[i]$ $(R(Place)[i])$. For ranking x in I_{PE}, it computes the plausibilities of the predicates *availableFor*, *reachable* and *at*. By Formula (3) to the power set $\mathcal{H} = 2^{\{availableFor, reachable\}}$ of the predicates in the precondition, we have $pl(availableFor, x)[p] = pl(reachable, x)[p] = 0.9$ and $pl(at, x)[e] = 1$. The pairs $(x, 0.9)$, $(x, 0.9)$ and $(x, 1)$ are inserted into the rankings $R(available)[p]$, $R(reachable)[p]$ and $R(at)[e]$, respectively.

<u>Semantic retrieval</u>: Given a query $q = \{"yellowcar"; \tau(YC, O_{req}) = Vehicle^P \sqcap Car^P \sqcap \forall canCarry. (Passenger^P \sqcap Goods^P); SS = \{[URI; haveFun;$ In(Passenger psg, Goods gds, TargetLocation tl); Out(); Prec(avaiableFor(psg)); Eff(at(psg,tl) $\wedge$ at(gds,tl))]}\}, two subqueries are processed in parallel based on I_{MOC} and I_{SS}:

- The subquery in MO concept aspect considers $\tau(YC, O_{req})$ as input, which is classified into a local concept $FamilyCar$, the ranking $R(FamilyCar)$ is subsequently retrieved as result $\mathcal{R}_{MOC}$ of MOC subquery processing.
- The subquery processing in semantic service aspect is done by two subprocesses over I_{IO} and I_{PE}. For the IO matching, $R(People)[i]$, $R(Goods)[i]$ and $R(Place)[i]$ are retrieved, which is followed by the execution of TA. That yields a ranked list $R(\text{haveFun})[io]$ of objects relevant to q in terms of the IO of haveFun. For the searching in I_{PE}, two rankings $R(availableFor)[p]$ and $R(at)[e]$ are retrieved. The merged ranked list $R(\text{haveFun})[pe]$ of them, in this context, is the final result of PE matching. The merge of $R(\text{haveFun})[io]$ and $R(\text{haveFun})[pe]$ results in the ranking $R(\text{haveFun})$ of objects relevant to q in terms of the service haveFun. Since q has only one service, no further merging is needed. Therefore, $\mathcal{R}_{SS} = R(\text{haveFun})$. Final aggregation is done by running TA on $\mathcal{R}_{SS}$ and $\mathcal{R}_{MOC}$.

4   Evaluation

We present and discuss the results of our comparative experimental evaluation of the retrieval performance of our prototyped SemI system in this section.

Experimental Settings. To the best of our knowledge, there does not exist a test collection with labeled query-answer sets in which the multimedia objects are annotated with both conceptual description and semantic services. Therefore, for the performance experiments, we manually annotated 616 media objects, in particular 591 3D scene graphs in X3D[5] and 25 3D scene graphs in XML3D[6], with MOCs which are defined in an ontology O in OWL2 which comprises 260 concepts, 48 roles and 7 role restrictions. These 3D scene graphs describe real-world objects like a car or computer which functionalities are described by references to 33 semantic services in OWL-S in total. The annotations are embedded into the XHTML files of the scene graphs with standard RDFa. For the labeled query answer set, we created a preliminary set Q of 20 queries each of which labeled with 10 relevant scene graphs that are scored from 1.0 to 0.1 ($rel \in \{1.0, 0.9, \dots, 0.1\}$), others are scored with 0. We set $N = 10$ for all $q \in Q$; $\theta = 0.5$; $\phi = \psi = 0.5$; and $a = 0.1$, $b = 0.9$ for importance function.

For the comparative experimental evaluation of our approach to SemI-enabled multimedia retrieval, we choose (a) the latent semantic analysis (LSA) approach proposed by [16] as its syntactic-based competitor and (b) the RDF triple index-based approach (RIR) [1] as its RDF-based competitor. Based on TML[7], we

---

[5] http://www.web3d.org/x3d/content/examples
[6] http://www.xml3d.org/
[7] http://tml-java.sourceforge.net/

apply the term-document analysis of LSA to the syntactic descriptions of our annotated 3D multimedia models. For each labeled query $q \in Q$, we generate the LSA query vector by counting the frequencies of the terms that appear in the syntactic, conceptual and semantic service descriptions of q. We create the RDF triples of the 3D models for the RIR approach by applying the Jena OWL analyzer[8] to the conceptual descriptions as well as service parameters of models. The precondition and effect of services are encoded in RDF plain literals. In addition, we employ the indexing facilities of MySQL database to index the generated RDF triples in terms of their subject, predicate and object, respectively. For querying the RDF triples, one SPARQL query is generated for each labeled query.

Metrics. We test the SemI search performance in terms of the following metrics:

- Macro-averaged precision (MAP_λ) at 11 recall levels (RE_λ) (MAP@RE) with equidistant steps of 0.1: $MAP_\lambda = \frac{1}{|Q|} \sum_{q \in Q} max\{pre|re \geq RE_\lambda$, for $\forall \langle pre, re \rangle \in PR_q\}$. A set PR_q of precision-recall $\langle pre, re \rangle$ pairs is observed for each query q when scanning the ranked 3D scenes in the returned answer set of q stepwise for true positive. Nearest-neighbor interpolation is used for estimation of missed precision values for some queries at some recall levels.
- Averaged precision (AP): $AP = \frac{1}{|Q|} \sum_{q \in Q} \frac{\text{\# of relevant objects}}{10}$.
- Averaged discounted cumulative gain at rank position 10 (DCG_{10}): $DCG_{10} = \frac{1}{|Q|} \sum_{q \in Q} (rel_1(q) + \sum_{i=2}^{10} \frac{rel_i(q)}{log_2 i})$ where $rel_i(q)$ is the labeled relevancy score of the scene ranked at i in the result list for a query q.
- Averaged query response time (AQRT) measured in milliseconds: $AQRT = \frac{1}{|Q|} T(Q)$ where $T(Q)$ is the total processing time of all queries in Q.

Experimental results. We experimentally compare the search performance of our SemI system with the LSA and RIR systems in terms of MAP @RE, AP, DCG_{10} and AQRT. All tested systems process the same set of 20 labeled queries. As shown in Fig. 2, SemI significantly outperforms its competitors in terms of MAP@RE, AP and DCG_{10}. and achieves, in particular, up to around 25 % and 43 % more MAP at lower recall levels than RIR and LSA, respectively. In addition, the SemI search based on semantic indexing appears to be more beneficial in terms of ranking than RIR and LSA since the resulting DCG_{10} value of SemI is about 1.58 and 1.96 times larger than the DCG_{10} values of RIR and LSA. Main reason for this is that SemI uses the approximated concept subsumption measure which effectively alleviates strict logic false negatives during the semantic indexing process. Syntactic matching of the LSA approach is sensitive to mismatches of exact string matching and the ambiguity of keywords. The same holds for the exact pattern matching of SPARQL query processing by the RIR system. However, the SemI approach is not as time efficient as its competitors.

---

[8] http://jena.apache.org/documentation/ontology/

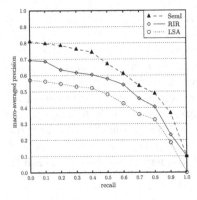

	SemI	RIR	LSA
AP	**0.721**	0.653	0.517
DCG_{10}	**2.087**	1.319	1.067
AQRT (sec)	0.169	0.091	**0.015**

Fig. 2. (left) MAP@RE of SemI, RIR and LSA; (right) AP, DCG_{10} and AQRT of SemI, RIR and LSA.

5   Related Works

Most approaches to the retrieval of multimedia objects rely on their non-logic-based semantic annotation, hence focus on the indexing of merely syntactic-based and domain-specific metadata and object properties. Leung et al. proposed a collaborative indexing approach [11] for image retrieval. The indices of image objects are created according to a specific pre-defined set of terms and adjusted with respect to observed user feedback. The quality of indexing of this approach is sensitive to the uncertainty and arbitrariness of user feedbacks. In [9], by the proposed multimodal descriptor, multimedia objects of multiple types like images, audio files and 3D models are packed if they describe the same real-world entity. Objects with different kinds of features, like geometry and audio frequency, are scored and ranked in different indices by means of various similarity measures. The drawback is that the scoring in different indices become incomparable which makes the ranking aggregation difficult. SemI avoids this, as it can employ the same similarity measure for different object features.

LSA-based approaches like [18] and [16] support the indexing of multimedia objects based on annotation term frequencies in their syntactic descriptions. Singular vector decomposition is used to reduce the number of involved descriptive documents while preserving the similarity structure among the terms. Similar with other syntactic tagging or labeling-based approaches like [15] these efforts are also error-prone when handling those terms with multiple meanings.

In case multimedia objects are annotated with RDF, there exist quite a few mature RDF stores (indices) with efficient SPARQL query processing which can be exploited for object retrieval. Alvez et al. proposed an efficient image retrieval approach [1] based on the indexing of RDF annotations. In [17], RDF annotations are extracted from MPEG7 video files and indices are created for the purpose of efficient retrieval. However, the query on RDF data in these methods is based on a boolean query model which is designed for exact matches. Besides, without a largely redundant indices based on the RDF(S) entailment regime, SPARQL is still prone to ignore the logical concept hierarchy during query processing. The work [8] focuses on the indexing and retrieval for distributed multimedia contents. In addition to the limitations mentioned above, the distributed RDF query answering in [8] closely relies on the prior knowledge of RDF sources and appears sensitive to unshared RDF schema namings. Logic-based annotation on multimedia objects offers more accurate comparison that is less prone to be affected by syntactic mismatching and strict SPARQL pattern matching.

Despite quite a few concept-based multimedia feature annotation/extraction approaches, the problem of how to best preserve the semantics of object annotations during the indexing and query processing remains open and has been rarely discussed so far. Yang et al proposed an approach [19] which establishes high-level content signatures of multimedia contents based on their logic-based representations. In addition, for handling imprecise queries a hierarchical metadata of objects is constructed on the basis of signatures and linguistic relationships. However, the constructed high-level content signature may have lost the formal semantics of the indexed multimedia objects. Besides, the linguistic extension can introduce more ambiguity to the remaining semantics. In contrast, SemI approach preserves as much as possible the semantics during the indexing.

6   Conclusion

In this paper, we presented and comparatively evaluated a new approach to semantic indexing of annotated multimedia. The main contribution is that SemI-enabled multimedia retrieval may achieve significantly better precision at recall, averaged precision and discounted cumulative gain than latent semantic analysis-based or RDF-based multimedia retrieval system available today.

References

1. Alvez, C., Vecchietti, A.: Efficiency analysis in content based image retrieval using RDF annotations. In: Batyrshin, I., Sidorov, G. (eds.) MICAI 2011, Part II. LNCS, vol. 7095, pp. 285–296. Springer, Heidelberg (2011)
2. Arndt, R., Troncy, R., Staab, S., Hardman, L., Vacura, M.: COMM: designing a well-founded multimedia ontology for the web. In: Aberer, K., Choi, K.-S., Noy, N., Allemang, D., Lee, K.-I., Nixon, L.J.B., Golbeck, J., Mika, P., Maynard, D., Mizoguchi, R., Schreiber, G., Cudré-Mauroux, P. (eds.) ASWC 2007 and ISWC 2007. LNCS, vol. 4825, pp. 30–43. Springer, Heidelberg (2007)
3. Bilasco, I., Gensel, J., Villanova-Oliver, M., Martin, H.: An MPEG-7 framework enhancing the reuse of 3D models. In: Proceedings of International Conference on 3D Web Technology, pp. 65–74. ACM (2006)
4. Di Noia, T., Di Sciascio, E., Donini, F.M.: A tableaux-based calculus for abduction in expressive description logics: preliminary results. In: Proceedings of DL Workshop, p. 477. CEUR-WS.org (2009)
5. Fagin, R.: Combining fuzzy information: an overview. ACM SIGMOD Rec. **31**(2), 109–118 (2002). ACM

6. Idrissi, N., Martinez, J., Aboutajdine, D.: Bridging the semantic gap for texture-based image retrieval and navigation. J. Multimedia **4**(5), 277–283 (2009)
7. Klusch, M.: Semantic web service coordination. In: Schumacher, M.,Schuldt, H., Helin, H. (eds.) CASCOM: Intelligent Service Coordination in the Semantic Web, pp. 59–104. Springer (2008)
8. Laborie, S., Manzat, A., Sedes, F.: Managing and querying efficiently distributed semantic multimedia metadata collections. IEEE Multimedia **99**(1), 1–9 (2009)
9. Lazaridis, M., Axenopoulos, A., Rafailidis, D., Daras, P.: Multimedia search and retrieval using multimodal annotation propagation and indexing techniques. Sig. Process.: Image Commun. **28**(4), 351–367 (2012). Elsevier
10. Lee, T.B., et al.: The semantic web. Sci. Am. **284**(5), 34–43 (2001)
11. Leung, C.H.C., Liu, J., Chan, A.W.S., Milani, A.: An architectural paradigm for collaborative semantic indexing of multimedia data objects. In: Sebillo, M., Vitiello, G., Schaefer, G. (eds.) VISUAL 2008. LNCS, vol. 5188, pp. 216–226. Springer, Heidelberg (2008)
12. Mezaris, V., Dimou, A., Kompatsiaris, I.: Local invariant feature tracks for high-level video feature extraction. In: Proceedings of International Workshop on Image Analysis for Multimedia Interactive Services, pp. 1–4. IEEE (2010)
13. Nesbigall, S., Warwas, S., Kapahnke, P., Schubotz, R., Klusch, M., Fischer, K., Slusallek, P.: ISReal: a platform for intelligent simulated realities. In: Filipe, J., Fred, A., Sharp, B. (eds.) ICAART 2010. CCIS, vol. 129, pp. 201–213. Springer, Heidelberg (2011)
14. Papadopoulos, G.T., Briassouli, A., Mezaris, V., Kompatsiaris, I., Strintzis, M.G.: Statistical motion information extraction and representation for semantic video analysis. IEEE Trans. Circuits Syst. Video Technol. **19**(10), 1513–1528 (2009). IEEE
15. Pereira, F., Alves, A., Oliveirinha, J., Biderman, A.: Perspectives on semantics of the place from online resources. In: Proceedings of International Conference on Semantic Computing, pp. 215–220. IEEE (2009)
16. Qi, G., Aggarwal, C., Tian, Q., Ji, H., Huang, T.: Exploring context and content links in social media: a latent space method. IEEE Trans. Pattern Anal. Mach. Intell. **99**(1), 850–862 (2011)
17. Sebastine, S.C., Thuraisingham, B., Prabhakaran, B.: Semantic web for content based video retrieval. In: Proceedings International Conference on Semantic Computing, pp. 103–108. IEEE (2009)
18. Souvannavong, F., Merialdo, B., Huet, B.: Latent semantic indexing for semantic content detection of video shots. In: Proceedings International Conference on Multimedia and Expo, pp. 1783–1786. IEEE (2004)
19. Yang, B.: DSI: A model for distributed multimedia semantic indexing and content integration. ACM Trans. Multimedia Comput. Commun. Appl. **6**(1), 3:1–3:21 (2010). ACM

A Proof-of-Concept for Orthographic Named Entity Correction in Spanish Voice Queries

Julián Moreno Schneider[1]([⊠]), José Luis Martínez Fernández[2], and Paloma Martínez[1]

[1] Computer Science Department, Universidad Carlos III de Madrid, Avda. Universidad 30, 28911 Leganés, Madrid, Spain
jmschnei@inf.uc3m.es, jmartinez@daedalus.es
[2] DAEDALUS – Data, Decisions and Language S.a., Avda. de La Albufera 321, 28031 Madrid, Spain
pmf@inf.uc3m.es

Abstract. Automatic speech recognition (ASR) systems are not able to recognize entities that are not present in its vocabulary. The problem considered in this paper is the misrecognition of named entities in Spanish voice queries introducing a proof-of-concept for named entity correction that provides alternative entities to the ones incorrectly recognized or misrecognized by retrieving entities phonetically similar from a dictionary. This system is domain-dependent, using sports news, especially football news, regardless of the automatic speech recognition system used. The correction process exploits the query structure and its semantic information to detect where a named entity appears. The system finds the most suitable alternative entity from a dictionary previously generated with the existing named entities.

Keywords: Automatic speech recognition · Audio transcription · Question answering · Phonetic representation · Named entity correction · Machine learning

1 Introduction

Automatic speech recognition (ASR) technology can be integrated in information access systems to allow searching on multimedia contents. But, in order to assure an adequate retrieval performance it is necessary to state the quality of the recognition phase, especially in speaker-independent and domain-independent environments.

Particularly important is the case in which named entities are not recognized because the information access system works with incomplete input data and is not able to find any useful information.

ASRs are not able to recognize entities that are not present in its vocabulary so the problem considered in this paper is the misrecognition of named entities in Spanish voice queries. Most works on this area try to modify the acoustic or language models of the ASR, but sometimes there is no possibility of make any change in the ASR system, e.g. if a real-time reaction is needed so there is no time to modify the acoustic model or if some predefined system (as Android or IPhone Speech Recognition) is integrated

© Springer International Publishing Switzerland 2014
A. Nürnberger et al. (Eds.): AMR 2012, LNCS 8382, pp. 181–190, 2014.
DOI: 10.1007/978-3-319-12093-5_10

into an application. In this case, the problem will be addressed from that point of view: **'there is no possibility of making any change in the ASR system'**.

As can be seen in the examples of Table 1, the main problem lies in the entities that are falsely recognized, i.e. the obtained entity is not the one that was said ('Woody Allen' - 'Raúl González'), or it is not even a named entity, i.e. getting a common noun when a named entity was said ('Kun Agüero' - 'una huelga').

Table 1. Examples of misrecognized named entities

Original query	Recognized query
¿Cuál fue la última película dirigida por **Woody Allen**? (What was the last film directed by **Woody Allen**?)	¿Cuál fue la última película dirigida por **Raúl González**? (What was the last film directed by **Raúl González**?)
¿En qué equipo juega **Kun Agüero**? (Which team does **Kun Agüero** play in?)	¿En qué equipo juega **una huelga**? (Which team does **una huelga** play in?)

2 Related Work

Entity correction has not been directly addressed as an independent phase after the recognition process but it has been traditionally managed within the error correction in speech recognition. There are no specific works for named entities in the state-of-the-art because in speech recognition an entity properly recognized is as important as any other word. Thus, most of the work developed in this sense tries to correct the whole transcribed query without focusing on named entities and few studies have addressed correcting only entities. The work described in [7] should be highlighted. It does not correct entities but it performs query expansion using phonetically similar words; [1] applies high-level lexical and syntactic information based on a syllables model for error correction (using also a thesaurus and domain specific dictionaries); [2] uses a rule-based system collecting error patterns and uses them to identify error in the query; [3] studies occurrence probabilities of words in dialogue; [4] detects and corrects errors using a context-based system; [5] makes a post-processing of the ASR output to correct transcription errors by offering alternatives (selected by estimating probabilities) that are not available at the output of the recognizer and [6] lets the user to select the alternatives by means of a confusion network with a large vocabulary.

Some related work can be extracted from the Spoken Term Detection Evaluation (STD) 2006 [12], but these works are related to the search of sequences of spoken words in big audio collections. They do not perform entity correction.

3 Proposal

This paper introduces a preliminary experiment for named entity correction that provides alternative entities to those incorrectly recognized or misrecognized by retrieving phonetically similar entities. This system is domain-dependent, using sports news, specifically football news, regardless of the automatic speech recognition system used. The correction process exploits the query structure and the semantic types of phrases to detect where a named entity appears (for instance, the query "Which team does Cristiano Ronaldo play for?" has the structure "which team does FOOTBALL PLAYER play for?" where the semantic type FOOTBALL PLAYER points a named entity susceptible of being reviewed. The detection of a misrecognized named entity is done by searching it in the previously defined dictionaries (if the dictionary does not contain the named entity then it is considered to be incorrectly recognized).

The treatment needed on these entities is essentially a correction assuming that in some cases the entity will not be correctly recognized or even is not an entity (see the previous example of "Football player").

As the main difference with the related work, it can be pointed out that this proposal is a dictionary-based system that works directly over named entities instead of trying to correct each word or the whole transcription. Considering that there is no specific work on named entity correction in Spanish voice queries the objective is to perform this correction through a post processing over transcribed queries with ASR-independence (considering that it is not possible to modify nor the ASR nor its models). The domain is limited to sports news to get the named entities dictionary.

The system must find the most suitable alternative to the entities received inside the input query. To search for these alternatives a phonetic comparison between the recognized entity (by the ASR) and the entities stored in the dictionary is used and the highest scored entity is obtained (by using string comparison measurements). This functionality (together with the system's architecture) is shown in Fig. 1.

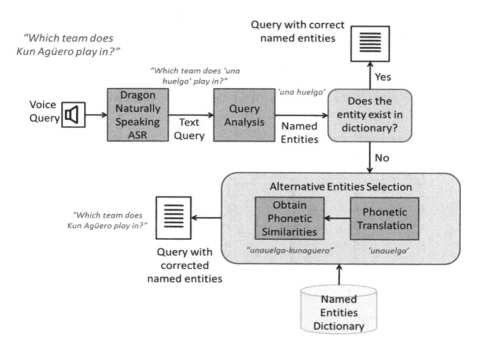

Fig. 1. Architecture of the system

The functionality of the system is structured in three main parts. Firstly, the ASR transcribes the input voice query providing a textual query. In this case two different ASR systems have been used: Dragon Naturally Speaking and Windows Speech Recognizer.

The second part of the architecture is composed by the 'entity extraction module'. It takes care of the query analysis and searches entities inside it. This search is performed by means of a rule-based system that considers five different query patterns. These patterns are shown in Table 2.

Table 2. Available query patterns

Query patterns
¿En qué equipo juega **##JUGADOR**? (What team does **##PLAYER** play for?)
¿Quién marcó el último gol en el estadio **##ESTADIO**? (Who scored the last goal in the stadium **##STADIUM**?)
¿Quién es el máximo goleador del **##EQUIPO**? (Who was the maximum scorer of **##TEAM**?)
¿Cuántos goles ha marcado **##JUGADOR** este año? (How much goals has **##PLAYER** scored this year?)
¿Cuántos penaltis se pitaron en el último partido que se jugó en **##ESTADIO**? (How many kick goals were dictated in the last game played in **##STADIUM**?)

To determine the corresponding pattern for the input query a direct comparison is not appropriate because it can contain transcription errors. Due to that, a bag of words approach is used. It counts the number of words of each pattern contained in the input query. Once the pattern has been determined, the entity is extracted by means of its position in the query.

The next step of the system is checking whether the extracted entity has been correctly recognized or not. This functionality is performed by determining the presence of the entity in the dictionaries. If so, the entity is considered to be properly recognized and no correction is done.

Table 3. Spanish phonetic letter correspondence.

Character	Phoneme	Character	Phoneme	Character	Phoneme
a	a	k	k	t	t
b	b	l	l	u	u
c	c	m	m	v	b
d	d	n	n	w	ui
e	e	ñ	N	x	ks
f	f	o	o	y	i
g	g/j	p	p	z	z
h	–	q	k	Blank	–
i	i	r	r/R		
j	j	s	s		

On the contrary, if the entity does not appear in the dictionaries, then alternative entities for that incorrectly recognized (or misrecognized) entity are provided. A phonetic representation of the input entity is generated using a rule-based system implemented as an adaptation of the work of J. Gil [8] and LivingSpanish [11] for Spanish phonetic letter correspondence. This representation is shown in Table 3.

The similarity between the phonetic representation of the recognized entity and the phonetic representation of the named entities of the dictionary is evaluated. Indeed, several measures have been tested, such as Euclidean, Monge-Elkan, Levenshtein, Needlemann-Wunsch, Smith-Waterman, Gotoh or Smith-Waterman-Gotoh, Jaro, Jaro-Winkler and Soundex distances. A complete description of these measurements can be found in [9, 10].

The implemented dictionary is composed by a set of named entities together with its associated information (in XML format). It structure can be seen in Table 4.

Table 4. Example of named entity stored in dictionary.

```
<dictionary>
        <properties>
                <totalentities>2000</totalentities>
                <totalFP>1900</totalFP>
                <totalFS>42</totalFS>
                <totalFT>42</totalFT>
                <searchedentities>2345</searchedentities>
                <searchedFP>2200</searchedFP>
                <searchedFS>85</searchedFS>
                <searchedFT>60</searchedFT>
        </properties>
        <entities>
                <entity>
                        <text>Lionel Messi</text>
                        <type>FootballPlayer</type>
                        <popularity>0.9</popularity>
                        <historic>4</historic>
                </entity>
                ...
        </entities>
</dictionary>
```

The dictionary is composed by a set of properties and 2004 different named entities. The properties are total number of stored entities, total number of each type of entity (players, stadiums and teams) and the times that each type of entity has been selected as a suitable alternative. The entities are divided into 1874 football player names, 42 stadium names and 88 team names.

Besides, each entity is composed by its associated text, the type it belongs to, a popularity score defined by an expert and the number of times it has been selected as alternative.

4 Preliminary Evaluation

The first evaluation was carried out using 168 Spanish voice queries read by 7 different users. These queries are uniformly distributed over the five query patterns. Some examples are shown in Table 5.

Table 5. Examples of input queries read by users.

Original query	Transcribed query (with one ASR)
¿En qué equipo juega juan josé collantes?	El equipo, Juan José Collantes
¿Quién marcó el último gol en el estadio los pajaritos?	Quien marcó lo temor en el estadio los pajaritos
¿Quién es el máximo goleador del valencia?	Quién es el máximo goleador del Valencia

The queries have been transcribed using both ASRs. The first ASR was used with four different acoustic models trained with videos of different length. The first model was not trained; the second was trained only with approx. 5 min of sport news videos; the third model was trained with 50 min of sport news videos; and the fourth was trained with 40 min of football news videos. The second ASR (WSR) was not trained and only its default model was used.

The first test is performed to validate the functionality and performance of the entity classification module. In order to do that, the entities were manually extracted from the transcribed queries and then classified into types for using them as reference. The five different ASR models are tested and four different classification techniques are shown. The first technique is a direct comparison between the transcribed query and the patterns; the second is a bag-of-words technique that uses all the words (including the entity tags (#footbalplayer)); the third improves the bag-of-word technique by eliminating the tags; and the last performs a phonetic comparison between the query and the patterns.

The results obtained by the entity type classifier and the entity extractor are shown on the next table (Table 6). As can be seen, the phonetic comparison classification is the better.

Table 6. Results of entity classification module validation

	WSR	DNS - 1	DNS - 2	DNS – 3	DNS – 4
Direct comparison	0	0	0	0	0
Complete BoW	78,57 % (132)	70,83 % (119)	72,62 % (122)	73,21 % (123)	58,33 % (98)
Limited BoW	82,74 % (139)	64,88 % (109)	72,02 % (121)	70,83 % (119)	54,17 % (91)
Phonetic comparison	88,69 % (149)	86,9 % (146)	90,48 % (152)	89,88 % (151)	77,38 % (130)

The second test was performed to validate the phonetic representation system. The phonetic representation that was finally used works properly as long as the entities are 'Spanish' entities while it fails with entities from other languages. Table 7 shows some examples.

Table 7. Examples of phonetic representation of named entities

Named entity	Phonetic translation	Expected correct translation?	Correct translation
Cristiano Ronaldo	kristianoronaldo	Yes	Yes
Schweinsteiger	scuieinsteijer	No	No
HamitAltintop	amitaltintop	No	Yes
Lionel Messi	lionelmessi	Yes	Yes

The corpus used for that task was composed by 168 entities. These entities were introduced into the phonetic representation system and the output of each entity was manually revised to determine if it was correctly represented. The amount of correct represented entities was 150 entities. That leads to an accuracy of 89,29 % (Table 8).

Table 8. Phonetic representation accuracy

Phonetic representation accuracy	89,29 %

It can be remarked that all the entities that were Spanish names were correctly represented while the errors occur when there is a foreign entity (*giorgioventurin-jiorjiobenturin, ilijanajdoski-ilijanajdoski*). This is a known problem of the phonetic representation module since it has been only implemented for entities in Spanish.

The last test validates the module that retrieves alternative entities using the same corpus of 168 entities. For this purpose some different phonetic distance measurements where used. The best comparison measurements for phonetic comparison are Levenshtein Distance and Monge-Elkan-Levenshtein Distance obtaining figures near to 56.55 % in Top@10 (Levenshtein) and 50.60 % in Top@1 (Monge-Elkan-Levenshtein) (Tables 9 and 10).

Table 9. Preliminary evaluation results for entity correction using WSR

Precision without entity correction	Precision with entity correction		
		Levenshtein	Monge-Elkan-Levenshtein
30.95 %	Top@1	49.40 %	50.60 %
	Top@3	52.38 %	53.57 %
	Top@5	52.98 %	54.17 %
	Top@10	56.55 %	55.36 %

Table 10. Preliminary evaluation results for entity correction using DNS

Precision without entity correction	Precision with entity correction		
		Levenshtein	Monge-Elkan-Levenshtein
19.05 %	Top@1	38.10 %	38.10 %
	Top@3	42.86 %	41.07 %
	Top@5	42.86 %	41.07 %
	Top@10	43.45 %	42.26 %

As can be seen in the previous results, the phonetic entity correction system increases the accuracy in both cases: using WSR it increases 19,65 % and with DNS the increment is a 19,05 % for the total amount of entities.

Table 11 shows the results of the entity correction module using a system with multiple dictionaries, i.e. it uses a different dictionary for each entity type. It depends on the performance of the entity type extraction but increases the entity correction in 3 % approximately.

Table 11. Results knowing the type of the entity to be corrected

	Precision without entity correction	Precision with entity correction	
		Levenshtein	Monge-Elkan-Levenshtein
WSR	30.95 %	52.98 %	51.79 %
DNS	19.05 %	41.67 %	40.48 %

These are promising results considering that it is an ongoing work.

5   Some Conclusions

The correction of named entities in IR systems accessed by voice is absolutely necessary. There are mainly two reasons for that; on the one hand the entities are an essential unit of information for IR systems, on the other hand in most scenarios the acoustic and language model of the ASR cannot be modified to improve the results, letting this to a post processing after the recognition process.

Some comparison distance measurements were tested and finally only two of them were selected for final tests. These measures have proved very useful when making phonetic comparison. Additionally, the results are even improved when the arithmetic mean between both measures is used as a new measure (Monge-Elkan-Levenshtein).

Storing the database entities in an XML file was a good decision covering both objectives, on the one hand having a structured design, on the other hand allowing an easy understanding.

After a preliminary evaluation, the 52 % (approx.) of entity alternatives are right choices, and after making a qualitative assessment, it can be said that whenever the entity has not been recognized, the system will be able to offer an appropriate alternative.

This work has got promising results but is still in an early development stage. There are some improvements that can be outcome to the system. The first improvement would be the adaptation of the phonetic representation system to take into account different pronunciations (accents) and especially words in other languages. Besides, some ASR's return acronyms and the phonetic expansion of these acronyms could be useful for the desired purpose.

The way the system uses to determine if the entity has been correctly recognized must be further studied. Now it considers a correctly recognized entity if it is present inside the dictionaries. A phonetic comparison together with a threshold could be better.

The number of input queries that the system recognizes is now limited to 6 and it must be increased as well as the amount of entity types (now limited to three). In general, making some tests on different domains or adding more than one domain at the same time is a crucial point in order to validate the system.

Acknowledgments.   This work has been partially supported by the Regional Government of Madrid under the Research Network MA2VICMR (S2009/TIC-1542) and by the Spanish Center for Industry Technological Development (CDTI, Ministry of Industry, Tourism and Trade) through the BUSCAMEDIA Project (CEN-20091026).

References

1. Jeong, M.: Using higher-level linguistic knowledge for speech recognition error correction in a spoken QA dialog. In: Proceedings of the HLT-NAACL Special Workshop on Higher-Level Linguistic Information for Speech Processing, pp. 48–55 (2004)
2. Kaki, S., Eiichiro Sumita, and Hitoshi Iida.: A Method for Correcting Speech Recognition Using the Statistical features of Character Co-occurrence, COLING-ACL'98, 653–657 (1998)
3. Ringger, E.K., Allen, J.F.: A fertility model for post correction of continuous speech recognition ICSLP'96, pp. 897–900 (1996)
4. Sarma, A., Palmer, D.: Context-based speech recognition error detection and correction. In: Proceedings of HLT-NAACL (2004)
5. Ringger, E.K., Allen, J.F.: Error correction via a post-processor for continuous speech recognition. In: Proceedings of the International Conference on Acoustics, Speech, and Signal Processing, vol. 1, pp. 427–430, Atlanta, GA (1996)
6. Ogata, J., Goto, M.: Speech repair: quick error correction just by using selection operation for speech input interfaces. In: Proceedings of Eurospeech'05, pp. 133–136 (2005)
7. Reyes-Barragán, A., Villaseñor-Pineda, L., Montes-y-Gómez, M.: Expansión fonética de la consulta para la recuperación de información en documentos hablados. Septiembre, 2011 Procesamiento del Lenguaje Natural, Revista n° 47, pp. 57–64 (2011)
8. Gil, J. Transcripción fonética: Representación escrita de los sonidos que pronunciamos. Fonética para profesores de español: De la teoría a la práctica. p. 547. Arco/Libros (2007)

9. Gusfield, D.: Algorithms on Strings, Trees, and Sequences: Computer Science and Computational Biology. Cambridge University Press, New York (1997)
10. Cohen, W.W., Ravikumar, P., Fienberg, S.E.: A comparison of string distance metrics for name-matching tasks. In: Proceedings of II Web 2003 – IJCAI Workshop on Information Integration on the Web, pp. 73–78 (2003)
11. LivingSpanish: Correspondencia de fonemas y grafías en español. http://www.livingspanish.com/correspondencia-fonetica-grafia.htm (2011)
12. Fiscus, J.G., Ajot, J., Garofolo, J.S., Doddington, G.: Results of the 2006 spoken term detection evaluation, pp. 45–50 (2007)

Music Retrieval

From Improved Auto-Taggers to Improved Music Similarity Measures

Klaus Seyerlehner[1], Markus Schedl[1(✉)], Reinhard Sonnleitner[1],
David Hauger[1], and Bogdan Ionescu[2]

[1] Department of Computational Perception,
Johannes Kepler University, Linz, Austria
markus.schedl@jku.at
[2] Image Processing and Analysis Laboratory,
University Politehnica of Bucharest, Bucharest, Romania

Abstract. This paper focuses on the relation between automatic tag prediction and music similarity. Intuitively music similarity measures based on auto-tags should profit from the improvement of the quality of the underlying audio tag predictors. We present classification experiments that verify this claim. Our results suggest a straight forward way to further improve content-based music similarity measures by improving the underlying auto-taggers.

Keywords: Music information retrieval · Music similarity · Auto-tagging · Music recommendation · Tag prediction

1 Introduction

Audio *tags* are semantic textual annotations (e.g., *"beat"*, *"fast"* or *"rock"*) that are used to describe songs. Typically, tags are collected by large online music platforms such as *Last.fm*[1] that allow users to annotate the songs they are listening to. However, there also exist several other methods to collect tag information [13]. Audio tags can also be obtained through surveys, music annotation games or web-mining. Another variant which is in the focus of this paper is to obtain tag information via *auto-tagging*. An auto-tagger is typically a purely content-based method (i.e. only based on a set of audio features extracted out of the audio signal) for predicting tags which might be associated with a song. Consequently, one can interpret an auto-tagger as a method that transforms *an audio feature space* into *a semantic space*, where music is described by words. This process is often referred to as *automatic tag prediction* or *automatic tag classification*.

While automatic tag classification can be viewed just as an interesting performance task that extends traditional genre classification to multi-label classification, there also exist several application scenarios where auto-tagging can be extremely beneficial. For example, auto-tags can be used to visualize and explore

---

[1] www.last.fm

© Springer International Publishing Switzerland 2014
A. Nürnberger et al. (Eds.): AMR 2012, LNCS 8382, pp. 193–202, 2014.
DOI: 10.1007/978-3-319-12093-5_11

music collections in a semantic space without relying on community data, which is typically incomplete or unavailable for some songs in a personal music collection. Another application field of auto-tags is to compute song similarities from automatically estimated tags. This approach to music similarity is of special interest in this paper, as we want to study how the quality of the estimated auto-tags influences the quality of a music similarity measure that is build on top of them. Intuitively, a tag based music similarity measure should profit from improving the quality of the underlying tag predictors. However, no empirical evidence of this assumption exists to the best of our knowledge. Thus, the main contribution of this paper is to fill this gap and provide experimental evidence of this relation. From a technical point of view, this relation is especially interesting, as it transforms the ill-defined task of improving a music similarity measure into the well-defined machine learning task of predicting audio tags and defines a straightforward way to improve content-based music similarity measures.

The outline of the remainder of this paper is as follows: In the subsequent section we give a brief introduction to automatic tag classification, as well as auto-tag based music similarity and discuss related work in these research areas. Then, in Sect. 3 we present the outline of the conducted experiment, which is structured into two successive experiments. We report on these sub-experiments in Sects. 4 and 5 respectively. Finally, in Sect. 6 we provide a brief summary and discuss the obtained results.

2   Related Work

While automatic tag prediction recently gained a lot of research attention and can be considered an emerging research area in Music Information Retrieval (MIR), the idea of predicting tags is relatively old. To the best of our knowledge Whitman et al. [17], Slaney [12] and Berenzweig et al. [1] were the first to introduce concepts related to auto-tags. While Slaney was working on animal sounds only, he already introduced the concept of an acoustic and a semantic space. In contrast, Whitman was already working on music and interpreted automatic tag prediction as a multi-label classification problem, while Berenzweig called the semantic tag space *"anchor space"* and was the first to compute similarities among songs based on tag information. However, it seems that in these early days the lack in computational resources and the unavailability of adequate tag sources limited further development in this research direction. Then driven by the general growing interest in tags in the MIR community around the year 2006 this idea was picked up again by West et al. [16], Eck et al. [4], Mandel and Ellis [7] and Turnbull et al. [14]. The latter introduced in [14] a first formal definition of the tag prediction task:

The task of predicting tags can be interpreted as a special case of multi-label classification and can be defined as follows: Given a set of tags $T = \{t_1, ..., t_A\}$ and a set of songs $S = \{s_1, ..., s_R\}$ predict for each song $s_j \in S$ the tag annotation vector $y = (y_1, ..., y_A)$, where $y_i > 0$ if tag t_i has been associated with the audio track by a number of users, and $y_i = 0$ otherwise. Thus, the y_i's describe the

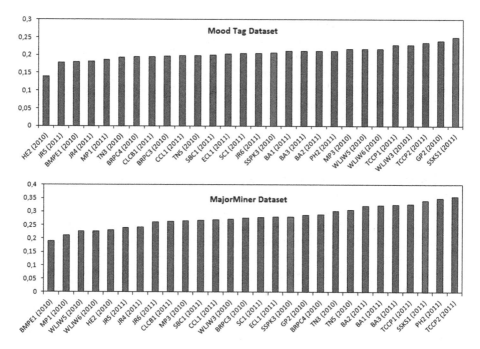

Fig. 1. Comparison of tag classification algorithms (**average per tag f-Score**) of the MIREX 2010 and MIREX 2011 campaigns.

strength of the semantic association between a tag t_i and a song s_j and are called *tag affinities, semantic weights* or *tag profiles*. If the semantic weights are mapped to $\{0, 1\}$, then they can be interpreted as class labels, which can be used for training and evaluating tag classifiers.

Recent research work in automatic tag prediction has mainly put the focus on the classification part. For instance in [2] and in [8] the authors propose to extend straightforward binary classification strategies by introducing a second layer of tag classifiers. The inputs of the second classification layer are the predictions of the binary tag classifiers from the first layer. This advanced approach allows to make use of inter-tag correlations in the second classification stage and is one way of improving the classification part of an auto-tagger.

Another recent trend in the context of automatic tag classification is to use auto-tags for music similarity estimation [3,15]. The common main idea behind *auto-tag based music similarity systems* is to first estimate a song's tag profile and then compare the estimated tag profiles of two songs. Interestingly, during the last two runs of the MIREX Audio Music Similarity and Retrieval task (2010 and 2011), hybrid (content- and tag-based) music similarity measure achieved the first rank [10]. This clearly indicates that auto-tag based similarity will become an important part of music similarity estimation.

3   Experiment: Outline

Intuitively, qualitative improvements in automatic tag classification should indirectly also lead to qualitative improvements of auto-tag based music similarity systems. To experimentally verify this assumption, we have conducted an experiment which is subdivided into two phases:

- In the first phase (*Train Auto-Taggers*) of the experiment, we train two sets of tag classifiers on the same tag classification datasets: A high quality set of tag classifiers (*high*) and a set of low quality tag classifiers (*low*). For both tag classification datasets we ensure that the *high* quality tag classifiers perform significantly better than the *low* quality tag classifiers.
- Then in the second phase (*Estimate Tag Profile Similarities*) of the experiment, we built two tag-based music similarity measure precisely in the same way. The only difference is that, one is based on the *high* quality tag-predictors and the other one is based on the *low* quality tag predictors. Both similarity measures are then evaluated via nearest-neighbor genre classification on six well-known datasets to identify qualitative differences between the two variants. In case the assumption about the relation between tag classification and tag-based music similarity is correct, it is expected that the music similarity measures based on the high quality tag predictors performs significantly better than the similarity measure based on the low quality tag predictors.

The following two sections present details and results of the execution of the two sub-experiments.

4   Phase 1: Train Auto-Taggers

To generate two sets of tag predictors of different quality we make use of the tag classification systems we have submitted to the MIREX tag classification tasks. The low quality tag predictors are generated by our 2010 submission [10], while the high quality tag predictors are generated by the improved submission in 2011 [9]. Both submissions are based on so-called *block-level* audio features [11]. In contrast to standard audio features these features allow to better capture local temporal information and together form a highly descriptive audio feature set. The descriptive power of this feature set has already been demonstrated during several evaluation campaigns (MIREX'2010 [10], MIREX'2011 [9] and MediaEval'2011 [5]).

The main differences between the 2010 submission and the 2011 submission are that two additional block-level features, the *Local Single Gaussian Model* (LSG) and the *George Tzanetakis Model* (GT), were added in 2011 and that the classification method was changed. In the 2010 submission the dimensionality of the high dimensional audio feature space was first reduced via a PCA (the extracted block-level forms a 9448 dimensions vector space), as it was not tractable to use the uncompressed feature set in combination with a *support vector machine* classifier. For our 2011 submission we decided to directly use the uncompressed feature set and replace the *support vector machine* classifier by a *random forest* classifier, since random forest classifiers can handle very high dimensional feature spaces and can make use of multi-core CPUs. For a more detailed description of the two algorithms we refer to [9,10].

Figure 1 visualizes the average per tag f-Score of all submissions in 2010 and 2011. This allows to compare our system to other tag classification approaches, but what is even more interesting in the context of this paper is that there is obviously a significant improvement of our submission in 2011 over our submission in 2010. Thus, this comparison suggests to use the 2010 submission to generate the low quality tags and the 2011 system to generate the high quality tags. It is, however, worth mentioning that the quality of our 2010 system, although it is used to generate the low quality tag predictors, is still quite competitive and not just a baseline system. In the following we introduce two tag classification datasets. These datasets are then used to first experimentally verify the qualitative difference of the two approaches and then to learn a pair of low and high quality tag predictors form each dataset.

4.1 Datasets

Magnatagatune. The first dataset in our evaluation is the Magnatagatune [6] dataset. This huge dataset contains 21642 songs annotated with 188 tags. The tags were collected by a music and sound annotation game, the TagATune[2] game. The dataset also includes 30 s audio excerpts of all songs that have been annotated by the players of the game. All the tags in the dataset have been verified (i.e. a tag is associated with an audio clip only if it is generated independently by more than 2 players, and only tags that are associated with more than 50 songs are included). From the tag distribution (Fig. 2) one can see that in terms of binary decisions (tag present/not present), the classification tasks are extremely skewed. So 110 out of the 188 tags apply to less than 1 % of all songs and the 87 most frequently used tags account for 89.86 % of all annotations.

RadioTagged. The second dataset in our evaluation is called RadioTagged, because the audio files in this dataset were recorded from internet radio streams. Audio fingerprinting was used to identify the recorded tracks and retrieve artist, album, song name and cover art. In a second step *Last.fm* was queried with artist and track name to obtain tags on track level. We only kept those songs for which we were able to retrieve tags. After this process we ended up with a total of 10557 full length songs and 1072 tags. From Fig. 3 one can see an effect similar to the one of the Magnatagatune dataset: most of the tags do only apply to a fraction of all songs. This is especially interesting as the two datasets originate from different annotation processes (annotation game and social tagging), but according to their summarization plots the general structure of both datasets is still very similar. The most frequently applied tag is *"rock"* and is set for 48.2 % of all songs, while 569 out of 1072 tags are applied to less than 1 %. Compared

---

[2] http://www.tagatune.org

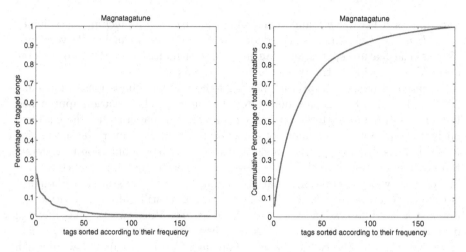

Fig. 2. Percentage of annotated songs per tag (left) and percentage of accumulated annotations of the first k most frequent tags (Magnatagatune).

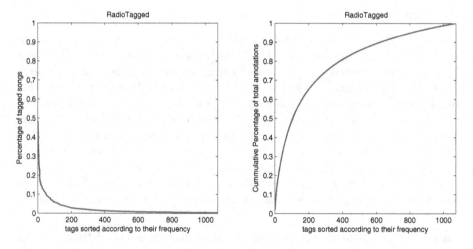

Fig. 3. Percentage of annotated songs per tag (left) and percentage of accumulated annotations of the first k most frequent tags (RadioTagged).

to the Magnatagatune dataset the number of tags in this dataset is far higher, while it does only contain about half the number of songs.

4.2   Evaluation Metrics

In our experiments all reported quality measures are first computed separately for each tag and are then averaged over all tags to come up with a global evaluation metric. For each tag we compute the following standard quality metrics: *f-Score*, *AUC-ROC* and the *accuracy*. Furthermore, we also report a modified

Table 1. Qualitative comparison of the high quality (*high*) and the low quality (*low*) tag classification systems. Marked results (*) indicate statistically significant differences.

	Magnatagatune		RadioTagged	
	low	*high*	*low*	*high*
avg f-Score	0.1575	**0.2225***	0.0490	**0.0878***
AUC-ROC	0.6951	**0.8615***	0.5691	**0.6622***
Acc.	0.9719	**0.9749**	0.9608	**0.9640**
p@50 AB	0.1947	**0.2661***	0.0349	**0.0959***

variant of the *precision @k (p@k)* quality metric, which is called *precision @k above baseline (p@k AB)*. For each tag we first estimate the baseline precision at k, which is the expected precision when k samples are randomly drawn from the ground truth population without replacement. Obviously the baseline precision of each tag clearly depends on the individual class distribution. To reduce the influence of the individual class distribution of the tags on the overall metric, we first compute the *p@k* and then subtract the estimated baseline, which then gives *p@k AB*. For all our experiments we choose a fixed value of $k = 50$.

4.3 Results

We have evaluated both tag classification systems (*high* and *low*) on both tag classification datasets via a two fold-cross-validation. Table 1 shows the results. For both datasets we could identify a significant difference between the *high* and the *low* quality system. So we have ensured for both datasets that the learned low and high quality auto-taggers differ significantly in terms of quality. In the next phase we will use these auto-taggers to predict tag profiles for songs of well-known music genre classification datasets to assess quality of a similarity measure based on these tag-predictors.

5   Phase 2: Estimate Tag Profile Similarities

In the previous phase we have ensured that the generated auto-taggers (*high* and *low*) differ significantly in terms of quality. Therefore, in case our assumption is correct we should end up with a higher quality music similarity measure for tag profiles estimated by the high quality tag classifiers compared to tag profiles estimated by the low quality tag classifiers. To actually estimate song similarities we follow the approach in [10] and compare the generated auto-tag profiles using the Manhattan distance.

Then, to assess the quality of both approaches (*low* and *high*) we evaluate them on six different genre classification datasets. In the following two subsections these datasets and the utilized quality measure are briefly introduced.

5.1 Datasets

To measure the quality of the resulting similarity estimates, we follow the standard approach in MIR and evaluate different approaches via nearest neighbour genre classification. In our experiments 6 different well-known genre classification datasets are used: *GTZAN, ISMIR 2004 Genre, ballroom, Homburg, Unique, 1517-Artist*. It is worth mentioning that all these datasets are publicly available.

5.2 Evaluation Metrics

To assess the quality of the music similarity estimates of an algorithm the resulting similarity matrix containing all estimated pairwise distances among all songs in a collection is analyzed. The percentage of genre matches in the top k most similar songs is computed for each query song. To obtain an overall quality measure the per song results are averaged over the whole dataset. This quality measure is one of the automatic statistics that is computed at the MIREX Audio Music Similarity and Retrieval Task. There it is called *Genre Neighbourhood Clustering*, but is named *precision @k* (*p@k*) here. Interestingly, the results of the human music similarity evaluations at MIREX are year by year highly correlated with the *p@k* quality metric. Thus, this measure is an excellent choice to automatically assess the quality of music similarity systems. In our evaluation we will report the *precision @10* (p@10). For datasets *1517-Artist, Homburg* and *Unique* artist filtered results are reported.

5.3 Results

The results of this experiment are summarized in Table 2. For both sets of classifiers, the one trained on the *Magnatagatune* dataset and the one trained on the *RadioTagged* dataset, the high quality auto-tag similarity measure outperforms the low quality version **on all six evaluation datasets**. Although this is inline with the intuitive expectations this result is a very encouraging, as the main

Table 2. Comparison of auto-tag based music similarity algorithms (*p@10*) based on *high* and *low* quality auto-taggers. Marked results (*) indicate statistically significant differences.

Dataset	Magnatagatune		RadioTagged	
	low	*high*	*low*	*high*
GTZAN	0.4569	**0.5765***	0.5459	**0.6167***
ISMIR 2004	0.7715	**0.7955**	0.7217	**0.7472**
ballroom	0.4089	**0.4645***	0.3983	**0.6143***
Homburg	0.3322	**0.4036***	0.3892	**0.4108***
Unique	0.5506	**0.6114***	0.5618	**0.6062***
1517-Artists	0.1839	**0.2355***	0.2128	**0.2475***

implications of this experiment is that any improvement in automatic tag classification, will also lead to an improved content-based music similarity measure. Consequently, improvements both on the feature side and on the classification part can easily be integrated into an existing audio similarity algorithm.

6 Conclusions

In this paper we have focused on the relation between automatic tag classification and auto-tag based music similarity. Intuitively, auto-tag based music similarity algorithms should directly profit from qualitative improvements in automatic tag classification. Based on the results of our experiment we conclude that this assumption is correct, which is a very encouraging result, as this defines a systematic and straightforward way to further improve content-based music similarity algorithms and content-based music recommender systems by improving the underlying automatic tag prediction systems.

Acknowledgements. This research is supported by the Austrian Science Funds (FWF): P22856-N23 and Z159.

References

1. Berenzweig, A., Ellis, P., Lawrence, S.: Anchor space for classification and similarity measurement of music. In: Proceedings of the 2003 International Conference on Multimedia and Expo (ICME-03) (2003)
2. Bertin-Mahieux, T., Eck, D., Maillet, F., Lamere, P.: Autotagger: a model for predicting social tags from acoustic features on large music databases. J. New Music Res. **37**(2), 115–135 (2008)
3. Bogdanov, D., Serrà, J., Wack, N., Herrera, P., Serra, X.: Unifying low-level and high-level music similarity measures. IEEE Trans. Multimedia (2010)
4. Eck, D., Lamere, P., Mahieux, T.B., Green, S.: Automatic generation of social tags for music recommendation. In: Proceedings of the 21st Conference on Neural Infromation Processing Systems (NIPS-07) (2007)
5. Ionescu, B., Seyerlehner, K., Rasche, C., Vertan, C., Lambert, P.: Content-based video description for automatic video genre categorization. In: Schoeffmann, K., Merialdo, B., Hauptmann, A.G., Ngo, C.-W., Andreopoulos, Y., Breiteneder, C. (eds.) MMM 2012. LNCS, vol. 7131, pp. 51–62. Springer, Heidelberg (2012)
6. Law, E., Ahn, L.: Input-agreement: a new mechanism for collecting data using human computation games. In: Proceedings of the 27th International Conference on Human Factors in Computing Systems (CHI-09) (2009)
7. Mandel, M.I., Ellis, D.P.W.: A web-based game for collecting music metadata. J. New Music Res. **37**(2), 151–165 (2008)
8. Ness, S., Theocharis, A., Tzanetakis, G., Martins, L.: Improving automatic music tag annotation using stacked generalization of probabilistic SVM outputs. In: Proceedings of the 17th ACM International Conference on Multimedia (2009)
9. Seyerlehner, K., Schedl, M., Knees, P., Sonnleitner, R.: A refined block-level feature set for classification, similarity and tag prediction. In: Online Proceedings of the 7th MIR Evaluation eXchange (MIREX-11) (2011)

10. Seyerlehner, K., Schedl, M., Pohle, T., Knees, P.: Using block-level features for genre classification, tag classification and music similarity estimation. In: Online Proceedings of the 6th MIR Evaluation eXchange (MIREX-10) (2010)
11. Seyerlehner, K., Widmer, G., Schedl, M., Knees, P.: Automatic music tag classification based on block-level features. In: Proceedings of the 7th Sound and Music Computing Conference (SMC-10) (2010)
12. Slaney, M.: Mixture of probability experts for audio retrieval and indexing. In: Proceedings of the IEEE International Conference on Multimedia and Expo (2002)
13. Turnbull, D., Barrington, L., Lanckriet, G.: Five approaches to collecting tags for music. In: Proceedings of the 9th International Conference on Music Information Retrieval (2008)
14. Turnbull, D., Barrington, L., Torres, D., Lanckriet, G.: Semantic annotation and retrieval of music and sound effects. IEEE Trans. Audio Speech Lang. Process. **16**(2), 467–476 (2008)
15. West, K., Cox, S.: Incorporating cultural representations of features into audio music similarity estimation. IEEE Trans. Audio Speech Lang. Process. **18**(3), 625–637 (2010)
16. West, K., Cox, S., Lamere, P.: Incorporating machine-learning into music similarity estimation. In: Proceedings of the 1st ACM Workshop on Audio and Music Computing Multimedia (AMCMM-06) (2006)
17. Whitman, B., Rifkin, R.: Musical query-by-description as a multiclass learning problem. In: IEEE Multimedia Signal Processing Conference (MMSP) (2002)

Ambiguity in Automatic Chord Transcription: Recognizing Major and Minor Chords

Antti Laaksonen[(✉)]

Department of Computer Science, University of Helsinki, Helsinki, Finland
ahslaaks@cs.helsinki.fi

Abstract. Automatic chord transcription is the process of transforming the harmonic content of a music signal into chord symbols. We use difficult chord transcription cases in the Beatles material to compare human performance to computer performance. Surprisingly, in many cases musically oriented participants are unable to determine whether the chord is major or minor. We further analyze ambiguous chords and find out that there are often no clear rules for chord interpretation. This suggests that the standard evaluation method in automatic chord transcription based on a single ground truth is inadequate.

Keywords: Automatic chord transcription · Signal processing · Musical context · Major and minor chords · Listening experiment

1 Introduction

Automatic chord transcription is the process of extracting the harmonic content from a music signal and representing it through chord symbols. The harmony of Western music is largely based on the twelve major and minor chords. These chords form the basis of the chord symbols. There are several different ways to express the chords. In this paper, for example, the symbol G denotes a G major chord and the symbol Gm denotes a G minor chord.

During the last decade, automatic chord transcription has been an active research field, and good results have been achieved in real-world transcription tasks. Currently, chord transcription algorithms are usually combinations of low-level signal processing methods and high-level probabilistic models. Section 2 presents an overview of the techniques used in automatic chord transcription.

So far, automatic chord transcription research has focused almost entirely on popular music. There are both traditional and practical reasons for this. First, although most classical music works can be marked with chord symbols, this is almost never done. The way the chords are played is strictly defined by the composer and simple chord symbols would not be sufficient for an acceptable performance.[1] Second, there is almost always a full score available. Therefore there is no need for a chord transcription or any other transcription.

---

[1] This has not always been the case. The figured bass notation used in J. S. Bach's time resembles today's chord symbols.

© Springer International Publishing Switzerland 2014
A. Nürnberger et al. (Eds.): AMR 2012, LNCS 8382, pp. 203–213, 2014.
DOI: 10.1007/978-3-319-12093-5_12

Fig. 1. A typical chord transcription of the first few bars of the Beatles' *Yesterday*.

Fig. 2. Another transcription which contains three 'incorrect' chords. However, this transcription sounds good and an average listener hardly notices any difference.

One aspect rarely mentioned in research papers concerning automatic chord transcription is that there are often several good transcriptions for a chord or a chord sequence. Consider the two transcriptions of the Beatles song *Yesterday* in Figs. 1 and 2. The first transcription is a particularly accurate transcription of the original studio recording. The second transcription contains three 'incorrect' chords: two G minor chords in bars 2 and 4 and the final A major chord. This transcription could have been done by a musician who has heard *Yesterday* many times in his or her life but cannot recall all the details about the chords.

Fig. 3. A third transcription with one 'incorrect' chord. This transcription does not sound good despite the fact that all the other chords are correct.

A natural evaluation metric in automatic chord transcription is the ratio of the number of correctly labeled audio segments compared to the total number of audio segments. In this case, the accuracy of the second transcription is 75 %

because there are 20 beats in the transcription and 15 of them are correctly transcribed. However, the transcription sounds good and a listener that is not very familiar with *Yesterday* will probably not notice any difference.

Now, consider the third transcription in Fig. 3. There is only one 'incorrect' chord, the F major chord in the third bar, and the transcription accuracy is 80 %. According to this evaluation metric, the third transcription seems to be better than the second transcription. Yet the third transcription does not sound good because the F major chord does not fit in the musical context. This is an important problem and we will come back to it at the end of the paper.

The organization of the rest of the paper is as follows: In Sect. 2, we present previous research concerning automatic chord transcription. In Sect. 3, we describe our experiment with human listeners, and in Sect. 4, we analyze typical errors in recognizing major and minor chords. Finally, in Sect. 5, we present our conclusions.

2   Current Systems

This section presents techniques used in automatic chord transcription. The transcription is based on chromagrams which are derived from the frequency spectrums of the audio data. The first paper describing this kind of approach to chord transcription is [3]. More detailed information of methods used in chord transcription algorithms is provided in [2, 11].

2.1   Chromagram

The first step in automatic chord transcription is to divide the input audio data into small segments and calculate frequency spectrums for each of the segments using the Fourier transform or a similar method. After this, the frequencies representing the same notes in different octaves are combined into chromagrams. A chromagram is a vector of 12 values assigned to each of the segments: each value corresponds to one note in the chromatic scale.

A simple method for finding a chord for each segment is to compare the chromagrams with chord patterns for all the possible chords. For example, a chord pattern for a C major chord could be [1, 0, 0, 0, 1, 0, 0, 1, 0, 0, 0, 0] where the ones denote the notes C, E and G. However, the frequencies in the signal that are not chord notes present some challenges.

First of all, a musical sound consists of several frequencies: the fundamental frequency is the frequency we actually hear, but in addition to this, there are harmonics which are multiples of the fundamental frequency. Thus the chromagram contains frequencies which do not correspond to chord notes. A promising approach to filtering harmonics from the chromagram is to produce an approximate transcription using chord patterns with harmonics [7].

Moreover, some frequencies in the audio signal have nothing to do with the chords. For example, the melody often contains notes that are not chord notes.

A common way to reduce this problem is to average the results of several consecutive segments. This method is based on the fact that a chord always lasts for several segments.

2.2 Musical Context

Every musician knows that the melody, the chords and the key of a musical work are connected to each other. These factors create the musical context that affects the probabilities of chords. The probabilities depend, of course, on the type of the music that is being transcribed. The following examples assume that we are working with Western music. The musical context has been used in several ways in automatic chord transcription:

Chord transitions: Certain chord transitions are more likely than others. For example, a G major chord is often followed by a C major chord but very rarely by an Eb minor chord. There have been various methods for deriving chord transition probabilities. A music theory approach [1,11] favors the transitions between chords that are close to each other in the circle of fifths. This works because their scales share a lot of common notes. In addition to this, several transcription algorithms [5,11,13] have used transition probabilities that are derived from existing chord transcriptions.

Key: Certain chords are typically used in specific keys. For example, C major and E minor chords are often used in the key of G major, whereas C# major and Bb minor chords are not typically used in the key of G major. These typical chord patterns have been used in some advanced chord probability models [5,8].

Structure: Recognizing the structure of the piece can help the chord transcription. For example, in popular music, there are usually repeating refrains with identical chord progressions. This information can be used to produce more consistent transcriptions [6].

While there are several ways to estimate chord probabilities, it is not clear how important the probabilities actually are in chord transcription. In an evaluation of chord transcription methods [2], the choice of chord transition probabilities had no real significance to the transcription result. The only important parameter was the probability of staying in the same chord.

3 Experiment

Major and minor chords have a central role in Western music. A common error in automatic chord transcription is that a major chord is recognized as a minor chord or vice versa. For human listeners, distinguishing between major and minor chords is, in principle, an easy task. However, the results of this section show that many cases that are difficult for computers are also difficult for humans.

There are different transcription situations where problems with major and minor chords arise. Here we consider the following three classes:

A. *Minor instead of Major:* A major chord is recognized as a minor chord with the same root note. E.g. a C major chord is recognized as a C minor chord.
B. *Major instead of Minor:* A minor chord is recognized as a major chord with the same root note. E.g. a C minor chord is recognized as a C major chord.
C. *Relative Chords:* A chord is recognized as its relative chord. E.g. a C major chord is recognized as an A minor chord or an A minor chord is recognized as a C major chord.

Note that class C covers both situations where the actual chord is major or minor. It was important to separate classes A and B because they have unique challenges with regards to chord transcription.

In our experiment, human listeners with a musical background were asked to listen to a series of chords and classify each chord as a major chord or a minor chord. We concentrated on major and minor chords in the above situations for two reasons. First, these transcription situations are important in practice. Second, the experiment is easy for the participants because there are only two alternatives to choose from for each chord.

3.1 Material

The standard dataset that is used in automatic chord transcription evaluation has been the Beatles dataset [12] which offers hand-made ground truths for a collection of songs from the Beatles studio albums. The Beatles dataset has also been used as the main dataset in the MIREX chord transcription task [10].

The best current chord transcription algorithms achieve a transcription accuracy of about 80 % when evaluated using the Beatles dataset. Most of the top algorithms use advanced probabilistic models. The drawback with these approaches is the possibility of overfitting the dataset. However, Mauch's MM1 algorithm [9] also achieves an accuracy of 80 % without machine learning. Our experiment is based on the results of the MM1 algorithm because we assume that it gives a representative picture of the challenges in automatic chord transcription.

Table 1. The distribution of difficult chords in our collection. A total of 202 chords (55.96 %) belong to one of the classes A, B and C.

Class	Amount	Percentage
A. Minor instead of Major	49	13.57 %
B. Major instead of Minor	85	23.55 %
C. Relative Chords	68	18.84 %
Other Transcription Errors	159	44.04 %
Total	361	100.00 %

The material in our experiment consists of chords from the Beatles dataset that the MM1 algorithm recognizes incorrectly. We call these chords difficult chords. We built a collection of segments where the duration of the chord in the ground truth is at least 1.5 s and the MM1 algorithm outputs a single incorrect chord. Furthermore, we ignored segments where the tuning substantially differs from the standard A (440 Hz) or where there is no clear chord content.

We ended up with a collection of 361 difficult chords shown in Table 1. A total of 202 chords (55.96 %) belong to one of the classes A, B and C. For the experiment, we randomly selected 10 chords from each of the three classes. Direct audio excerpts from the Beatles songs were used in the experiment.

3.2 Arrangements

A total of 81 people participated in our experiment. All of the participants had some kind of musical background. The participants ranged from self-taught amateur musicians to professionals with a university degree. Most of the participants had attended transcription courses in music schools or universities as part of their musical education.

The experiment was carried out through a web page. The participants were asked to listen to 30 chords of 1.5 s in random order. For each chord, the participants were asked to determine whether the chord was a major or a minor chord. The root notes of the two options were given. The participants were allowed to listen to each chord as many times as they wanted and they could use any type of musical instrument to help with the transcription. A reference note (A at 440 Hz) was also given. There were no time limitations on the experiment.

3.3 Results

Our initial hypothesis was that recognizing major and minor chords would be a rather easy task for musically oriented human listeners. However, the task proved to be surprisingly difficult.

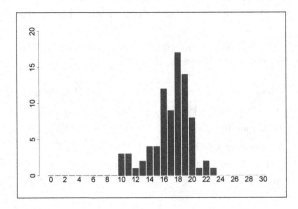

Fig. 4. The distribution of correctly recognized chords. The x-axis denotes the number of correctly recognized chords, and the y-axis denotes the number of participants with the corresponding result in the experiment.

Figure 4 shows the distribution of correctly recognized chords in the experiment. The minimum and maximum results were 10/30 and 23/30 chords and the average result was 17.06/30 chords. The number of correctly recognized chords was very low: this result was only slightly higher than randomly choosing one of the two chord options.

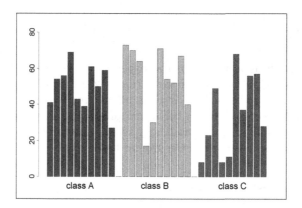

Fig. 5. The number of correct answers for each chord. The y axis denotes the number of participants who answered correctly for the corresponding chord.

However, there are notable differences in the results when individual chords are examined. Figure 5 shows the number of correct answers for each chord in the experiment grouped by the three classes. There were chords where almost all the participants chose the correct answer and chords where almost all the participants chose the wrong answer. Finally, in some cases the participants had no general opinion about the quality of the chord.

Table 2. The number of correctly recognized, unclear and incorrectly recognized chords in the experiment.

Class	Correct	Unclear	Incorrect
A. Minor instead of Major	6	3	1
B. Major instead of Minor	7	1	2
C. Relative Chords	4	1	5
Total	17	5	8

We divided the segments into three groups shown in Table 2: correctly recognized chords where more than 48 participants chose the correct chord, unclear chords where 33–48 participants chose the correct chord, and incorrectly recognized chords where less than 33 participants chose the correct chord. The thresholds were chosen so that the probability that more than 48 participants (or less than 33 participants) recognize a certain chord correctly is less than 5 % if the choices are random.

In total, there were 17 correctly recognized chords, 5 unclear chords and 8 incorrectly recognized chords. Class C was the only class with more incorrectly recognized chords than correctly recognized chords. In the following section we analyze the reasons why some chords are so difficult to recognize.

4   Analysis

The following analysis is based on the experiment described in the previous section and a thorough study of our collection of 202 difficult chords. There are three main reasons why some chords were difficult to recognize. First, in many cases the problem seems to be on the signal processing level. Second, in many cases the chord quality is ambiguous and several interpretations are possible. Third, in some cases the chord in the ground truth seems to be incorrect.

4.1   Signal Processing Problems

There are many chords in the material where human listeners can rather easily identify whether they are major or minor. However, frequencies produced by harmonics and notes outside the chord cause problems for automatic transcription methods. A typical problem is that the fifth harmonic of a tone is the major third and there can be a strong major component in a pure minor chord. This problem has been addressed in [7].

Fig. 6. A segment from *Sun King*. The reference chord in the second bar is F#m^7 defined by the bass, but the F# note is played softly.

Sometimes the problem is that the note defining the chord is played softly. Figure 6 shows one example of this situation. This segment is taken from *Sun King*, the guitar part is on the upper staff and the bass part is on the lower staff. The chord in the second bar is F#m^7 because there is a F# note in the bass and an A major chord in the guitar. However, the F# note is difficult to recognize because it is played softly which easily leads to an A major interpretation.

These types of transcription problems are 'easy' because there is no ambiguity in the chord and the problem seems to be on the signal processing level.

4.2   Ambiguous Chords

Most of the chords that are difficult for human listeners are ambiguous: either there are both major and minor elements or both are missing. The following examples illustrate typical situations where the chord content is ambiguous.

Fig. 7. A segment from *Ticket to Ride*. The reference chord in the second bar is E major but there is also a strong G note in the melody.

The segment in Fig. 7 is taken from *Ticket to Ride*. The chord in the second bar in the ground truth is E major probably because of the accompaniment. However, in the melody, there is a strong G note which is the third of an E minor chord. This gives a strong feeling of an E minor harmony.

Fig. 8. A segment from *Come Together*. The reference chord in the second bar is B minor but only the D note in the backing voice supports this interpretation.

Figure 8 shows an example of a missing chord note in the song *Come Together*. In the accompaniment of the second bar, only the notes B and F# are present which leaves the quality of the chord open. The ground truth interpretation of the chord is B minor, possibly because there is a D in the backing voice at the beginning of the second bar. However, the D note can hardly be heard and the fifth harmonic of the root note B supports the perception of a major chord.

Ambiguous chords are problematic because the quality of the chord cannot be directly derived from the audio signal. Musical context can help the transcription but there are no clear interpretation rules for difficult situations. Even transcriptions created by experts contain differences. The problem is that there are several valid interpretations but only one chord in the ground truth.

4.3   Ground Truth Differences

During the analysis we compared the reference chords in [12], the chords in *The Beatles – Complete Scores* [4] and the chords in our hand-made transcriptions and found several cases where we disagree with one or both of the sources. Of course, the possibilities for interpretation are based on subjective decision.

There are two different levels of ambiguity: (1) what the performers actually play and sing and (2) how this should be denoted with chord symbols. One possibility would be to use datasets where the original scores are available. This would remove the first level of ambiguity. However, this requirement would force to use mainly classical music datasets.

5   Conclusions

Our work suggests that many problems in automatic chord transcription originate from the inherent ambiguity of chords. We suggest that our results concerning major and minor chords can be generalized to more complex cases, where the differences may be even more subtle. Therefore it seems to be an impossible goal to design an algorithm which produces perfect chord transcriptions when a single ground truth is used as the evaluation method.

In practice, the most important property of a chord transcription is that it sounds good as a whole. Different interpretations and rearrangements are acceptable as long as the performers and the audience are content with the transcription. This is a very different goal than maximizing the number of chords matching the ground truth. Our future work aims to define more precisely the goodness of a chord transcription and find new ways to build more flexible ground truths that take chord ambiguity into account.

Acknowledgements. This work has been supported by the Helsinki Doctoral Programme in Computer Science and the Academy of Finland (grant number 118653).

References

1. Bello, J., Pickens, J.: A robust mid-level representation for harmonic content in music signals. In: Proceedings of the 6th International Conference on Music Information Retrieval (2005)
2. Cho, T., Weiss, R., Bello, J.: Exploring common variations in state of the art chord recognition systems. In: Proceedings of the 7th Sound and Music Computing Conference (2010)
3. Fujishima, T.: Realtime chord recognition of musical sound: a system using common Lisp music. In: Proceedings of the 25th International Computer Music Conference (1999)
4. Fujita, T., Hagino, Y., Kubo, H., Sato, G.: The Beatles - Complete Scores. Hal Leonard Corporation, London (1993)
5. Lee, K., Slaney, M.: Acoustic chord transcription and key extraction from audio using key-dependent HMMs trained on synthesized audio. IEEE Trans. Audio Speech Lang. Process. **16**(2), 291–301 (2008)
6. Mauch, M., Noland, K., Dixon, S.: Using musical structure to enhance automatic chord transcription. In: Proceedings of the 10th International Conference on Music Information Retrieval (2009)

7. Mauch, M., Dixon, S.: Approximate note transcription for the improved identification of difficult chords. In: Proceedings of the 11th International Conference on Music Information Retrieval (2010)
8. Mauch, M.: Automatic chord transcription from audio using computational models of musical context. Ph.D. thesis (2010)
9. Mauch, M.: Simple Chord Estimate: Submission to the MIREX Chord Estimation Task (2010)
10. MIREX Wiki. http://www.music-ir.org/mirex/wiki/
11. Papadopoulos, H., Peeters, G.: Large-scale study of chord estimation algorithms based on chroma representation and HMM. In: Proceedings of the 5th International Workshop on Content-Based Multimedia Indexing (2007)
12. Reference Annotations: The Beatles. Centre for Digital Music at Queen Mary, University of London (2010). http://isophonics.net/content/reference-annotations-beatles
13. Ryynänen, M., Klapuri, A.: Automatic transcription of melody, bass line, and chords in polyphonic music. Comput. Music J. **32**(3), 72–86 (2008)

Capturing the Temporal Domain in Echonest Features for Improved Classification Effectiveness

Alexander Schindler[1,2]([✉]) and Andreas Rauber[1]

[1] Department of Software Technology and Interactive Systems,
Vienna University of Technology, Vienna, Austria
{schindler,rauber}@ifs.tuwien.ac.at
[2] Intelligent Vision Systems, AIT Austrian Institute of Technology,
Vienna, Austria

Abstract. This paper proposes Temporal Echonest Features to harness the information available from the beat-aligned vector sequences of the features provided by The Echo Nest. Rather than aggregating them via simple averaging approaches, the statistics of temporal variations are analyzed and used to represent the audio content. We evaluate the performance on four traditional music genre classification test collections and compare them to state of the art audio descriptors. Experiments reveal, that the exploitation of temporal variability from beat-aligned vector sequences and combinations of different descriptors leads to an improvement of classification accuracy. Comparing the results of Temporal Echonest Features to those of approved conventional audio descriptors used as benchmarks, these approaches perform well, often significantly outperforming their predecessors, and can be effectively used for large scale music genre classification.

1   Introduction

Music genre classification is one of the most prominent tasks in the domain of Music Information Retrieval (MIR). Although we have seen remarkable progress in the last two decades [5, 12], the achieved results are evaluated against relative small benchmark datasets. While commercial music services like Amazon[1], Last.fm[2] or Spotify[3] maintain large libraries of more than 10 million music pieces, the most popular datasets used in MIR genre classification research - GTZAN, ISMIR Genre, ISMIR Rhythm and Latin Music Database - range from 698 to 3227 songs - which is less than 0.1‰ of the volume provided by on-line services.

Recent efforts of the Laboratory for the Recognition and Organization of Speech and Audio (LabROSA)[4] of Columbia University lead to the compilation

---

[1] http://www.amazon.com/music
[2] http://www.last.fm
[3] http://www.spotify.com/
[4] http://labrosa.ee.columbia.edu

© Springer International Publishing Switzerland 2014
A. Nürnberger et al. (Eds.): AMR 2012, LNCS 8382, pp. 214–227, 2014.
DOI: 10.1007/978-3-319-12093-5_13

of the Million Song Dataset (MSD) [1] - a large collection consisting of music meta-data and audio features. This freely available dataset gives researchers the opportunity to test algorithms on a large-scale collection that corresponds to a real-world like environment. The provided data was extracted from one million audio tracks using services of The Echo Nest[5]. Meta-data consists of e.g. author, album, title, year, length. There are two major sets of audio features that are described as 'Mel Frequency Cepstral Coefficients (MFCC) like' and 'Chroma like' and a number of additional descriptors including tempo, loudness, key and some high level features e.g. danceability, hotttnesss.

Unfortunately, due to copyright restriction, the source audio files cannot be distributed. Only an identifier is provided that can be used to download short audio samples from 7digital[6] for small evaluations and prototyping. Again, these audio snippets are not easily obtained due to access restrictions of the 7digital-API. Consequently, the set of features provided so-far constitutes the only way to utilize this dataset. Although the two main audio feature sets are described as similar to MFCC [11] and Chroma, the absence of accurate documentation of the extraction algorithms makes such a statement unreliable. Specifically no experiments are reported that verify that the Echo Nest features perform equivalent or at least similar to MFCC and Chroma features from conventional state-of-the-art MIR tools as Marsyas [19] or Jmir [13]. Further, several audio descriptors (e.g. MFCCs, Chroma, loudness information, etc.) are not provided as a single descriptive feature vector. Using an onset detection algorithm, the Echonest's feature extractor returns a vector sequence of variable length where each vector is aligned to a music event. To apply these features to standard machine learning algorithms a preprocessing step is required. The sequences need to be transformed into fixed length representations using a proper aggregation method. Approaches proposed so far include simply calculating the average over all vectors of a song [3], as well as using the average and covariance of the timbre vectors for each song [1]. An explicit evaluation of which method provides best results has not been reported, yet.

This paper provides a performance evaluation of the Echonest audio descriptors. Different feature set combinations as well as different vector sequence aggregation methods are compared and recommendations towards optimal combinations are presented. The evaluations are based on four traditional MIR genre classification test sets to make the results comparable to conventional feature sets, which are currently not available for the MSD. This approach further offers benchmarks for succeeding experiments on the Million Song Dataset.

The remainder of this paper is organized as follows: In Sect. 2 a detailed description of the Echonest features is provided. Section 3 lays out the evaluation environment. In Sect. 4 the conducted experiments are described and results are discussed. Finally, in Sect. 5 we draw conclusions and point out possible future research directions.

---

[5] http://the.echonest.com/

[6] http://us.7digital.com/

2 Echonest Features

The Echonest Analyzer [7] is a music audio analysis tool available as a free Web service which is accessible over the Echonest API[7]. In a first step of the analysis audio fingerprinting is used to locate tracks in the Echonest's music metadata repository. Music metadata returned by the Analyzer includes artist information (name, user applied tags including weights and term frequencies, a list of similar artists), album information (name, year) and song information (title). Additionally a set of identifiers is provided that can be used to access complimentary metadata repositories (e.g. musicbrainz[8], playme[9], 7digital).

Further information provided by the Analyzer is based on audio signal analysis. Two major sets of audio features are provided describing timbre and pitch information of the corresponding music track. Unlike conventional MIR feature extraction frameworks, the Analyzer does not return a single feature vector per track and feature. The Analyzer implements an onset detector which is used to localize music events called *Segments*. These *Segments* are described as sound entities that are relative uniform in timbre and harmony and are the basis for further feature extraction. For each *Segment* the following features are derived from musical audio signals:

Segments Timbre are casually described as MFCC-like features. A 12 dimensional vector with unbounded values centered around 0 representing a high level abstraction of the spectral surface (see Fig. 1).

Segments Pitches are casually described as Chroma-like features. A normalized 12 dimensional vector ranging from 0 to 1 corresponding to the 12 pitch classes C, C#, to B.

Segments Loudness Max represents the peak loudness value within each segment.

Segments Loudness Max Time describes the offset within the segment of the point of maximum loudness.

Segments Start provide start time information of each segment/onset.

Onset detection is further used to locate perceived musical events within a *Segment* called *Tatums*. *Beats* are described as multiple of *Tatums* and each first *Beat* of a measure is marked as a *Bar*. Contrary to *Segments*, that are usually

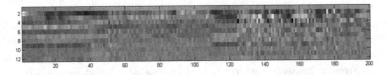

Fig. 1. First 200 timbre vectors of 'With a little help from my friends' by 'Joe Cocker'

---

[7] http://developer.echonest.com

[8] http://musicbrainz.org

[9] http://www.playme.com

shorter than a second, the Analyzer also detects *Sections* which define larger blocks within a track (e.g. chorus, verse, etc.). From these low-level features some mid- and high-level audio descriptors are derived (e.g. tempo, key, time signature, etc.). Additionally, a confidence value between 0 and 1 is provided indicating the reliability of the extracted or derived values - except for a confidence value of '−1' which indicates that this value was not properly calculated and should be discarded. Based on the audio segmentation and additional audio descriptors the following features provide locational informations about music events within the analyzed track:

Bars/Beats/Tatums start the onsets for each of the detected audio segments.
Sections start the onsets of each section.
Fadein stop the estimated end of the fade-in.
Fadeout start the estimated start of the fade-out.

Additionally a set of high-level features derived from previously described audio descriptors is returned by the Analyzer:

Key the key of the track $(C, C\#, \ldots, B)$.
Mode the mode of the track (major/minor).
Tempo measured in beats per minute.
Time Signature three or four quarter stroke.
Danceability a value between 0 and 1 measuring of how danceable this song is.
Energy a value between 0 and 1 measuring the perceived energy of a song.
Song Hotttnesss a numerical description of how hot a song is (from 0 to 1).

3   Evaluation

This section gives a description of the evaluation environment used in the experiments described in Sect. 4. The Echonest features are compared against the two conventional feature sets Marsyas and Rhythm Patterns. The evaluation is performed on four datasets that have been widely used in music genre classification tasks. The performance of the different features is measured and compared by classification accuracy that has been retrieved from five commonly used classifiers.

3.1   Feature Sets

The following feature sets are used in the experiments to evaluate the performance of features provided by the Echnoest Analyzer.

Echonest Features: Echonest features of all four datasets were extracted using the Echonest's open source Python library Pyechonest[10]. This library provides methods for accessing the Echonest API. Python code provided by the MSD Web page[11] was used to store the retrieved results in the same HDF5[12] format which is also used by the MSD.

---

[10] https://github.com/echonest/pyechonest/
[11] https://github.com/tb2332/MSongsDB/tree/master/PythonSrc
[12] http://www.hdfgroup.org/HDF5/

Marsyas Features: The Marsyas framework [19] is an open source framework for audio processing. It implements the original feature sets proposed by Tzanetakis and Cook [20]. The Marsyas features are well known, thus only brief description of the features included in this evaluation are provided. For further details we refer to [19,20].

Marsyas features were extracted using a version of the Marsyas framework[13] that has been compiled for the Microsoft Windows operating system. Using the default settings of bextract the complete audio file was analyzed using a window size of 512 samples without overlap, offset, audio normalization, stereo information or downsampling. For the following features mean and standard deviation values were calculated (the number of dimensions provided corresponds to the total length of the feature vector):

Chroma Features (chro) corresponding to the 12 pitch classes C, C#, to B.

Spectral Features (spfe) is a set of features containing Spectral Centroid, Spectral Flux and Spectral Rolloff.

Timbral Features (timb) is a set of features containing Time ZeroCrossings, Spectral Flux and Spectral Rolloff, and Mel-Frequency Cepstral Coefficients (MFCC).

Mel-Frequency Cepstral Coefficients (mfcc).

Psychoacoustic Features: Psychoacoustics feature sets deal with the relationship of physical sounds and the human brains interpretation of them. The features were extracted using the Matlab implementation of rp_extract[14] - version 0.6411.

Rhythm Patterns (RP) also called fluctuation patterns [14], are a set of audio features representing fluctuations per modulation frequency on 24 frequency bands according to human perception. The features are based on spectral audio analysis incorporating psychoacoustic phenomenons. A detailed description of the algorithm is given in [15].

Rhythm Histograms (RH) features are capturing rhythmical characteristics of an audio track by aggregating the modulation values of the critical bands computed in a Rhythm Pattern [9].

Statistical Spectrum Descriptors (SSD) describe fluctuations on the critical bands and capture both timbral and rhythmic information. They are based on the first part of the Rhythm Pattern computation and calculate substantially statistical values (mean, median, variance, skewness, kurtosis, min, max) for each segment per critical band [9].

Temporal Statistical Spectrum Descriptor (TSSD) features describe variations over time by including a temporal dimension to incorporate time series aspects. Statistical Spectrum Descriptors are extracted from segments of a musical track at different time positions. Thus, TSSDs are able to reflect

---

[13] http://marsyas.info
[14] http://www.ifs.tuwien.ac.at/mir/downloads.html

rhythmical, instrumental, etc. changes timbral by capturing variations and changes of the audio spectrum over time [10].

Temporal Rhythm Histogram (TRH) capture change and variation of rhythmic aspects in time. Similar to the Temporal Statistical Spectrum Descriptor statistical measures of the Rhythm Histograms of individual 6-second segments in a musical track are computed [10].

3.2   Data Sets

For the evaluation four data sets that have been extensively used in music genre classification over the past decade have been used.

GTZAN. This data set was compiled by George Tzanetakis [18] in 2000–2001 and consists of 1000 audio tracks equally distributed over the 10 music gen-res: blues, classical, country, disco, hiphop, pop, jazz, metal, reggae, and rock.

ISMIR Genre. This data set has been assembled for training and development in the ISMIR 2004 Genre Classification contest [2]. It contains 1458 full length audio recordings from Magnatune.com distributed across the 6 genre classes: Classical, Electronic, JazzBlues, MetalPunk, RockPop, World.

ISMIR Rhythm. The ISMIR Rhythm data set was used in the ISMIR 2004 Rhythm classification contest [2]. It contains 698 excerpts of typical ballroom and Latin American dance music, covering the genres Slow Waltz, Viennese Waltz, Tango, Quick Step, Rumba, Cha Cha Cha, Samba, and Jive.

Latin Music Database (LMD) [17] contains 3227 songs, categorized into the 10 Latin music genres Axé, Bachata, Bolero, Forró, Gaúcha, Merengue, Pagode, Salsa, Sertaneja and Tango.

3.3   Classifiers

The classifiers used in this evaluation represent a selection of machine learning algorithms frequently used in MIR research. We evaluated the classifiers using their implementations in Weka [6] version 3.7 with 10 runs of 10-fold cross-validation.

K-Nearest Neighbors (KNN) the nonparametric classifier has been applied to various music classification experiments and has been chosen for its pop-ularity. Because the results of this classifier rely mostly on the choice of an adequate distance function it was tested with Euclidean (L2) and Manhattan (L1) distance as well as $k = 1$.

Support Vector Machines have shown remarkable performance in supervised music classification tasks. SVMs were tested with different kernel methods. Linear PolyKernel and RBFKernel (RBF) are used in this evaluation, both with standard parameters: penalty parameter set to 1, RBF Gamma set to 0.01 and $c = 1.0$.

J48. The C4.5 decision tree is not as widely used as KNN or SVM, but it has the advantage of being relatively quick to train, which might be a concern processing one million tracks. J48 was tested with a confidence factor used for pruning from 0.25 and a minimum of two instances per leaf.

RandomForest. The ensemble classification algorithm is inherently slower than J48, but is superior in precision. It was tested with unlimited depth of the trees, ten generated trees and the number of attributes to be used in random selection set to 0.

NaiveBayes. The probabilistic classifier is efficient and robust to noisy data and has several advantages due to its simple structure.

4   Experiments and Results

This section describes the experiments that were conducted in this study.

4.1   Comparing Echonest Features with Conventional Implementations

The features *Segments Timbre* and *Segments Pitches* provided by the Echonest's Analyzer are described as MFCC and Chroma 'like'. Unfortunately no further explanation is given to substantiate this statement. The documentation [7] gives a brief overview of the characteristics described by these feature sets, but an extensive description of the algorithms used in the implementation is missing.

Compared to conventional implementations of MFCC and Chroma features the most obvious difference is the vector length of *Segments Timbre* - which is supposed to be a MFCC like feature. Most of the available MFCC implementations in the domain of MIR are using 13 cepstral coefficients as described in [11] whereas the Echonest Analyzer only outputs vectors with dimension 12. Although the number of coefficients is not strictly defined and the use of 12 or 13 dimensions seems to be more due to historical reasons, this makes a direct comparison using audio calibration/benchmark testsets impossible.

To test the assumption, that the Echonest features are similar to conventional implementations of MFCC and Chroma features, the audio descriptors are evaluated on four different datasets using a set of common classifiers as described in the evaluation description (see Sect. 3.3). Echonest Segments Timbre were extracted as described in Sect. 3.1. The beat aligned vector sequence was aggregated by calculating mean and standard deviation for each dimension. The Marsyas framework was used as reference. Mean and standard deviations of the MFCC features were extracted using `bextract`.

Table 1 shows the accuracies of genre classification for MFCC and Chroma features from the Echonest Analyzer and Marsyas. Significance testing with a significance level $\alpha = 0.05$ is used to compare the two different features. Significant differences are highlighted in bold letters. According to these results the assumption that *Segments Timbre* are similar to MFCC does not hold. There are significant differences on most of the cases and except for the GTZAN dataset

Table 1. Comparing MFCC and Chroma implementations of the Echonest Analyzer (EN) and Marsyas (MAR) by their classification accuracy on the GTZAN, ISMIR Genre (ISMIR-G), ISMIR Rhythm (ISMIR-R) and Latin Music Dataset (LMD) datasets. Significant differences ($\alpha = 0.05$) between EN and MAR are highlighted in bold letters.

Segments Timbre / MFCC

Dataset	GTZAN		ISMIR-G		ISMIR-R		LMD	
	EN1	MAR	EN1	MAR	EN1	MAR	EN1	MAR
SVM Poly	**61.1**	69.0	**75.1**	62.1	**63.1**	57.1	**78.4**	60.4
SVM RBF	35.1	39.3	**46.8**	44.1	30.3	31.0	**41.2**	38.0
KNN K1 L2	**58.1**	63.4	**77.0**	64.2	**49.2**	43.3	**78.7**	58.4
KNN K3 L2	57.6	61.4	**77.0**	63.0	51.8	46.8	**79.4**	56.9
KNN K1 L1	**56.6**	63.0	**77.9**	63.0	**49.4**	44.0	**79.1**	57.9
KNN K3 L1	**56.4**	62.3	**76.6**	61.5	50.0	47.1	**79.9**	57.4
J48	**44.7**	49.7	**69.4**	52.9	40.4	37.4	**62.5**	44.4
Rand-Forest	**54.7**	59.1	**75.8**	60.8	50.8	45.3	**74.7**	54.0
NaiveBayes	**50.5**	55.9	**63.2**	49.6	**53.3**	38.3	**68.4**	46.7

Segments Pitches / Chroma

Dataset	GTZAN		ISMIR-G		ISMIR-R		LMD	
	EN2	MAR	EN2	MAR	EN2	MAR	EN2	MAR
SVM Poly	**37.0**	41.2	**64.3**	50.3	38.7	38.6	**54.1**	39.4
SVM RBF	**26.1**	22.0	**50.2**	46.7	22.6	24.9	**32.6**	26.1
KNN K1 L2	**38.0**	42.7	**62.1**	46.0	31.8	28.9	**57.1**	37.3
KNN K3 L2	**35.0**	41.1	**63.0**	50.6	28.9	28.9	**54.7**	36.0
KNN K1 L1	**38.2**	44.1	**62.8**	45.4	32.5	27.4	**56.4**	37.3
KNN K3 L1	**36.5**	42.5	**62.7**	51.3	32.5	28.4	**53.8**	36.7
J48	**27.8**	40.1	**53.7**	43.8	29.0	26.9	**41.5**	33.6
Rand-Forest	**37.0**	48.5	**62.1**	50.5	35.2	30.6	**53.3**	39.1
NaiveBayes	**34.1**	28.6	**59.7**	46.8	**39.7**	22.7	**47.0**	26.9

the Echonest features outperform the Marsyas MFCC implementation. Even more drastic are the differences between *Segments Pitches* and Marsyas Chroma features except for the ISMIR Rhythm dataset. Similar to *Segments Timbre Segments Pitches* perform better except for the GTZAN dataset.

4.2 Feature Selection and Proper Aggregation of Beat Aligned Vector Sequences

The second part of the experiments conducted in this study deals with the huge amount of information provided by the by the MSD respectively the Echonest Analyzer. Currently no evaluations have been reported that give reliable benchmarks on how to achieve maximum performance on these features sets.

Scope of Selected Features: Due to number of features provided by the MSD only a subset of them was selected for the experiments. A comprehensive comparison of all possible feature combinations is beyond the scope of this publication. The focus was set on the beat aligned vector sequences *Segments Timbre*, *Segments Pitches*, *Segments Loudness Max* and *Segments Loudness Max Time*. Further *Segments Start* was used to calculate the length of a segment by subtracting the onsets of two consecutive vectors.

Aggregation of Echonest Vector Sequences: A further focus has been set on the feature sets that are provided as beat aligned vector sequences. Such sequences represent time series of feature data that can be exploited for various MIR scenarios (e.g. audio segmentation, chord analysis). Many classification tasks in turn require a fixed-length single vector representation of feature data. Consequently, the corresponding Echonest features need to be preprocessed. A straight forward approach would be to simply calculate an average of all vectors resulting in a single vector, but this implies discarding valuable information. Lidy et al. [8,10] demonstrated how to effectively exploit temporal information of sequentially retrieved feature data by calculating statistical measures. The temporal variants of Rhythm Patterns (RP), Rhythm Histograms (RH) and Statistical Spectrum Descriptor (SSD) describe variations over time reflecting rhythmical, instrumental, etc. changes of the audio spectrum and have previously shown excellent performance on conventional MIR classification benchmark sets as well as non-western music datasets.

For this evaluation the vector sequences provided by the Echonest Analyzer were aggregated by calculating the statistical measures mean, median, variance, skewness, kurtosis, min and max.

Temporal Echonest Features: Temporal Echonest Features (TEN) follow the approach of temporal features by Lidy et al. [10], where statistical moments are calculated from Rhythm Pattern features. To compute Temporal Rhythm Patterns (TRP) a track is segmented into sequences of 6 s and features are extracted for each consecutive time frame. This approach can be compared to the vector sequences retrieved by the Echonest Analyzer, except for the varying time frames caused by the onset detection based segmentation. To capture temporal variations of the underlying feature space, statistical moments (mean, median, variance, min, max, value range, skewness, kurtosis) are calculated from each dimension.

We experimented with different combinations of Echonest features and statistical measures. The combinations were evaluated by their effectiveness in classification experiments using accuracy as measure. The experiments conclude with a recommendation of a featureset-combination that achieves maximum performance on most of the testsets and classifiers used in the evaluation.

Multiple combinations of Echonest features have been tested in the experiments. Due to space constraints only a representative overview is given as well as the most effective combinations.

EN0. This represents the trivial approach of simply calculating the average of all *Segments Timbre* descriptors (12 dimensions).

EN1. This combination is similar to EN0 including variance information of the beat aligned *Segments Timbre* vectors already capturing timbral variances of the track (24 dimensions).

EN2. Mean and variance of *Segments Pitches* are calculated (24 dimensions).

EN3. According to the year prediction benchmark task presented in [1] mean and the non-redundant values of the covariance matrix are calculated (90 dimensions).

EN4. All statistical moments (mean, median, variance, min, max, value range, skewness, kurtosis) for *Segments Timbre* are calculated (96 dimensions)

EN5. All statistical moments of *Segments Pitches* and *Segments Timbre* are calculated (192 dimensions).

Temporal Echonest Features (TEN). All statistical moments of *Segments Pitches*, *Segments Timbre*, *Segments Loudness Max*, *Segments Loudness Max Time* and lengths of segments calculated from *Segments Start* are calculated (216 dimension).

4.3   Results

Table 2 shows the results of the evaluations for each dataset. Echonest features are located to the right side of the tables. Only EN0 and EN3-TEN are displayed, because EN1 and EN2 are already presented in Table 1. Bold letters mark best results of the Echonest features. If a classifier shows no bold entries, EN1 or EN2 provide best results for it. Conventional feature sets on the left side of the tables provide an extensive overview of how the Echonest features perform in general.

Good Results with Simple but Short Feature Sets: The trivial approach of simply averaging all segments (EN0) provides expectedly the lowest precision results of the evaluated combinations. As depicted in Table 2, the values range between Marsyas MFCC and Timbre features. On the other hand, taking the low dimensionality of the feature space into account, this approach constitutes a good choice for implementations focusing on runtime behavior and performance. Especially the non-parametric K-Nearest-Neighbors classifier provides good results. Adding additional variance information (EN1) provides enhanced classification results on *Segments Timbre* features. Specifically Support Vector Machines gain from the extra information provided. As pointed out in Table 3, this combination already provides top or second best results for K-Nearest Neighbors and Decision Tree classifiers. Again, addressing performance issues, the combinations EN0 and EN1 with only 12 or 26 dimensions may be a good compromise between computational efficiency and precision of classification results.

Chroma features are reported to show inferior music classification performance compared to MFCC [4]. This behavior was reproduced. Marsyas Chroma features as well as Echonest *Segments Pitches* (EN2) provide the lowest results for their frameworks.

Table 2. Comparing Echonest, Marsyas and Rhythm Pattern features by their classification accuracy. Best performing Echonest feature combinations are highlighted in bold letters.

Classifiers	chro	spfe	timb	mfcc	rp	rh	trh	ssd	tssd	EN0	EN3	EN4	EN5	TEN
ISMIR Genre Dataset														
SVM Poly	50.3	54.9	67.7	62.1	75.1	64.0	66.5	78.8	80.9	67.0	67.2	78.5	80.4	**81.1**
SVM RBF	46.6	44.2	50.0	44.1	69.0	55.5	64.5	64.1	72.0	44.3	49.1	64.9	69.4	**70.9**
KNN K1 L2	46.0	56.3	65.8	64.2	72.9	60.7	63.3	77.8	76.6	76.8	64.0	75.5	75.9	**77.8**
KNN K1 L1	45.4	56.5	65.9	63.0	71.5	60.8	63.3	78.5	77.6	77.1	60.8	77.6	78.3	**81.3**
J48	43.8	53.3	56.5	52.9	61.9	56.9	56.7	69.6	68.3	68.5	64.5	67.4	66.5	68.0
Rand-Forest	51.5	60.4	62.3	60.8	69.8	65.2	65.4	75.7	74.6	74.3	65.9	74.7	73.2	74.4
NaiveBayes	46.8	53.2	52.3	49.6	63.5	56.7	60.2	61.0	40.2	66.1	45.5	**63.8**	56.0	63.3
Latin Music Database														
SVM Poly	39.4	38.2	68.6	60.4	86.3	59.9	62.8	86.2	87.3	70.5	69.6	82.9	87.1	**89.0**
SVM RBF	26.1	19.1	51.0	38.0	79.9	36.6	53.2	71.6	83.3	29.2	40.9	69.4	76.6	**79.3**
KNN K1 L2	37.3	42.5	62.7	58.4	74.3	58.7	49.5	83.1	78.4	73.5	52.2	77.3	79.0	**80.9**
KNN K1 L1	37.3	43.2	61.5	57.9	73.8	59.0	53.1	83.8	81.7	72.6	49.8	79.8	81.6	**83.0**
J48	33.6	38.4	48.8	44.3	57.1	43.3	43.8	64.7	64.4	58.7	53.9	60.5	61.7	**64.8**
Rand-Forest	39.4	46.4	58.1	53.6	58.8	50.3	47.5	76.3	73.0	69.9	54.9	74.1	73.5	**75.9**
NaiveBayes	26.9	35.7	43.5	46.7	66.0	47.0	49.9	64.1	67.8	66.5	40.4	70.8	71.1	**73.3**
GTZAN														
SVM Poly	41.1	43.1	75.2	67.8	64.9	45.5	38.9	73.2	66.2	56.4	53.6	63.9	65.2	**66.9**
SVM RBF	22.0	27.1	52.1	37.7	56.7	31.4	39.9	53.1	63.3	36.7	22.3	46.6	56.3	**56.5**
KNN K1 L2	41.9	42.1	67.8	61.8	51.5	40.2	32.7	63.7	53.4	56.3	39.9	56.8	56.1	**58.2**
KNN K1 L1	43.6	43.0	68.2	61.7	53.4	39.8	35.8	64.1	60.6	55.1	36.4	56.9	56.3	**58.7**
J48	38.6	39.2	53.6	48.9	38.3	32.6	31.6	52.0	50.6	**45.0**	39.1	44.3	43.6	44.1
Rand-Forest	48.0	47.2	64.2	57.9	45.9	39.6	38.0	63.4	59.3	54.7	41.1	54.0	53.2	**55.0**
NaiveBayes	28.1	40.0	52.2	54.9	46.3	36.2	35.6	52.4	53.0	53.1	29.5	**53.6**	52.5	53.3
ISMIR Rhythm														
SVM Poly	38.1	41.4	60.7	54.5	88.0	82.6	73.7	58.6	56.0	55.1	51.7	62.7	63.7	**67.3**
SVM RBF	25.1	27.9	36.4	29.7	79.6	36.6	63.2	42.1	55.3	24.7	26.6	37.1	46.5	**53.1**
KNN K1 L2	28.3	34.8	43.9	37.3	73.7	77.7	51.5	45.5	39.8	43.5	34.6	44.5	43.0	**45.7**
KNN K1 L1	26.8	35.8	44.3	38.9	71.4	73.9	60.3	43.4	42.1	44.0	32.9	46.9	44.7	**49.2**
J48	26.9	33.7	37.6	37.1	64.3	67.6	65.9	37.6	35.8	38.5	34.0	38.5	40.5	**48.0**
Rand-Forest	31.0	38.1	44.4	43.8	64.9	71.6	68.2	46.6	44.1	47.5	37.1	47.9	48.8	**53.5**
NaiveBayes	23.3	37.0	37.7	36.5	75.9	69.0	69.3	44.4	46.8	52.8	25.1	52.8	49.9	**55.1**

Better Results with Complex Feature Sets: Providing more information to the classifier expectedly results in better performance. Adding more statistical measures to simple feature sets (EN4) provides no significant performance gain but increases the length of the vector by a factor of 4. Also combining *Segments Timbre* with *Segments Pitches* and calculating the statistical moments (EN5) only provides slightly better results. The 192 dimensions of this combination may alleviate this result when performance issues are taken into consideration. Only the initially as benchmark proposed approach by [1] (EN3) provides inferior results.

Table 3. Overview of which Echonest feature combination performs best for a certain classifier on the datasets (a) GTZAN, (b) ISMIR Genre, (c) ISMIR Rhythm and (d) LMD

Dataset	EN0	EN1	EN2	EN3	EN4	EN5	TEN
SVM Poly							a, b, c, d
SVM RBF							a, b, c, d
KNN K1 L2		c					a, b, d
KNN K3 L2		a, b, c					d
KNN K1 L1		c					a, b, d
KNN K3 L1		c					a, b, d
J48	a	b					c, d
Rand-Forest		a, b					c, d
NaiveBayes	b				a		c, d

Recommendation: Temporal Echonest Features: Including additional information of loudness distribution and the varying lengths of segments in the feature set (TEN), enhances performance for all classifiers and provides the best results of the experiments (see Table 3). For many testset-classifier combinations the Temporal Echonest Features provide best results for all feature sets. Compared to similar performing features like TSSD - which have a dimension of 1176 - TMEs outperform on precision and computational efficiency belongings. Table 3 summarizes the best performing Echonest feature combinations.

5   Conclusion and Future Work

In this paper, we presented a comparison of Echonest features - as provided by the Million Song Dataset - with feature sets from conventionally available feature extractors. Due to the absence of audio samples, researcher solely rely on these Echonest features. Thus, the aim was to provide empirically determined reference values for further experiments based on the Million Song Dataset. We used six different combinations of Echonest features and calculated statistical moments to capture their temporal domain. Experiments show that Temporal Echonest Features - a combination of MFCC and Chroma features combined with loudness information as well as the distribution of segment lengths - complimented by all calculated statistical moments - outperforms almost all datasets and classifiers - even conventional feature sets, with a prediction rate of up to 89 %. Although higher percentages have been reported on these datasets based on other feature sets or hybrid combinations of different feature sets, these additional audio descriptions are not available on the MSD. Additionally it was observed, that top results can already be obtained calculating average and variance of *Segments Timbre* features. This short representation of audio content favors the development of performance focused systems.

Further research will focus on the remaining features provided by the Million Song Dataset. Since these descriptors provide an already highly aggregated representation of the extracted audio content, harnessing this information may lead to shorter feature vectors. Also large scale evaluations on the Million Song Dataset - that were not performed in this paper due to the absence of consolidated genre classification subsets - are needed.

6   Distribution of Data

All feature sets described in Sect. 4, including Temporal Echonest Features (TEN) and the different aggregated feature combinations EN0 - EN5, are provided for download on the Million Song Dataset Benchmarking platform [16]:

http://www.ifs.tuwien.ac.at/mir/msd/

This Web page provides a wide range of complementary audio features for the Million Song Dataset. Additional features have been extracted from nearly one million corresponding audio samples that have been downloaded from 7digital. The aggregated Echonest features are provided as single files containing all vectors for the tracks of the MSD and are stored in the WEKA Attribute-Relation File Format (ARFF) [21]. Additionally different benchmark partitions based on different genre label assignments are provided for instant use and comparability.

References

1. Bertin-Mahieux, T., Ellis, D.P.W., Whitman, B., Lamere, P.: The million song dataset. In: Proceedings of the 12th International Conference on Music Information Retrieval (ISMIR 2011) (2011)
2. Cano, P., Gómez, E., Gouyon, F., Herrera, P., Koppenberger, M., Ong, B., Serra, X., Streich, S., Wack, N.: ISMIR 2004 audio description contest. Technical report (2006)
3. Dieleman, S., Schrauwen, B.: Audio-based music classification with a pretrained convolutional network. In: Proceedings of the 12th International Conference on Music Information Retrieval (ISMIR 2011) (2011)
4. Ellis, D.P.W.: Classifying music audio with timbral and chroma features. In: Proceedings of the 8th International Conference on Music Information Retrieval (ISMIR 2007) (2007)
5. Fu, Z., Lu, G., Ting, K.M., Zhang, D.: A survey of audio-based music classification and annotation. IEEE Trans. Multimed. **13**(2), 303–319 (2011)
6. Hall, Mark, Frank, Eibe, Holmes, Geoffrey, Pfahringer, Bernhard, Reutemann, Peter, Witten, Ian H.: The WEKA data mining software: an update. SIGKDD Explor. **11**(1), 10–18 (2009)
7. Jehan, T., DesRoches, D.: Analyzer documentation (analyzer version 3.08). Website (2011). http://developer.echonest.com/docs/v4/_static/AnalyzeDocumentation.pdf. Accessed 17 Apr 2012
8. Lidy, T., Mayer, R., Rauber, A., Pertusa, A., Inesta, J.M.: A cartesian ensemble of feature subspace classifiers for music categorization. In: Proceedings of the 11th International Conference on Music Information Retrieval (ISMIR 2010) (2010)

9. Lidy, T., Rauber, A.: In: Proceedings of the 6th International Society for Music Information Retrieval Conference (ISMIR 2005) (2005)

10. Lidy, T., Silla Jr., C.N., Cornelis, O., Gouyon, F., Rauber, A., Kaestner, Caa, Koerich, A.L.: On the suitability of state-of-the-art music information retrieval methods for analyzing, categorizing and accessing non-Western and ethnic music collections. Signal Process. **90**(4), 1032–1048 (2010)

11. Logan, B.: Mel frequency cepstral coefficients for music modeling. In: International Symposium on Music Information Retrieval (2000)

12. McKay, C., Fujinaga, I.: Musical genre classification: is it worth pursuing and how can it be improved. In: Proceedings of the 7th International Conference on Music Information Retrieval (ISMIR 2006), pp.101–106 (2006)

13. McKay, C., Fujinaga, I.: jMIR: tools for automatic music classification. In: Proceedings of the International Computer Music Conference, pp. 65–68 (2009)

14. Pampalk, E., Rauber, A., Merkl, D.: Content-based organization and visualization of music archives. In: Proceedings of the 10th ACM International Conference on Multimedia, p. 570 (2002)

15. Rauber, A., Pampalk, E., Merkl, D.: The SOM-enhanced JukeBox: organization and visualization of music collections based on perceptual models. J. New Music Res. **32**(2), 193–210 (2003)

16. Schindler, A., Mayer, R., Rauber, A.: Facilitating comprehensive benchmarking experiments on the million song dataset. In: Proceedings of the 13th International Conference on Music Information Retrieval (ISMIR 2012) (2012)

17. Silla, Jr., C.N., Koerich, A.L., Catholic, P., Kaestner, C.A.A.: The Latin music database. In: Proceedings of the 9th International Conference of Music Information Retrieval, p. 451. Lulu. com (2008)

18. Tzanetakis, G.: Manipulation, analysis and retrieval systems for audio signals. Ph.D. thesis (2002)

19. Tzanetakis, George, Cook, Perry: Marsyas: a framework for audio analysis. Organised Sound **4**(3), 169–175 (2000)

20. Tzanetakis, George, Cook, Perry: Musical genre classification of audio signals. IEEE Trans. Speech Audio Process. **10**(5), 293–302 (2002)

21. Witten, I.H., Frank, E., Trigg, L., Hall, M., Holmes, G., Cunningham, S.J.: Weka: practical machine learning tools and techniques with Java implementations (1999)

Adaptation and HCI

SKETCHify – An Adaptive Prominent Edge Detection Algorithm for Optimized Query-by-Sketch Image Retrieval

Ihab Al Kabary[(⊠)] and Heiko Schuldt

Department of Mathematics and Computer Science,
Databases and Information Systems Group, University of Basel, Basel, Switzerland
{ihab.alkabary,heiko.schuldt}@unibas.ch

Abstract. Query-by-Sketch image retrieval, unlike content based image retrieval following a Query-by-Example approach, uses human-drawn binary sketches as query objects, thereby eliminating the need for an initial query image close enough to the users' information need. This is particularly important when the user is looking for a known image, i.e., an image that has been seen before. So far, Query-by-Sketch has suffered from two main limiting factors. First, users tend to focus on the objects' main contours when drawing binary sketches, while ignoring any texture or edges inside the object(s) and in the background. Second, the users' limited ability to sketch the known item being searched for in the correct position, scale and/or orientation. Thus, effective Query-by-Sketch systems need to allow users to concentrate on the main contours of the main object(s) they are searching for and, at the same time, tolerate such inaccuracies. In this paper, we present SKETCHify, an adaptive algorithm that is able to identify and isolate the prominent objects within an image. This is achieved by applying heuristics to detect the best edge map thresholds for each image by monitoring the intensity, spatial distribution and sudden spike increase of edges with the intention of generating edge maps that are as close as possible to human-drawn sketches. We have integrated SKETCHify into QbS, our system for Query-by-Sketch image retrieval, and the results show a significant improvement in both retrieval rank and retrieval time when exploiting the prominent edges for retrieval, compared to Query-by-Sketch relying on normal edge maps. Depending on the quality of the query sketch, SKETCHify even allows to provide invariances with regard to position, scale and rotation in the retrieval process. For the evaluation, we have used images from the MIRFLICKR-25K dataset and a free clip art collection of similar size.

1 Introduction

Users of digital cameras are faced with the challenge of managing increasingly large image collections. Typical tasks in such collections include the search for known images, i.e., images that have been looked at before. Due to the sheer size of collections, individual objects are rarely annotated with metadata describing

© Springer International Publishing Switzerland 2014
A. Nürnberger et al. (Eds.): AMR 2012, LNCS 8382, pp. 231–247, 2014.
DOI: 10.1007/978-3-319-12093-5_14

their content – as this activity can hardly be automated. This lack of metadata renders image retrieval inherently difficult. Therefore, content-based image retrieval (CBIR) is becoming increasingly important as it frees users from manually providing such metadata. Most existing CBIR systems provide two fundamentally different modes of operation. Query-by-Example follows an approach where a query image, which is close enough to the users' information need is required in order to obtain acceptable results. Without such a query image, it is difficult to achieve good retrieval quality, even if sophisticated relevance feedback mechanisms are available as search is done by comparing the similarity of inherent image features such as color, shape, texture or prominent regions of interest. Query-by-Sketch, in turn, is another approach to CBIR that addresses this problem by using human generated binary sketches as query objects. This eliminates the need for finding a query image and allows users to focus on the most important details of the image(s) they are looking for. This is particularly important when a user is looking for a specific seen-before image (a.k.a. "known item search"). However, one of the main challenges faced in Query-by-Sketch is the lack of the users' drawing ability to accurately sketch the known item being searched for. The user might forget the spatial location(s) of the main object(s) within the known image or forget the exact scale or orientation. Moreover, users should be able to focus on the objects' main contours without having to worry about clutter in the background or the objects' fine texture. Hence, a major requirement for effective Query-by-Sketch systems is to tolerate such inaccuracies and to allow users to focus on drawing only the prominent edges of the known item being searched for and yet achieve satisfactory results.

Image feature descriptors commonly used in Query-by-Sketch systems are already designed to provide at least some basic support for different types of invariances. In this paper, we present SKETCHify, an adaptive algorithm for extracting image features that significantly extends translation and scale invariances for Query-by-Sketch and, at the same time, leads to more compact feature descriptors which reduce retrieval time. SKETCHify, based on the Canny edge detector [3], allows to identify prominent objects within an image by automatically adjusting the threshold values of the edge detector parameters, independently for each image. It monitors the intensity, spatial distribution and sudden spike increase of edges pixels at various thresholds, and results in an edge map that is as close as possible to the user-drawn binary sketch. This enhances similarity search, as it enables the identification of bounding box(es) wrapped around the main object(s) within an image, which in turn, enables partial image search, and in certain cases enables complete scale and translation invariances. SKETCHify has been integrated in QbS, our Query-by-Sketch system [16] and we report on evaluations that have been performed using the MIRFLICKR-25K dataset with 25,000 images and a clipart collection of similar size.

The paper is organized as follows: Sect. 2 discusses related work. Section 3 introduces QbS and Sect. 4 presents the SKETCHify approach to extend search invariances and to improve Query-by-Sketch image retrieval. Section 5 summarizes the results of the experimental evaluation of SKETCHify. Section 6 concludes the paper.

2   Related Work

Systems that support Query-by-Sketch like [4,7,9,11,12,20] aim at providing tolerance for a variety of inaccuracies in the sketches. This is mainly needed to assist users with limited sketching skills or users who do not precisely remember the location, scale and/or orientation of the object(s) within the known item being searched for. For this purpose, a variety of visual feature descriptors have been designed or modified to provide invariances to tolerate such inaccuracies.

The MPEG-7 Visual Standard [6] defines a large set of feature descriptors that can be used in image retrieval, out of which the Edge Histogram Descriptor (EHD) [1] is frequently used for sketch-based retrieval. It represents the distribution of 5 types of edges partitioned into 4×4 non-overlapping blocks. Angular Radial Partitioning (ARP) [5] also uses spatial distribution of edges by means of which both sketch and image are partitioned into subregions according to an angular-radial segmentation. The image distortion model (IDM) [14] which has been used for handwritten character recognition and in medical automatic annotation tasks is yet another descriptor that has proven to work well as a distance function. It evaluates displacements of individual pixels between images within a so-called warp-range and takes patches of surrounding pixels (local context) into account for more detailed comparisons. However, it is computationally expensive when used with large warp-ranges and local context. The more recent Tensor [8] descriptor also subdivides the image and sketch respectively into rectangular blocks. It determines the orientation of large gradients within each block and correlates them with the edge directions in the user sketch.

All these algorithms tolerate a limited degree of inaccuracies in the sketches. In contrast, the SKETCHify approach presented in this paper provides heuristics that not only significantly extends the tolerance range of search invariances – in some cases, it even tolerates absolute invariance for scale and translation of objects and thus enables partial image search. It builds on the idea that the majority of feature descriptors used in Query-by-Sketch systems rely on some edge detector [22]. A fundamental aspect of using these edge detectors is deciding on the thresholds for identifying local changes in intensity. Because the accuracy of the resulting edge maps is highly sensitive to the input values chosen for these single or multi-variable thresholds, no specific fixed threshold values can obtain optimal results for every individual image in any image collection. SKETCHify attempts to find the optimal threshold for every individual image even in a heterogeneous image collection. Figure 1 shows three different images from the MIRFLICKR-25K collection and the resulting edge maps obtained from running the Canny edge detector [5] at various thresholds. For the remainder of this paper we will summarize the two thresholds of the Canny edge detector (T_H high threshold and T_L low threshold), with β, where $\beta = T_H$ and $T_L = 0.4 \times T_H$. It is obvious that there is no fixed threshold that would give reasonable edge maps for all three images. Essentially, it is the goal of SKETCHify to solve this problem and to automatically select the most suitable threshold. This main weakness of needing to set the threshold values manually for each image within a collection has led to attempts for automating the calculation of these thresholds.

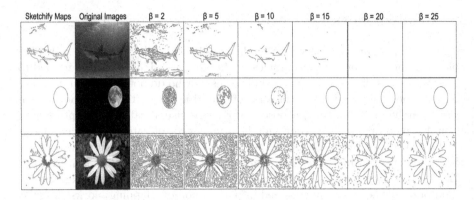

Fig. 1. Edge maps generated with the Canny edge detector at various thresholds

The Otsu method, with computations based entirely on the grayscale histogram, has been used to obtain the adaptable thresholds for the Canny edge detector as shown in [10]. Examples of more scope specific adaptations can be shown in [18] to detect the sky-sea line in infrared images for identifying targets for missile guidance systems and in [17] to extract coastline from satellite imagery. The Berkeley detector [19] is another successful approach that uses the assistance of supervised machine learning to detect and localize boundaries in natural scenes using local image measurements like brightness, color and texture. It has been used to create the contours needed by Edgel [4] that creates an index for large-scale sketch-based image search systems.

Another set of algorithms that focuses on the locations that attract the gaze of an observer are human visual attention (VA) algorithms. These algorithms are built on the fact that main objects within an image are often distinguished when they stand out from the rest of the image. VA saliency map algorithms [15] like in [2,13,21] take into account color, texture and luminance to produce saliency maps that emulate which regions within an image (in order of importance) a human would notice most, and as a result would most probably sketch. Our proposed unsupervised approach focuses on setting the threshold values of edge detectors automatically on generic, natural, unrestrained image collections by applying heuristics to detect the best thresholds for each image.

3   QbS: A Query-by-Sketch System

QbS [20] provides an interactive application that is used to retrieve known images based on user-provided sketches as query input. The system runs on a Tablet PC to facilitate sketch input but also supports interactive paper and smart pen technology as query frontend. For similarity search, two visual features are exploited: Angular Radial Partitioning (ARP) [5] is used as a rapid method to retrieve images, when the user draws a rough sketch, and places the edges in a spatial distribution that is sufficient to distinguish the known item from the rest of the images within the collection. The Image Distortion Model (IDM) [14] is used as a more accurate but computationally more expensive approach, when users require a more detailed comparison between their sketches and the images within the collection. A textual filter, based on Apache Lucene[1], can be used beforehand to limit the search space if images have attached annotations.

3.1 Angular Radial Partitioning (ARP)

QbS supports ARP at various resolutions. By default ARP is used with 8 angular and 4 radial partitions in which the number of edge pixels inside sketch and edge map are counted. This leads to a 32 dimensional feature vector. The user is given the possibility to distinguish areas that shall remain free of edges (e.g., a plain area) or areas where the content is unknown (e.g., the user does not remember the content). The user can easily select these regions of interest using input devices such as the mouse, a stylus-pen of a tablet, or even interactive paper. Internally this is handled by comparing the vectors using a weighted Manhattan distance: For areas which should get treated as "unknown", a weight of zero is assigned; all non-empty partitions have a weight of one – as well as areas which are intentionally left empty. The user can choose between the two cases, and therefore has the freedom to either sketch the full image or focus on drawing the parts of the image the user remembers well (treating all other areas as unknown). ARP can handle small misplacements and rotation, as long as not many edge pixels are moved from one partition to another. If the user anticipates even more rotation, invariance to rotation is provided by applying the 1D FFT on the features as proposed in [5]. For translations that are further off the position of the sketch, a heuristic can be applied: The original image is cut into several regions that still cover enough pixels to give meaningful search results. In addition to the image as a whole and a bounding box covering all the area with edge pixels, we also extract slightly smaller regions shifted towards each of the image corners. During search, the sketch is compared with all regions, thus compensating to a certain extent for translation. As most of these regions are smaller than the full image, comparing the sketch or parts of the sketch can also allow for some invariance of scale. Features of these multiple sub-images with and without 1D FFT are stored separately for a selection of edge threshold values. For simple, regular searches that do not use any search invariance option, the query feature vector from the user's sketch will be compared to exactly one feature vector for each image in the collection. Invariant searches will compare the distances with all corresponding representations, and only select the most similar version of the image features.

3.2 Image Distortion Model (IDM)

IDM can be used when a more detailed sketch-to-edge map comparison is required. Initially, the edge map image is scaled down to a smaller size (e.g.,

---

[1] http://lucene.apache.org/

32 pixels for the longer side), and comparisons on the scaled down sketch and edge maps are performed pixel-by-pixel. To allow for invariances, a warp range can be defined at query time, allowing the user to specify the tolerance allowed for pixels to be misplaced by translation, scaling or rotation. Larger warp ranges allow for higher invariance but at a higher computational cost. Another parameter, the local context, allows for patches, instead of individual pixels to be matched. Larger values request more detailed comparisons, and less tolerance to drawing inaccuracies is achieved as bigger patches must match in order to achieve low distances in the similarity search process.

4   SKETCHify

SKETCHify has three main goals: First, to tune the edge detector thresholds automatically so as to obtain an edge map that is as close as possible to a human-drawn binary sketch, thus freeing the user from manually inputting the thresholds. Second, to encapsulate the detected object(s) in bounding box(es) so as to enable partial image search and complete scale and translation invariance when required by the user. Third, to generate compact feature representations which will result in an increased query runtime performance. This will enable Query-by-Sketch systems to avoid saving excessive feature files at different threshold levels but rather save concise files obtained from more accurate edge maps. In doing so, it removes the need for multi-feature comparisons between a sketch and the various edge maps for every image, which increases retrieval speed and decreases storage requirements in addition to the aforementioned benefits. It is worth noting that SKETCHify is entirely performed in the offline phase of the feature generation. The following subsections provide details on the algorithm.

4.1   Prominent Area Matrix

SKETCHify starts by generating the so-called *Prominent Area Matrix* (PAM) which, in brief, is a result of iterating through the various edge maps of a single image at various β values of the Canny edge detector and deducing the pixels that are considered prominent. The PAM size is equivalent to the image size, i.e., every cell corresponds to a pixel in the image. In these iterations, starting from the largest β value (showing only the most prominent edges), detecting an edge pixel triggers an increment of the score of the PAM at the relative cell. Incrementing can either be by a fixed value or by a value proportional to the importance of the β threshold. We refer to these options as *simple* and *weighted* increment methods. In the latter case, higher values of β get higher weights, since they only detect the very prominent edges with very sharp contrasts. Finally, after calculating the PAM, the cells scoring relatively low values are removed (less than 5 % of the maximum PAM cell score). Experiments showed that the weighted increment method outperformed the simple method. A visual representation of these steps for the weighted PAM can be seen in Fig. 2(b), (c).

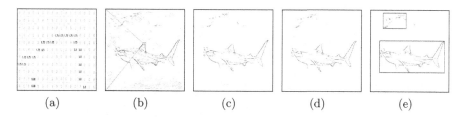

Fig. 2. The various steps of Sketchify: (a) Zoom-in on a PAM region, (b) PAM, (c) Remove low-valued cells, (d) Remove isolated cells, (e) Minimum bounding regions (im16102.jpg – MIRFLICKR-25K)

4.2   Clutter Spike Detection

From our observations of SKETCHify on various images, we noticed that very small fragmented edges appeared suddenly in large quantities at different thresholds with different images. Consequently, if we stop the PAM calculations at these β values, which are different from one image to another (among others, due to the difference in focal length chosen when taking the picture), the quality of detection of prominent edges can be further enhanced. We also noticed that with very high contrast images in which the main contours within an image are easily extracted with a high β value, avoiding the use of edge maps with lower (more sensitive) β values would generate better results. *Clutter spike detection*, which is the name we gave to this heuristic, relies on measuring the increase in edge pixels at each β value and detecting the maximum amount of sudden edge pixel increase. This heuristic assists SKETCHify to avoid using these edge maps in the PAM selection process. Note that we cannot use the percentage increase instead of the absolute increase of edge pixels to measure the maximum surge, since this can sometimes lead to premature detection. Instead, we count the edge pixels in the edge map at each β and measure the increase. Early termination is triggered when the initial three edge maps (with the largest β values) have nearly identical edge pixels, i.e., having the increase of edge pixels at subsequent edge maps less than 5 %. This indicates that the image is of very high contrast and prominent edges have successfully been segmented. Figure 3 shows the heuristic at work with $\beta=6$ being detected as the edge map with the largest spike in clutter. Hence, PAM calculation should be stopped before this stage and edge maps with β values 6, 4 and 2 should not be taken into account in SKETCHify's calculations.

4.3   Isolated Cell Removal

SKETCHify cannot guarantee the complete removal of background noise from each image, especially when the foreground is well blended with the background. However, further enhancements can be applied on the PAM to remove edge pixels that appear to be isolated so as to further improve the quality of the results, and make the edge representation even more similar to the user-drawn sketch

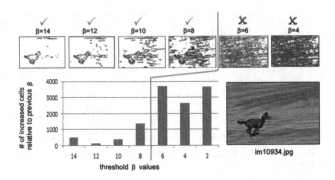

Fig. 3. Clutter spike detection enables SKETCHify to not take edge maps with β values 6, 4 and 2 into account (β=8 identified as the final edge map to be used in SKETCHify)

(which focuses on the main contours). This improvement will not only assist in removing clutter but will also help obtain more accurate minimum bounding box(es) around the main object(s), which consequently will provide the ability to perform partial image search in Query-by-Sketch systems with total scale and translation invariance. The initial approach for the *Isolated Cell Removal* filter is to count the number of neighborhood cells within a specific range. If that number is less than a reasonable threshold then the cell gets removed. This approach has proven not to be accurate enough because it does not consider the importance of each cell within the PAM. Therefore, an effective approach has been developed that takes into account the PAM scores by summing up the values within neighborhood cells of each cell and compares the sum to the maximum obtainable PAM score. In this way, if a pixel is surrounded by multiple low-valued edge pixels, it has a higher probability of getting removed, contrary to a pixel that has a few prominent neighborhood pixels. The window size for the neighbors taken into account was chosen to be 10 % of the size of the image after empirical observations. In an image of width and height equal to w and h, respectively, and with w' and h' symbolizing the width and height of half the window size respectively; a pixel (x, y) is removed if it meets the following condition:

$$\sum_{i=x-w'}^{x+w'} \sum_{j=y-h'}^{y+h'} PAM[i, j] < maxPAMscore$$

The isolated cell removal algorithm is not applied just once, but works in an iterative manner until it converges and no more pixels are further removed. The results of the isolated cells algorithm are illustrated in Fig. 2(d) followed by the bounding box detection in Fig. 2(e). The pseudo-code of the SKETCHify algorithm is shown in Algorithm 1.

Algorithm 1. SKETCHify

1: $start\beta \leftarrow 50$
2: $max\beta \leftarrow 0$
3: $step\beta \leftarrow 2$
4: $stop\beta \leftarrow 2$
5: $i \leftarrow 0$
6: $EdgePixPerMap[25] \leftarrow \{0, 0..0, 0\}$
7: $PAMs[25] \leftarrow \{null, null..null, null\}$
8: $PAM[imageWidth][imageHeight] \leftarrow \{0, 0..0, 0\}$
9: **for** $\beta = start\beta$ to $stop\beta$ **do**
10:    $edgemap \leftarrow getEdgeMap(image, \beta)$
11:    **for** $x = 0$ to $imageWidth$ **do**
12:      **for** $y = 0$ to $imageHeight$ **do**
13:        $c \leftarrow edgePixelsInCell(edgemap, x, y)$
14:        **if** $(c > 0)$ **then**
15:          $EdgePixPerMap[i] \leftarrow EdgePixPerMap[i] + 1$
16:          **if** $max\beta = 0$ **then**
17:            $max\beta = \beta$
18:          **end if**
19:          **if** (Weighted Scoring Scheme) **then**
20:            $PAM[x][y] \leftarrow PAM[x][y] + \frac{\beta}{max\beta}$
21:          **else**
22:            $PAM[x][y] \leftarrow PAM[x][y] + 1$
23:          **end if**
24:        **end if**
25:      **end for**
26:    **end for**
27:    $PAMs[i] \leftarrow PAM$
28:    $\beta \leftarrow \beta + step\beta$
29:    **if** $(i = 3)$ **then**
30:      **if** $(EdgePixPerMap[0] \approx EdgePixPerMap[1] \approx EdgePixPerMap[2])$ //
        change less than 5 % **then**
31:        $breakloop$
32:      **end if**
33:    **end if**
34:    $i \leftarrow i + 1$
35: **end for**
36: **if** $(i = 3)$ **then**
37:    $PAM \leftarrow PAMs[i]$
38: **else**
39:    $j \leftarrow getMaxEdgeSpike(EdgePixPerMap)$
40:    $PAM \leftarrow PAMs[j]$
41:    $PAM \leftarrow removeLowValuePixels(PAM)$
42:    $PAM \leftarrow removeIsolatedPixels(PAM)$
43: **end if**
44: $sketchifyMap \leftarrow edgeMapFromPAM(PAM)$

5   Evaluation

We have divided the evaluation of SKETCHify into two parts. The first part involves an evaluation of SKETCHify against semi-automatic ground truth edge maps generated with human feedback on what is perceived to be important edges, and an evaluation against visual attention saliency map algorithms. The second part focuses on the retrieval performance of QbS when SKETCHify is integrated in the feature generation phase (i.e., when sketch-based CBIR takes place on the basis of the compact features provided by PAM with Clutter Spike Detection and Isolated Cell Removal). For the evaluations we used the MIRFLICKR-25K[2] dataset comprising 25,000 images obtained from the popular image sharing site Flickr and containing photographs selected based on the flickr measure of "interestingness" which takes into account how many people watched the image, commented on it, tagged it, picked it as favorite. We also used a clipart collection of similar size, assembled from a variety of clipart websites. Both datasets include only Creative Commons-licensed pictures.

5.1   Sensing Prominence

In the first part of our evaluation, we compare SKETCHify against semi-automatic ground truth edge maps and then against well-known visual attention saliency map algorithms.

Comparing Against the Perfect Edge Map. In a first step, we involved 8 users in a semi-automatic task of collectively finding the ground truth edge maps for 12 images divided into three categories: (i) images with one object and a simple background, (ii) multiple objects with a simple background, and (iii) images where the background is blended and hard to extract from the foreground. The following steps of the heuristic are used to obtain the ground truth edge maps:

Step 1: The user sketches the 12 images.
Step 2: The user chooses for each sketch the closest edge map, from the set of edge maps obtained using the Canny edge detector at various thresholds.
Step 3: The user optionally deletes extra edge pixels (from the edge map of step 2) to create edge maps that are as close as possible to the sketch.
Step 4: Repeat step 1 to 3 by the 8 users.
Step 5: Superimpose for each image the resulting sketch-like edge maps, and if an edge pixel is found in 6 out of 8 of these sketch-like edge maps, then regard it as a ground truth edge pixel.

After obtaining ground truth edge maps, we compared SKETCHify against the canny edge detector at various thresholds by using a weighted hamming distance function. This function gives a penalty when an edge pixel is found

---

[2] http://press.liacs.nl/mirflickr/

in the ground truth edge map and not in the edge map being compared with. Figure 4 shows that SKETCHify outperforms Canny at all threshold values and is closest to the human perception of important edges used within a sketch. It is also worth noting that SKETCHify with the weighted PAM scoring performs slightly better than the simple PAM scoring scheme. The removal of isolated pixels, when applied, slightly improves the results although its main advantage is evident when applied before finding the bounding box(es) around main object(s) within the image. This can be crucial for image retrieval as it can assist in partial image search. Figure 4 also shows that the Canny operator at thresholds 8, 10 and 12 performs well in this set of images. However, they could be unreliable for other image collections and this is why fixing thresholds is not the solution (except in homogeneous collections).

SKETCHify and Saliency Maps. We have also compared SKETCHify against Saliency algorithms like the ones developed by Itti et al. [13], Bruce

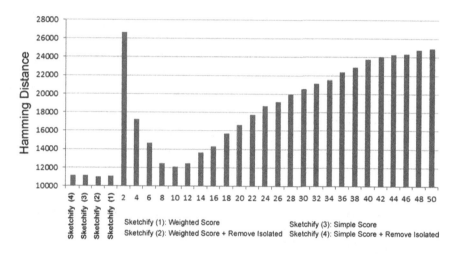

Fig. 4. SKETCHify vs. Canny operator at various thresholds

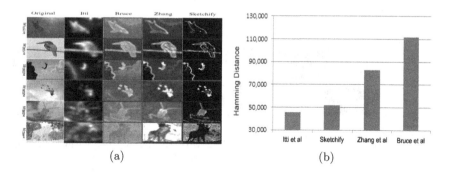

(a)   (b)

Fig. 5. Evaluation of SKETCHify against saliency maps

et al. [2] and Zhang et al. [21] in order to test their ability of identifying prominent areas within an image. We handpicked 105 images (the intention was to represent the same three categories as in Sect. 5.1) from the MIRFLICKER-25K collection. Figure 5(a) shows two images from each category of images and the resulting saliency maps and edge maps generated from the various algorithms. It is evident that for some images like the last image of the wolf, un-useful results are obtained (even misleading) with all algorithms including SKETCHify due to the unusual color composition of the image and in this case, the majority of edges are found for the background and not the foreground. The ground truth for a quantitative evaluation to compare SKETCHify with Saliency maps has been obtained using the following heuristics:

Step 1: User *traces* prominent areas (by drawing closed edges) for all images.
Step 2: Repeat step 1 by the 10 users.
Step 3: Fill areas within the objects detected to signify important areas.
Step 4: Superimposed segmented areas from step 3 (for each image) to generate the ground truth maps of the important areas. If a marked pixel is found in 7 out of 10 of these segmented areas, mark it as a ground truth pixel.

Having obtained the ground truth maps, we apply several filters on the outputs of the SKETCHify and saliency maps to make the evaluation fair. Initially, we resize all images to 512×512 pixels. Then we apply a filling filter, that checks if a pixel is surrounded in 15 out of 16 directions within a reasonable radius (64 pixels) with prominent pixels, in which case it is marked as prominent as well. This step fills up the majority of small empty spaces within objects that mostly appear in Zhang et al.'s saliency map and SKETCHify. We also apply histogram equalization to the output of all algorithms before applying a threshold that will remove very low intensity valued pixels. This step especially assists the saliency maps of Bruce et al. and Zhang et al. to reduce clutter.

Comparing SKETCHify with the saliency maps algorithms, as shown in Fig. 5(b), we discovered that the saliency map of Itti et al. was the best performer followed closely by SKETCHify. However, SKETCHify is concerned more on finding prominent edges, as opposed to prominent regions, and therefore it was encouraging to see that it also stands out in the competition against saliency map algorithms.

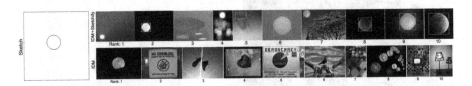

Fig. 6. Results for a simple sketch using IDM with (first row) and without (second row) SKETCHify (MIRFLICKR-25K)

5.2   Evaluating SKETCHify in QbS

Finally, we put SKETCHify to the test in the retrieval arena. SKETCHify was used in the feature generation phase to generate an edge map for each image, instead of the various edge maps at various thresholds. If bounding box(es) were obtainable around the object(s) in the image, another edge map for each object(s) was saved separately. The following subsection starts off with a qualitative evaluation followed by a quantitative evaluation.

Retrieval Quality. In order to show the potential of SKETCHify, we start with a few simple examples. Figure 6 shows a sketch, which has a circle, drawn towards the center of the sketching area using the QbS drawing tools. IDM is used for the similarity search with and without applying SKETCHify in the feature generation. SKETCHify enables what can be percieved as absolute scale and translation invariance as shown in images with rank 1, 2, 3, 4, 5 and 7. It also enables partial image search as shown in images with rank 3, 4, 5 and 7. The aspect ratio is not respected, like in the image with rank 3, due to the fact that objects detected during feature generation are treated separately and have been resized to equal width and height. This can be avoided, if required, by extending the canvas of the shorter side of the detected object and filling it with an empty background to match the size requirement. It is worth mentioning that the green boxes drawn on top of the image results show the part of the image that matches the query object when SKETCHify is used. On the other hand, using IDM without applying SKETCHify in the similarity search does return a few images that contain a circular-like object placed approximately at the same place as the circle in the sketch. However, even with a large wrap range of 4 (tolerance to invariance of close to a third of the 32×32 canvas), results were not as accurate as when using SKETCHify.

The second example considers the clipart collection that contains around 25,000 images. In this example, we show an incremental sketch drawn by a user, where the final objective was to sketch-and-search for a "no bike allowed" sign. Figure 7 shows the incremental sketching steps and results obtained when pausing and searching at each of these steps using IDM as a feature vector and applying SKETCHify. The scale and position of the sketches did not alter the results and appropriate images were retrieved.

SKETCHify also allows to lighten-up the graphical user interface of QbS by removing a control that was used to manually select a range of β values to use during runtime similarity search. Higher values indicated that the user only sketched very prominent edges and lower values meant that the user spent some time in sketching the known item in more details. We noticed in various user trials of the system, that this control was not understood/used by the majority. However, without SKETCHify, this control cannot be removed because comparing with the whole range of β edge maps at runtime would decrease the speed of the system dramatically. As SKETCHify enables us to remove this control, it also increased the speed of the system to a great extent. This is mainly due to the fact that less feature vectors are stored on disk and less comparisons

Fig. 7. Results for incremental sketches using IDM with SKETCHify (Clipart-25K)

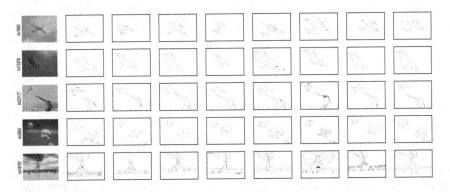

Fig. 8. Images used and sketches drawn for the evaluation

are needed at runtime. The MIRFLICKR-25K collection now needs only 0.83 GB in feature descriptors instead of 11.2 GB (7.4 % of previously required space).

Query Performance. We have chosen 5 different images to assess the performance of SKETCHify when used with QbS. We have collected sketches from 8 different users and asked them to sketch the 5 images in a very fast manner. Both images and sketches can be seen in Fig. 8. The chosen images include both images with single and multiple objects and with various degrees of complexity regarding the level of details in the background. We avoided using a text filter in the retrieval process to focus on the performance of the visual feature descriptors with and without applying SKETCHify.

The evaluation results illustrated at Fig. 9 show that SKETCHify was not only beneficial in creating more compact features files and a faster retrieval system, but its performance has even outperformed that of using IDM alone in a subset of the images. In image im18797, we noticed that SKETCHify's performance decreased due to the fact that sketch number 6 (for image im18797) in Fig. 8 was drawn by the user in great detail, including the clouds and even filling a portion of the Eiffel tower instead of just drawing edges, which led the system to give poorer ranks for this sketch when comparing with the edge map

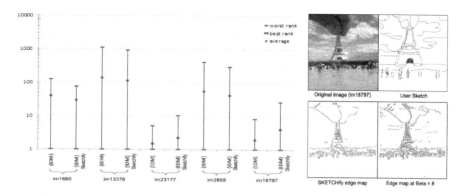

Fig. 9. SKETCHify evaluation

generated from SKETCHify and giving IDM (using a low valued β threshold) a higher rank. Figure 9 shows this and explains the rank drop from 8 when using IDM to 26 when using IDM alongside with SKETCHify. Taking the same image (im18797) into account and measuring the average response time of the retrieval system, we found that it decreased by 65 % on average (from close to 2 m to 41 s using an Intel Core i7 CPU @ 2.93 GHz equipped with 4 GB of RAM). The other images gave very similar response time improvements. Enabling full translation and scale invariance was noticed to be beneficial only when users submitted sketches that were reasonably close in structure and orientation to the generated edge maps (even if the sketches were of incorrect scale or placement).

6   Conclusions and Outlook

We have presented SKETCHify, an unsupervised adaptive algorithm that focuses on automatically setting the thresholds of edge detectors on generic, natural, unrestrained image collections by applying heuristics to detect the best threshold value for each image. This is done by monitoring the intensity, spatial distribution and sudden spike increase of edges with the intension of generating edge maps that are as close as possible to human-drawn sketches. Compared to existing approaches, it generates improved feature vectors and extends the ability of search invariance to reach, in some cases, full scale and translation invariance. Essentially, this enables partial image search in situations where bounding box(es) successfully wrap the detected main object(s) within images. The algorithm has been integrated into the QbS system and a detailed performance evaluation of the system has been done using the MIRFLICKR-25k benchmark and a clipart collection of similar size. Results have shown good results, especially when reasonable sketches are drawn, even if they are displaced or have a different scale. On a different angle, massive decrease in space requirements for storing the feature descriptors was achieved while, at the same time, query performance has increased. In addition, the user-friendliness of the complete Query-by-Sketch process has been significantly improved as users are freed from specifying a range of β edgemap values to search within at runtime. In future work, we will integrate additional filters to create sketch-like edge maps. Moreover, sketch-based retrieval will also be applied to video collections.

Acknowledgments. The work has been partly supported by the Swiss National Science Foundation, projects PAD-IR (No. 200020_126829) and MM-DocTable (No. 200020_137944).

References

1. Bober, M.: MPEG-7 visual shape descriptors. IEEE Trans. Circ. Syst. Video Technol. **11**(6), 716–719 (2001)
2. Bruce, N., Tsotsos, J.: Attention based on information maximization. J. Vis. **7**(9), 950–950 (2007)
3. Canny, J.: A computational approach to edge detection. IEEE Trans. PAMI **8**, 679–698 (1986)
4. Cao, Y., Wang, C., Zhang, L., Zhang, L.: Edgel inverted index for large-scale sketch-based image search. In: CVPR (2011)
5. Chalechale, A., Naghdy, G., Mertins, A.: Sketch-based image matching using angular partitioning. IEEE Trans. Syst. Man Cybern. **35**(1), 28–41 (2005)
6. Chang, S., Sikora, T., Puri, A.: Overview of the MPEG-7 standard. IEEE Trans. Circ. Syst. Video Technol. **11**(6), 688–695 (2001)
7. Del Bimbo, A., Pala, P.: Visual image retrieval by elastic matching of user sketches. IEEE Trans. PAMI **19**(2), 121–132 (1997)
8. Eitz, M., Hildebrand, K., Boubekeur, T., Alexa, M.: A descriptor for large scale image retrieval based on sketched feature lines. In: ACM SBIM, August 2009
9. Eitz, M., Hildebrand, K., Boubekeur, T., Alexa, M.: PhotoSketch: a sketch based image query and compositing system. In: SIGGRAPH (2009)
10. Fang, M., Yue, G.X., Yu, Q.C.: The study on an application of Otsu method in canny operator. In: Proceedings of the International Symposium on Information Processing (ISIP'09), vol. 2, pp. 109–112 (2009)
11. Flickner, M., et al.: Query by image and video content: the QBIC system. Computer **28**(9), 23–32 (1995)
12. Hirata, K., Kato, T.: Query by visual example. In: Pirotte, A., Delobel, C., Gottlob, G. (eds.) EDBT 1992. LNCS, vol. 580, pp. 56–71. Springer, Heidelberg (1992)
13. Itti, L., Koch, C., Niebur, E.: A model of saliency-based visual attention for rapid scene analysis. IEEE Trans. PAMI **20**, 1254–1259 (1998)
14. Keysers, D., Deselaers, T., Gollan, C., Ney, H.: Deformation models for image recognition. IEEE Trans. PAMI **29**(8), 1422–1435 (2007)
15. Koch, C., Ullman, S.: Shifts in selective visual attention: towards the underlying neural circuitry. Hum. Neurobiol. **4**(4), 219–227 (1985)
16. Kreuzer, R., Springmann, M., Kabary, I.A., Schuldt, H.: An interactive paper and digital pen interface for query-by-sketch image retrieval. In: Baeza-Yates, R., de Vries, A.P., Zaragoza, H., Cambazoglu, B.B., Murdock, V., Lempel, R., Silvestri, F. (eds.) ECIR 2012. LNCS, vol. 7224, pp. 317–328. Springer, Heidelberg (2012)
17. Liu, H., Jezek, K.: Automated extraction of coastline from satellite imagery by integrating canny edge detection and locally adaptive thresholding methods. Int. J. Remote Sens. **25**, 937–958 (2004)

18. Lu, J., Ren, J., Lu, Y., Yuan, X., Wang, C.: A modified canny algorithm for detecting sky-sea line in infrared images. Int. Conf. Intell. Syst. Des. Appl. **2**, 289–294 (2006)
19. Martin, D., Fowlkes, C., Malik, J.: Learning to detect natural image boundaries using local brightness, color, and texture cues. IEEE Trans. PAMI **26**, 530–549 (2004)
20. Springmann, M., Kopp, D., Schuldt, H.: QbS - searching for known images using user-drawn sketches. In: Proceedings of MIR 2010, pp. 417–420 (2010)
21. Zhang, L., Marks, T., Tong, M., Shan, H., Cottrell, G.: SUN: a Bayesian framework for saliency using natural statistics. J. Vis. **8**(7), 1–20 (2008)
22. Ziou, D., Tabbone, S.: Edge detection techniques - an overview. Int. J. Pattern Recogn. Image Anal. **8**, 537–559 (1998)

Adaptive Temporal Modeling of Audio Features in the Context of Music Structure Segmentation

Florian Kaiser[✉] and Geoffroy Peeters

STMS IRCAM-CNRS-UPMC, 1 Place Igor Stravinsky,
75004 Paris, France
{kaiser.florian,peeters.geoffroy}@ircam.fr

Abstract. This paper describes a method for automatically adapting the
length of the temporal modeling applied to audio features in the context
of music structure segmentation. By detecting regions of homogeneous
acoustical content and abrupt changes in the audio feature sequence, we
show that we can consequently adapt temporal modeling to capture both
fast- and slow- varying structural information in the audio signal. Evaluation of the method shows that temporal modeling is consistently adapted
to different musical contexts, allowing for robust music structure segmentation while gaining independence regarding parameter tuning.

Keywords: Music structure segmentation · Audio features temporal
modeling · Adaptivity

1 Introduction

The problem of music structure segmentation is concerned with the estimation
of the largest structural entities that compose a music piece. A verse in Popular
music, a bridge in Jazz music or a movement in Classical music constitute such
structural entities. As a front-end processing for audio indexing applications
such as audio browsing, summarization or annotation, the task knows a growing
interest in the Music Information Retrieval research community.

Given the large musical spectrum, research in this field is limited to a very few
tangible assumptions on the characteristics of musical structures. Formalization
of the task has been however especially active since the introduction in [2] of the
audio self-similarity matrix in which the pairwise similarity of an audio feature
sequence is computed. Such matrices give a rather understandable visualization
of the audio signal's content in terms of self-similarity and repetitions. The similarity matrix largely inspired the two main hypothesis that are made on the sections of a musical structure [9]. For the first hypothesis, i.e. the state hypothesis,
a section is characterized by a strong inner-homogeneity within its acoustical content. Feature frames within that section thus activate a single state and methods
such as HMM or clustering techniques can be applied to estimate the states in the
feature sequence. The state hypothesis closely relates to the concepts of verse and

© Springer International Publishing Switzerland 2014
A. Nürnberger et al. (Eds.): AMR 2012, LNCS 8382, pp. 248–261, 2014.
DOI: 10.1007/978-3-319-12093-5_15

chorus in popular music. In the second hypothesis, i.e. the sequence hypothesis, sections are solely defined by their repetitions and are composed of sequences of unrelated feature frames. Such sequences can be illustrated by the repetition of a melody. The musical structure is then visualized as stripes on the off-diagonals of the similarity matrix highlighting the repeated patterns in the audio features. A comprehensive overview of music structure segmentation methods that were proposed under both hypothesis can be found in [9].

Musical structures are thus characterized by rather long-term musical patterns that should be captured in the signal description. While the temporal scale of such structural patterns can hardly be extracted by means of low-level audio features solely, part of the research in music structure segmentation has focused on the integration of context in the signal description. In [10] a dynamic feature that models the spectral envelope over a short period of time is proposed to embed contextual timbre information. In [1] Dynamic Texture is applied to model timbral and rhythmical properties of sounds. In [8] a contextual measure of similarity that considers sequences of feature frames instead of single frames in the similarity matrix computation allows to strengthen the visualization of repetitive patterns in the audio features. In [6], the evolution of the tonal context in the audio signal is described by concatenating mid-term chroma sequences in Multi-Probe Histograms [12].

Including that contextual information and bringing the signal description to the temporal level of musical sections thus means that temporal modeling of audio features is applied at some stage. However, the temporal scale of structural sections varies between and within music pieces and the choice of an adequate window length to apply this temporal modeling is crucial. Moreover, though allowing for a better characterization of musical sections, temporal modeling nevertheless significantly damages the detection of boundaries between these. See for example the similarity matrices proposed in [6,8]. Modeling of feature sequences of a few seconds is indeed in contradiction with abrupt changes in the audio signal that characterize boundaries between sections. Ideally, the length of temporal modeling should thus be chosen as large for the description of a section, and reduced at the border of sections.

In this paper we propose a method to automatically adapt the window length on which to apply temporal modeling over the audio signal. Detecting both regions of relative stability and strong variance in the audio features, the length of modeling is increased while in the middle of a section and reduced when a boundary might be encountered. Doing so, we show that we can increase the temporal segmentation between sections while keeping the benefit of temporal modeling for the characterization of sections.

We first introduce in Sect. 2 a system for music structure segmentation that applies temporal modeling with Multi-Probe Histograms and illustrate the need for adaptivity in this context. Our approach for adaptive temporal modeling for the introduced model is then presented in Sect. 3 and the benefit of the approach with regard to the task is highlighted. Finally an evaluation of the impact of the adaptive temporal modeling on the music structure segmentation system is proposed in Sect. 4.

2   Adaptivity and Music Structure Segmentation

Adaptive windowing for temporal modeling is introduced in this paper in the context of the music structure segmentation system introduced in [6]. Temporal modeling is applied in this system in the sense that mid-term chroma features sequences are modeled by means of Multi-Probe Histograms in order to characterize tonal context variations. After a general introduction of the system, we present in this section the components of the system that may benefit from adaptivity in the length of the applied temporal modeling. The presented algorithm will serve in Sect. 4 to evaluate the impact of automatic temporal modeling length selection for the task.

2.1   System Overview

The general architecture of the system is a rather standard music structure segmentation architecture and is presented in Fig. 1. A signal description that relates to the harmonic content is first estimated at the features extraction stage with the calculation of the chroma features.

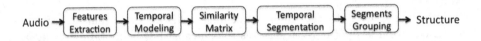

Fig. 1. Music structure segmentation overview

Tonal context is then derived from this description with the modeling of local chroma frames sequences with Multi-Probe Histograms (MPH). Therefore the chroma frames are split in subsequences of a given length, usually a couple of seconds, and each sequence is concatenated in a MPH. The MPH is a fixed-size representation of a chroma sequence which is determined by the dominant pitch classes transition between all adjacent frames of the sequence. A more detailed description can be found in [12]. Such histograms reflect the tonal structure of local portions of the audio and allow to model the evolution of the tonal context through time and capture slow varying harmonic patterns that relate to musical patterns. This description thus provides a good characterization of the inner-homogeneity of structural sections and also discriminates unrelated sections. Embedding the MPHs in a similarity matrix [2] strengthens the structure visualization. This is well illustrated with the similarity matrix extracted on the song *Things we said today* by the Beatles. Blocks of high similarity indicate long-term homogeneous tonal sections and a clear structural representation can be visualized (Fig. 2).

Based on this visualization of the structural information, music structure segmentation of a music piece consists in detecting the boundaries between its sections and group these segments together according to the musical structure. These two steps and their potential benefit in adaptive windowing are now presented.

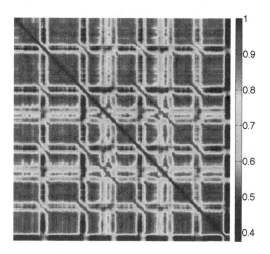

Fig. 2. Similarity matrix for the song *Things we said today* by the Beatles

2.2   Temporal Segmentation

A popular method based on similarity matrices for temporal segmentation was introduced in [3]. Arguing that section transitions are characterized by an abrupt change from one homogeneous acoustical content to another homogeneous acoustical content, the assumption behind the method is that boundaries in an audio signal are visualized in similarity matrices as 2-dimensional checkerboards such as the one represented in Fig. 3. Such a kernel is actually the ideal template of a transition between two different states of a musical structure. Correlating this gaussian checkerboard kernel along with the main diagonal of the similarity matrix, a novelty curve can be computed and serve for the detection of boundaries in the audio signal.

As illustrated by the boundary detection kernel, temporal segmentation supposes that our signal description allows to visualize both the inner-homogeneity of sections and the precise transition time instants. The use of temporal modeling of chroma frames has however the drawback that it necessarily smoothens the boundaries between sections. In that sense, adapting the length of the modeling in order to increase it while in the middle of a section and reduce it when narrowing a boundary may improve the performance of the temporal segmentation.

2.3   Segment Grouping

Once a temporal segmentation of the audio signal is estimated, grouping of segments according to the musical structure is obtained with the algorithm proposed in [5]. The method is based on the state hypothesis and uses the fact that information in similarity matrices is highly redundant over time in this hypothesis. Indeed, in the case of ideal states, structure is represented in the similarity

Fig. 3. Novelty detection kernel in [3]

matrix as uniform blocks. A single row or column of the similarity matrix thus contains the information of the whole state. Intuitively, the similarity matrix is thus ideally spanned by much lower dimensional basis, with a dimensionality that relates to the number of states in the musical structure. The method thus proposes to perform the dimension reduction of the similarity matrix by means of its Non-negative Matrix Factorization (NMF), and allows to group segments according to the structure by applying hierarchical clustering on the segments projected on this new basis.

The introduction of temporal modeling in the signal description is fully coherent with this algorithm in the sense that it strengthens the redundancy in the description of sections. While different sections of a same music piece may have various temporal scales, adaptive windowing can potentially enhance the description of the homogeneity in each section and consequently improve the segment grouping.

3   Adaptive Windowing

We present in this section our method for automatically adapting the length of temporal modeling with Multi-Probe Histograms over audio signals. After briefly introducing the approach, we describe the novelty score computation that allows to discriminate between homogeneous regions and abrupt changes in the audio features sequence. A method to consequently select the length of temporal modeling is then introduced.

3.1   Approach

Temporal modeling is generally applied by first splitting the audio feature sequence into a set of overlapping sub-sequences of a given length of N feature frames. Computing the Multi-Probe Histogram on the sub-sequences, one then captures information of slower variation than the one contained in the original audio feature sequence and thus embeds what we call contextual information. The core of the problem that is addressed in this paper now resides in the fact that the length of the subsequences should not be chosen as constant over the whole signal in order to capture both slow- and fast- varying structural information when it is needed.

One can easily illustrate the problem with a feature sub-sequence that would overlap two sections of the music piece. In that case, the goal of window adaptation consists in detecting in the sub-sequence the change point induced by the boundary in order to determine a stop criterion for the modeling. This way, frames of the sub-sequence that belong to the new section can be excluded from the current model. The consistency of each section modeling is consequently increased and the discrimination between the sections is preserved and not smoothened in the histogram modeling.

The first step of our method thus consists in the detection of potential break points within the feature subsequence that is currently modeled in a novelty score manner. This is discussed in the next subsection. Doing so over the whole signal, one can then consequently adapt the length of the sub-sequences over the audio signal. This part will be discussed in the second subsection.

3.2   Novelty Score Computation

Local Novelty Detection. We consider the whole feature sequence extracted on a music piece and divide it into K overlapping sub-sequences of length N, N being the maximum chroma subsequence length that may be considered. Regardless of the musical structure, the dynamics of music usually implies that the original feature sequence is composed of an alternation of relatively stable regions and abrupt changes. Therefore, any frame in any sub-sequences potentially constitutes a border between these regions. To estimate these borders, our approach first consists in considering each feature sub-sequence independently, and probe significant changes within its elements.

Considering the kth feature sub-sequence, we consider each of its frames as the potential division point of the sub-sequence. A novelty measure between all divisions in 2 of the subsequence then allows to estimate for all division point whether or not the subsequence is better modeled by a single histogram or by two histograms. The novelty score nov_k is computed for the kth feature sub-sequence as follows:

$$nov_k(i) = distance(MPH_1^i, MPH_{i+1}^N) \tag{1}$$

with MPH_a^b the Multi-Probe Histogram computed on the sequence from frame a to frame b, and with the frame $i \in [\tau, N - \tau]$, τ being the minimum sequence length for the modeling. It is to be noted that the subsets are not necessarily of equal lengths and one should thus choose a normalized model or use a dynamics-independent distance. In our case, the novelty is thus computed by means of the cosine distance.

The approach is illustrated in Fig. 4 with the example of two regions of different harmonic contexts in a feature sequence. Peaks that arise in the local novelty curve indicate likely changes in the audio signal with respect to the chosen audio features and temporal modeling. In contrast, regions of low novelty indicate temporal sequences of homogeneous acoustical content.

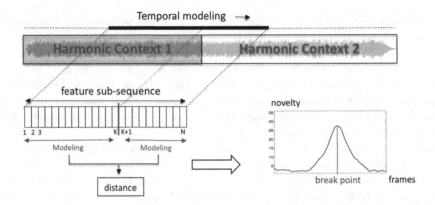

Fig. 4. Local novelty score computation within an audio signal

Sustainable Novelty Score. Calculating a novelty score within each subsequence of N frames, we obtain K local novelty scores within the feature sequence. Say feature sub-sequences overlap with a hop size of one frame, each feature frame is thus probed $N - 2\tau$ times as a potential border between two contexts, each time as part of a different sub-sequence. In order to ensure that the borders that will be selected for the window adaptation are sustainable over all local novelty scores, we fuse all novelty scores in a single one that covers the whole audio signal. The novelty for each frame is therefore defined as the sum of the $N - 2\tau$ novelty values computed for it in each sub-sequence it is part of.

Considering feature frames as break points between different context ensures the reliability of the novelty score. The benefit of estimating boundaries as part of different contexts was already highlighted in [11].

3.3   Window Adaptation

Peaks and Troughs Detection. Computing the audio features novelty score described above, one extracts break points in the feature sequence. In practice, these breakpoints are used to adapt the length of the temporal modeling applied to the audio features over time.

The novelty score is therefore used to discriminate between regions of rather homogeneous acoustical content, i.e. troughs in the novelty score, and regions of potential changes, i.e. peaks in the novelty score. Peaks and troughs are detected in the novelty score by means of the adaptive threshold estimation described in [4]. Therefore, a peak P in the novelty score is defined as the local maxima between two consecutive local minima, and a trough T as a local minima between two local maxima. Peaks-troughs are then selected as local maxima-minima values that exceed neighboring local minima-maxima values of at least a threshold δ as described in Eq. 2 and illustrated in Fig. 5.

$$P_i \quad \equiv \quad T_i + \delta \leq P_i \cap T_{i+1} + \delta \leq P_i \tag{2}$$

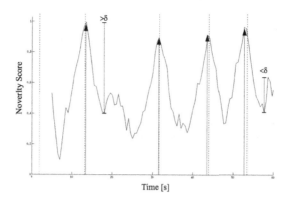

Fig. 5. Peaks and trough detection in the novelty curve

The threshold δ for the peak detection is automatically selected by clustering the novelty score into the classes of lowest and highest values, and setting δ to the mean value of the largest cluster. Supposing that there are more troughs than peaks in the novelty score, i.e. changes should not constantly occur in the music piece, we use a lower threshold value for the troughs detection in order to be less discriminative in their selection.

Window Length Selection. Peaks and troughs in the novelty score constitute our reference points for adapting the window length N of the temporal modeling. Indeed both fast-varying and slow-varying structural information can be captured by setting N to a minimal value at peaks and to a maximal value at troughs. To extend the window adaptation to the remaining time instants and thus the whole signal, only the time information of peaks and troughs is kept and their values set to N_{min} for peaks and N_{max} for troughs, with N_{min} and N_{max} respectively the minimum and maximum modeling lengths. Cubic interpolation of these points over time then allows to estimate an adapted window length for all time instant of the signal. Such an interpolation ensures a smooth variation of the temporal modeling length over time.

The window selection method is illustrated in Fig. 6. A novelty score and the peaks and troughs detected are represented. The interpolation of time points of maximal and minimal window lengths shows that the method allows to ideally select the window length according to the acoustical content. Examples of similarity matrices based on this adaptive temporal modeling concept shown in the next section of the paper illustrate that the method is indeed efficient to prevent smoothing effects between sections.

3.4   Application to Music Structure Visualization

A good example of the smoothing effect implied by temporal modeling is shown in Fig. 7.a. The music piece used for this similarity matrix example is a 30 s

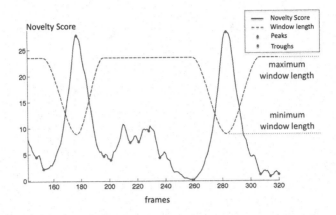

Fig. 6. Novelty score based window adaptation

excerpt of the *Mazurka op.63 n.3* composed by Chopin that is centered on a transition between two sections of different tonal contexts. The chroma vectors sampled at 10 Hz are extracted on the audio file and Multi-Probe Histograms computed on a sliding window of 5s, i.e. 50 frames, are embedded in a similarity matrix. While the discrimination between sections is globally clearly visible, one can hardly determine the exact boundary between the two sections.

We present in Fig. 7.b a similarity matrix computed on the same audio example but this time using the adaptive temporal modeling for the computation of Multi-Probe Histograms. The maximum window length is set to 80 frames and the minimum window length to 10 frames. Doing so, the inner-homogeneity of the two sections is still well captured by our modeling and the boundary visualization between the two sections is considerably increased. Indeed, the visualization highlights two clear boundaries in the transition between the two sections, suggesting a progressive evolution between the two tonal contexts.

This thus suggests that our method may increase the precision of the temporal segmentation of similarity matrices that use temporal modeling of audio features while keeping the benefit for the characterization of structural sections.

4   Evaluation

We now present an evaluation of the impact of the adaptive temporal modeling introduced in this paper on the music structure segmentation system presented in Sect. 2. Therefore the system is evaluated using both a constant modeling length and adaptive modeling for the Multi-Probe Histograms computation. After introducing the evaluation test set and the evaluation metrics, results will be presented and discussed.

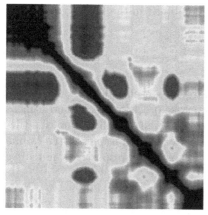 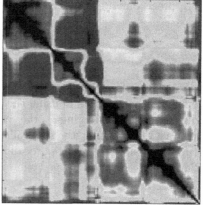

(a) 7.a MPHs over sequences of 50 (b) 7.b: MPHs over adapted chroma se-
chroma frames   quences

Fig. 7. MPH similarity matrices, excerpt of Chopin's *Mazurka op.63 n.3*

4.1   Test Set

The music structure segmentation system is evaluated on two different test sets:

- 18 pieces of the RWC Classic[1] annotated within the SALAMI Project[2]
- 180 songs of the Beatles annotated within the Isophonics corpus[3]

Besides the annotation of temporal boundaries, the structural information is annotated in the state hypothesis. All feature frames of a same section are thus labeled under a same state. Note that no semantic or functional labeling of the structure will be evaluated here. Only the retrieval of temporal boundaries and the grouping of segments under the correct structural states are evaluated.

4.2   Evaluation Metrics

Temporal Segmentation Evaluation. The temporal segmentation step produces a set of boundaries between sections. The following terms are defined for its evaluation: the True Positives (TP) as the number of correctly retrieved boundaries within a given tolerance range, the False Positives (FP) as the number of unexpected retrieved boundaries, and the False Negatives (FN) as the number of missing boundaries. The precision P and recall R are then defined as in Eqs. 3 and 4.

$$P = \frac{TP}{TP + FP} \tag{3}$$

$$R = \frac{TP}{TP + FN} \tag{4}$$

---

[1] http://staff.aist.go.jp/m.goto/RWC-MDB/AIST-Annotation/
[2] http://ddmal.music.mcgill.ca/salami
[3] http://isophonics.net/

Both precision and recall can then be combined in the F-Measure defined in Eq. 5.

$$F = \frac{2PR}{P + R} \tag{5}$$

Segment Grouping Evaluation. A largely consensual method to evaluate the frame labeling consists in using the pairwise precision, recall and F-measure introduced in [7]. Considering F_a the set of identically labelled frames in the reference annotation, and F_e the set of identically labelled frames in the estimated structure, the pairwise precision, recall and F-measure, respectively noted pw_P, pw_R and pw_F are then defined as:

$$pw\_P = \frac{|F_e \bigcap F_a|}{|F_e|} \tag{6}$$

$$pw\_R = \frac{|F_e \bigcap F_a|}{|F_a|} \tag{7}$$

$$pw\_F = \frac{2pw\_Ppw\_R}{pw\_P + pw\_R} \tag{8}$$

4.3 Set up

The "Constant Modeling" system is run with similarity matrices computed on chroma sequences Multi-Probe Histograms of three different lengths, i.e. 4, 6 and 8 s. The "Adaptive Modeling" system applies our adaptive temporal modeling length method with a maximal window of 8 s and a minimum window of 2s. The segment grouping is set to form 4 structural sections for each piece. The temporal segmentation evaluation is considered with a tolerance range of 2 s.

4.4 Results and Discussion

Structure segmentation performance evaluation for the RWC Classic and Isophonics Beatles datasets are reported in Tables 1 and 2 respectively.

The results that holds for both datasets is that the performance of temporal segmentation and segment grouping varies with the length of the modeling for the "Constant Modeling" method. Moreover, the lengths of modeling that provide the best temporal modeling results are not the same as the one that provide the best segment grouping performance. Indeed, for the RWC Classic dataset, reducing the length of modeling strongly increases the temporal segmentation performance. However, the homogeneity of structural sections is better captured with large temporal modeling, and hence, segment grouping is better handled with large windows as illustrated by the results. In that context, our adaptive modeling acts coherently with the musical context and allows for both a good temporal segmentation and the best segment grouping performance.

Table 1. Structure segmentation evaluation [%] - **RWC Classic Set**

	F	P	R
Constant Modeling - 2s	35,54	34,87	39,47
Constant Modeling - 4s	31,27	31,04	35,35
Constant Modeling - 6s	29,55	30,86	31,99
Constant Modeling - 8s	26,27	28,15	26,27
Adaptive Modeling	32,26	35,86	32,91

(a) 1.a: Temporal Segmentation [%]

	pw_F	pw_P	pw_R
Constant Modeling - 2s	47,33	56,74	45,04
Constant Modeling - 4s	48,54	50,90	52,34
Constant Modeling - 6s	49,36	52,32	53.09
Constant Modeling - 8s	51,48	55,30	53,69
Adaptive Modeling	52.45	54,81	55.14

(b) 1.b: Segment Grouping [%]

Table 2. Structure segmentation evaluation - **Isophonics Beatles Set**

	F	P	R
Constant Modeling - 2s	43,17	40,39	50,38
Constant Modeling - 4s	43,64	41,54	49,73
Constant Modeling - 6s	44,05	43,06	48,95
Constant Modeling - 8s	41,48	41,55	44,91
Adaptive Modeling	43,36	42,77	47,47

(a) 2.a: Temporal Segmentation [%]

	pw_F	pw_P	pw_R
Constant Modeling - 2s	59.37	52.68	73.94
Constant Modeling - 4s	60.34	54.28	73.53
Constant Modeling - 6s	59.29	52.47	74.16
Constant Modeling - 8s	57.54	51.60	70.36
Adaptive Modeling	61.17	55.59	73.51

(b) 2.b: Segment Grouping [%]

Concerning the Isophonics Beatles set, the good temporal segmentation and segment grouping obtained with the 4 s constant modeling suggests that harmonic patterns in this dataset are of a lower temporal scale than in the RWC Classic dataset. There again, the adaptive method consistently adapted modeling windows to this musical context and shows the best segment grouping performance and a good temporal segmentation performance. This all suggests that the approach captures both long term musical patterns and abrupt changes in the audio content without any strict assumption on the temporal scale of the sections.

It is to be noted that the impact of adaptive modeling is sensitively higher on the segment grouping performance. Indeed, temporal segmentation performances did match but not overcome the performance of the best constant modeling algorithm. In contrast, adaptive modeling did outperform for both datasets the segment grouping performance given by fixed-length temporal modeling. While we ideally could consider the two tasks separately, i.e. segment the music piece with a small temporal modeling length and increase it for the segment grouping, the method introduced in this paper gives a single performant and computationally effective solution for both problems.

5   Conclusion

We proposed in this paper a method for the automatic adaptation of the temporal modeling of audio features for music segmentation purposes. Modeling chroma features by means of Multi-Probe Histograms we detect regions of homogeneous acoustical content where the length of the modeling can be increased, and changing contexts where the length should be reduced. We may in that manner describe both fast- and slow- varying musical patterns in music signals. The evaluation in the context of music structure segmentation indeed showed that our method adapts the temporal modeling length to the acoustical context and allows to capture both long-term harmonic patterns and precise breaks between these patterns. While parameter tuning for the modeling length is hardly generalizable for large datasets, our method allows us to be independent on any strict assumption on the temporal scale of structural sections. We believe that the approach could be efficiently extended to other temporal modeling and for example allow to enhance the description of timbral contexts. More generally, adaptivity in music signals description is a challenging research for music information retrieval. As an example, music interaction systems that face different kinds of musical events through time could benefit from this research.

Acknowledgments. This work was partly supported by the Quaero Program funded by Oseo French agency.

References

1. Barrington, L., Chan, A.B., Lanckriet, G.: Modeling music as a dynamic texture. IEEE Trans. Audio Speech Lang. Process. **18**(3), 602–612 (2010)
2. Foote, J.: Visualizing music and audio using self-similarity. In: Proceedings of the ACM Multimedia, pp. 77–80 (1999)
3. Foote, J.: Automatic audio segmentation using a measure of audio novelty. In: Proceedings of the IEEE International Conference on Multimedia and Expo (2000)

4. Jacobson, A.: Auto-threshold peak detection in physiological signals. In: Proceedings of the 23rd Annual International Conference of the IEEE Engineering in Medicine and Biology Society, 2001, vol. 3, pp. 2194–2195 (2001)
5. Kaiser, F., Sikora, T.: Music structure discovery in popular music using nonnegative matrix factorization. In: Proceedings of the 11th International Society for Music Information Retrieval Conference (ISMIR), August 2010
6. Kaiser, F., Sikora, T.: Multi-probe histograms: a mid-level harmonic feature for music structure segmentation. In: Proceedings of the 14th International Conference on Digital Audio Effects (DAFx), Paris, France, September 2011
7. Levy, M., Sandler, M.: Structural segmentation of musical audio by constrained clustering. IEEE Trans. Audio Speech Lang. Process. **16**(2), 318–326 (2008)
8. Mueller, M., Kurth, F.: Enhancing similarity matrices for music audio analysis. In: Proceedings of IEEE International Conference on Acoustics, Speech, and Signal Processing (ICASSP) (2006)
9. Paulus, J., Müller, M., Klapuri, A.: Audio-based music structure analysis. In: Proceedings of the 11th International Society for Music Information Retrieval Conference (ISMIR) (2010)
10. Peeters, G.: Toward automatic music audio summary generation from signal analysis. In: Proceedings of the International Conference on Music Information Retrieval (ISMIR), pp. 94–100 (2002)
11. Sargent, G., Bimbot, F., Vincent, E.: A regularity-constrained viterbi agorithm and its application to the structural segmentation of songs. In: Proceedings of the International Society for Music Information Retrieval Conference (ISMIR) (2011)
12. Yu, Y., Crucianu, M., Oria, V., Damiani, E.: Combining multi-probe histogram and order-statistics based lsh for scalable audio content retrieval. In: ACM Multimedia (2010)

Detector Performance Prediction
Using Set Annotations

Robin Aly[1]([✉]) and Martha Larson[2]

[1] University of Twente, 7500AE Enschede, The Netherlands
r.aly@ewi.utwente.nl
[2] Delft University of Technology, 2628CD Delft, The Netherlands
m.a.larson@tudelft.nl

Abstract. Content-based videos search engines often use the output of concept detectors to answer queries. The improvement of detectors requires computational power and human labor. It is therefore important to predict detector performance economically and improve detectors adaptively. Detector performance prediction, however, has not received much research attention so far. In this paper, we propose a prediction approach that uses human annotators. The annotators estimate the number of images in a grid in which a concept is present, a task that can be performed efficiently. Using these estimations, we define a model for the posterior probability of a concept being present given its confidence score. We then use the model to predict the average precision of a detector. We evaluate our approach using a TRECVid collection of Internet archive videos, comparing it to an approach that labels individual images. Our approach requires fewer resources while achieving good prediction quality.

1 Introduction

Concept-based video search engines combine the output of concept detectors to answer queries. The detectors' performance therefore strongly influences search performance [11,13]. If the search performance is poor, a widely used approach is to modify all detectors of the considered target vocabulary. For some concepts, however, the original detector might have better performance than the modified version and such modifications might create unnecessary costs in terms of computational power and human labor. It is therefore important to modify detectors adaptively depending on their performance. On the other hand, *measuring* detector performance using large numbers of annotations is again very labor intensive. As a result, detector performance *prediction* that estimates, rather than measures, performance is therefore an important research direction, which has not received a lot of research attention. In this paper, we propose a method for detector performance prediction that employs human annotators efficiently.

Detector performance prediction is similar to the highly active field of query performance prediction in text retrieval [7]. Here, the aim is to predict the performance of a query with the aim to adapt the retrieval algorithm. Most approaches

© Springer International Publishing Switzerland 2014
A. Nürnberger et al. (Eds.): AMR 2012, LNCS 8382, pp. 262–275, 2014.
DOI: 10.1007/978-3-319-12093-5_16

use collection statistics of query terms to predict the query performance. A straightforward approach to detector performance prediction would therefore be to use statistics of the features employed by the detector. The approaches in query performance prediction, however, still have to improve to be applicable to real-world scenarios [6]. Additionally, we expect that statistic-based approaches are more challenging for detector performance prediction due to the high dimensionality of feature vectors. Motivated by the recent popularity of crowdsourcing applications, which replace complex computing tasks with human labor, we propose that a promising alternative direction exploits human judgments, which is feasible since concept detection is performed off-line.

A straightforward manual approach to use annotations for detector performance is to annotate a sample of images and to measure detector performance based on these annotations. This approach is used, for example, for detector performance evaluation[1] in the popular benchmarking workshop TRECVid [10]. A disadvantage of this approach, however, is that the costs of annotating individual images are high, especially because current concept vocabularies can contain several thousand concepts; a number that is likely to grow in the future [12]. Additionally, currently an increasing amount of research effort has been devoted to developing unsupervised techniques for training concept detectors using images that are automatically collected from the Web. Such detectors are trained "on the fly" using images returned by querying an image search engine and can change very frequently. In such a case, it is not possible to evaluate the performance of concept detectors using reference concept labels, since potentially imperfect search engine results rather than manually annotated data is used for training. In this paper, we propose another approach, where annotators estimate for a set of images how many contain a concept. We expect and show in experiments that set annotations require less time compared to the annotation of individual images.

Although here our main interest in set annotation is improving the effectiveness with which detector performance can be predicted, set annotations possibly also improve the annotation quality. In particular, presenting human annotators with sets allows them to quickly converge on an understanding of what exactly constitutes an occurrence of a concept and what does not. Even long before digital images came into widespread use, psychologists recommended presenting humans with multiple images at once in order to support development of the understanding of concepts [8]. The presence of multiple images is assumed to promote comparison and contrast behavior, thus allowing humans to more easily acquire concepts. Therefore, if presenting human annotators with image sets supports them in quickly stabilizing their understanding of what constitutes concepts, ultimately, set annotation may provide concept annotations that are more consistent across images and that reflect subtle but important differences in the images of a particular collection.

---

[1] Performance evaluation and performance prediction can be performed similar but differ in their aim: performance evaluation aims at comparing detectors and performance prediction aims at deriving actions (e.g. change of detector technique).

Using set annotations directly, however, limits the performance measures that we can consider. For example, while we can predict the precision, which is the ratio of top-ranked images that contain a concept, we cannot predict the average precision because it is defined on the annotations of individual images. To overcome this problem we define a probabilistic model of the concept labels for individual images and estimate the performance measures using this model. The model is based on the common assumption that the probability of a concept label increases with increasing detector confidence scores.[2] Using these probabilities, we can calculate the expected performance measure, which we use as a predictor for the detector performance.

The rest of this paper is structured as follows. Section 2 describes related work to this paper. Section 3 explains our approach to set annotation. Section 4 describes the experiments we used to investigate set annotations and presents their results. Section 5 concludes this paper.

2   Related Work

In this section, we present the work most related to our own. The work falls into three areas: evaluation of concept detectors, query performance prediction and multi-image presentations. We discuss each of the areas in turn.

As mentioned above, the evaluation of detector performance is similar to the prediction of detector performance: both try to assign a performance number to a detector. The motivation of both research directions, however, is different. Detector performance evaluation, on the one hand, is interested in the performance of a *set* of detectors in order to measure certain aspects of a method to build detectors in order to compare it with other measures. Detector performance prediction, on the other hand, is interested in the performance of *individual* detectors to decide whether they should be modified or not. The de-facto detector performance measure is the average precision, which can be seen from its use in the well-known evaluation workshop TRECVid [10]. The top-ranked images of several search engines are put into a pool. In this pool, each image is individually annotated or judged. Exhaustive annotations are, however, too expensive for the evaluation in today's collections. The workshop therefore currently uses methods which only assess only samples of this pool [14,15]. Our method instead annotates sets rather than individual images, expecting a higher annotation speed. The idea of only judging a sample could also be incorporated in our approach, which we propose for future work.

Detector performance prediction is similar to query performance prediction because they share similar aims. In query performance prediction, a predictor estimates the performance of a search engine for a particular query to allow the search engine to adapt its search algorithm. Query prediction methods can be divided into pre execution prediction, where the performance is predicted before any search algorithm is executed, and post execution prediction, where

---

[2] Detector confidence scores indicates the belief of a detector that an image contains a concept.

the prediction uses the ranking of an initial search result. Despite recent accuracy improvement, as previously mentioned, query performance prediction is still in its infancy [6]. We expect that detector performance prediction is more difficult since the amount of features and search/detector mechanisms are mathematically more complex. We therefore decide on an approach to detector performance prediction that uses manual annotations.

Psychologists have long been interested in the psychological and perceptual phenomena involved with presenting humans with sets of multiple images [4]. As already mentioned above, [8] investigates in the ways in which multi-image presentations support human acquisition of the understanding of (visual) concepts. Presentation formats containing multiple images are used in other image judgment settings. For example, the 'Real Time Photo Monitoring' service (http://crowdflower.com/rtfm) offered by the Crowdflower crowdsourcing platform elicits image appropriateness judgments from workers by presenting them with a grid of images. Another example, is implicit image tagging using gaze tracking. In [5], a system is described that presents users with a grid of images and tracks their gaze. Using features related to eye movement, the system is able to automatically determine which photos in the grid are associated with the target class that has been presented to the user. Our work differs from these examples in that we are interested in going as directly and efficiently as possible from a set of images to a performance prediction, without the intervening step of explicitly labeling individual images. Although eye tracking might ultimately be used to improve our approach, here we leave it out of consideration in order to implement an annotation system that can be used in a standard web browser without specialized equipment.

3   Detector Performance Prediction

In this section, we describe our method to predict detector performance through set annotations. We take the following steps: first, we describe factors that play a role in designing the annotation system, second, we describe our method to derive a probabilistic label of an image from the annotations, and finally, we detail how we use these probabilistic labels to estimate the average precision of a concept detector.

3.1   Dimensions of Set-based Annotations

Similarly to annotation tasks of individual images, set annotation tasks can be designed in many ways. Many dimensions play an important role in these design decisions. However, the scope of this initial study only allows us to test a single instance in this design space. In the following, we therefore discuss the dimensions that should be explored in set retrieval.

The first dimension is layout of the screen displayed to the annotator. For example, the images could be displayed in a grid or a circle form with various numbers of images per screen. Additionally to the general layout, the number of displayed images will be important. Although increasing numbers of images make the annotation process more efficient, they can also overload the annotator, which reduces efficiency. The decision on the screen layout should therefore be guided by results from human perception research. In this study we chose a grid of six times four images ($= 24$), which fits on the screens we considered.

The next dimension is the choice of images to display in general, possibly over multiple screens. In the general case, the number of images is bounded by the available annotation resources and more effort means better estimations. Since images with high confidence scores are the most influential in most evaluation measures, a straightforward choice is to use a certain number of the top ranked images. Note, however, that there might be other choices: in our study on the correlation of detector performance in [2] we showed that the variance of a detector's confidence scores strongly influences the search performance. Although this finding has not been turned into an evaluation measure, it hints the importance of also annotating images at lower ranks. The assignment of images to screens does not need to follow the ranking of images and varying the assignment of images to screens could improve results: for example, if the screen with the highest ranked images often contains the concept it is easy for annotators to overlook the few images for which this is not the case. It could therefore be desirable to randomize the assignment of images to screens within the images that should be annotated. We chose the simplest point of departure to annotate the 96 ($= 4 \cdot 24$) top ranked images and display them in their ranked order.

The kind of instructions that are given to the annotators is another dimension. We can instruct the annotators to produce a quick estimate of the number of images or to consider every image individually and count the results. This balances annotation speed and quality. Note that counting images still might carry advantages over individual annotations because the annotator is presented with context of this image. Furthermore, we need to instruct the annotator what the concept is he/she has to annotate. The obvious choice is to display the concept name and textual definition (usually one sentence) which the creators of the detectors used. However, we could also let the annotator decide on what concept this detector is mainly based on the displayed images. This would have the advantage that the annotator is free to decide what constitutes the concept for him/her, which is important because the annotator could be a searcher as well. For example, for the concept doorway with the definition "An opening you can walk through into a room or building" a detector could display often castle gates. According to this definition, the annotator objectively has to count the images of a castle gate as containing the images. However, in his/her perception this might not be the case and many users of the search engine could agree. These users will therefore be dissatisfied if presented with castle gates when searching for doorways. We regard this dimension as important and very relevant for future research. Due to the scope of this study, we did not instruct the annotators to estimate or count and displayed them the concept definitions that came with the detectors.

The last dimension, which we deem important, describes the ways to motivate annotators to do good work. This is especially important if the annotations are performed by workers on crowdsourcing platforms such as Amazon Mechanical Turk (www.mturk.com), whose motivation is usually considered to be extrinsic and not intrinsic in nature. For annotations of individual images, the literature has proposed many alternatives. For example, the annotations can be modeled as a game [1]. Although we agree on the importance of this aspect, we choose a basic design for the task. This allows us to study the general advantage over individual annotations.

Given decisions about the above dimensions, we can execute a set annotation task whose results are numbers of images that contain a concept within a larger set of images. Although we could for example already predict the precision of this larger set, other evaluation measures require individual annotations.

3.2   Probabilistic Individual Labels

Using the estimated number of images in a *set* that contain a concept, we define a probabilistic model of whether a *particular* image contains a concept or not. We will use this model to predict the average precision, which is based on individual image annotations. For each image we assume that we have a score that represents the confidence of a detector about an image containing a certain concept. Following our earlier work in [3], we assume that the posterior probability that the image contains the concept given its confidence score follows a logistic regression function:

$$P(C_i|\mathbf{w}, s_i) = \frac{1}{1 + \exp(-w_1 - w_2 s_i)} \tag{1}$$

where C_i is the random variable that image i contains the currently considered concept, $\mathbf{w} = (w_1, w_2)$ are unknown regression weights, and s_i is the detector score for the concept in the ith image. To simplify the notation we substitute in the following $\sigma_i(\mathbf{w}) = P(C_i|\mathbf{w}, s_i)$. Note that one usually varies the features (in this case s) in a logistic regression. In our setting, however, we have a set of fixed features but the weights $\mathbf{w}$ are unknown. Furthermore, let $M = \sum_i^N C_i$ be a random variable that denotes the number presented images N that contain the concept. The expected number of images is defined as follows:

$$E[M] = \sum_{i=1}^{N} E[C_i] = \sum_{i=1}^{N} \sigma_i(\mathbf{w}) \tag{2}$$

If we consider a large enough set, for a good probability measure, we will have $E[M] \simeq h$ because of the central limit theorem. To simplify the notation we set in the following $x(\mathbf{w}) = E[M]$. Note that the values are not necessarily equal because annotators can make mistakes in their estimation and the central limit has weaker applicability for smaller N. We determine the weights $\mathbf{w}_{ml}$ through their maximum likelihood estimate of the weights given the annotation h:

$$\mathbf{w}_{ml} = \underset{\mathbf{w}}{\mathrm{argmax}}\ p(H = h|\mathbf{w}) \tag{3}$$

We assume that $x(\mathbf{w})$ is Gaussian distributed around a mean μ_h with variance v_h, where both depend on the annotator and his/her estimate. We therefore have to maximize the likelihood $p(H = h|\mathbf{w}) = \mathcal{N}(x(\mathbf{w}), \mu_h, v_h)$ where $\mathcal{N}$ is the Gaussian probability density function, μ_h is the Gaussian's mean, and v_h is the Gaussian's variance (which are both derived from the annotation). We provide an explanation that is more elaborate and an algorithm to solve this maximization problem in the Appendix A. Using the optimized weights $\mathbf{w}_{ml}$, we can now calculate the probabilistic labels $P(C_i|s_i)$ from (1) for a concept given its confidence scores.

3.3   From Probabilistic Labels to Average Precision

We now turn to the prediction of the detector performance using the annotations and the probabilistic labeling above. Detector performance is usually measured in a conventional IR evaluation measure. Measures that only depend on the contained concepts in a set, for example the precision at a cut-off point, can simply be predicted from the set annotations (if the annotator estimates that 12 images of the 24 first images in a ranking contain a concept the precision at 24 is 0.5). Other evaluation measures, however, require individual labels that set annotations cannot provide directly. For example, the average precision measure is defined as follows.

$$AP(C) = \frac{1}{\sum_{i=1}^{N} C_i} \sum_{r=1}^{N} \frac{\sum_{i=1}^{r} C_i}{r} C_r \qquad (4)$$

where r is the currently considered rank, N is the number of images, C_x for $1 \leq x \leq N$ are binary indicator variables that assume the value 1 if concept C occurs in image x, and otherwise assume the value 0. We therefore cannot calculate the average precision directly, because the individual concept labels C_i are unknown. From (4) and the probabilistic labeling in Sect. 3.2, we can derive, however, the expected average precision (AP) given this labeling:

$$E[AP] = \frac{1}{\sum_{i=1}^{N} P(C_i|s_i)} \sum_{r=1}^{N} \frac{\sum_{i=1}^{r} P(C_i|s_i)}{r} P(C_r|s_r)$$

where we assume independence between the individual expressions. The expected average precision is the average precision that we expect from this probabilistic model and is therefore a good estimator for the actual average precision. In this paper, we focus on the average precision measure and leave the exploration of other evaluation measures that are based on individual labels for future work.

4   Experiments

In this section, we describe the experiments that we conducted to evaluate our set annotation approach for detector performance prediction. Note that, due to the early stage of detector performance prediction as a research discipline, the reported experiments can only be regarded as an initial study and further, more in-depth experimentation are needed.

Aspect	Value
Dataset	iacc.1.b [9]
Concepts	Light(23) [9]
TOP-N	$96(= 4 \cdot 24)$
Measures	$P@96, AP$
Detector systems	**AP**
ITI-CERTH_1	0.0253
CMU1_1	0.0767
ecl_liris_I_1	0.0563
Marburg3_2	0.0912
TokyoTech_Canon_1_1	0.1488

(a)

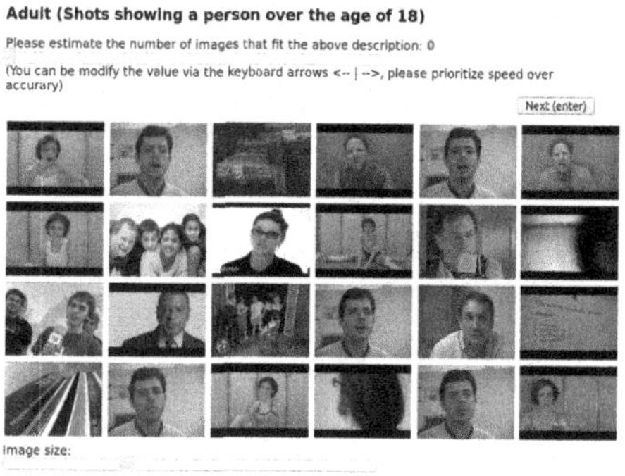

(b)

Fig. 1. (a) Information on the experimental set up. The two numbers reported for annotators and annotations refer to the base-line and set annotation study (see the text). The system names are stripped from the common prefix $F\_A\_$. (b) The interface used for set annotations.

4.1 Setup

Figure 1 (a) summarizes the facts of our experimental setup. We chose the Internet Archive collection part b that was used by the TRECVid workshop in 2011 [9]. The workshop provides ground truth for 23 concepts. We chose the runs with the strongest and weakest performance, whose performance should be easy to differentiate, and three detector runs around the median, making the prediction more difficult. In order to evaluate set annotations we conducted two studies with real annotators: first, a set annotations where 12 persons with a background in computer science in an age range of 25–50 annotated the TOP-96 images in screens of 24 images at a time, second, a study where we asked a 32 year-old person to annotate individual TOP-96 images of the run ecl_liris_I_1 to obtain a baseline about the annotation accuracy of this method. Note that to the best of our knowledge there is no other set annotation method to compare our results with, which would be clearly desirable.

Figure 1 (a) shows a screen shot of the user interface that we used for set annotations. At the top, the screen shows a concept definition that gives the guidelines for his/her annotation. Below there is a slider control through which the annotator indicates the estimated number of images that contain the described concept. The slider can also be controlled by the arrow keys of the keyboard for quicker access. The set of images to annotate are shown below. The annotator was able to scale the images according to his/her choice. The interface for individual images showed only one image and allowed for more efficient navigation (using the right and left arrow key for binary annotation decisions). We ensured that each screen of 24 images was annotated at least twice.

To evaluate the quality of the set annotations, we compare the annotation time, the compliance to the ground truth, and the feedback of annotators about the task to the individual annotations. Furthermore, because detector

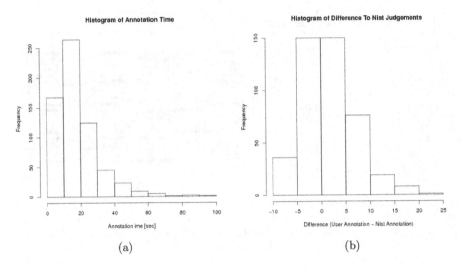

Fig. 2. (a) Distribution of the annotation time for a single image screen (24 images), (b) distribution to the annotations from NIST.

performance prediction is a relatively new research direction, there is no standard evaluation procedure of predictors. Ideally, an evaluation procedure for performance prediction compares the accuracy of a performance predictor with the actual performance measured by the ground truth. The evaluation of query performance predictors, however, has shown that the prediction of the ranking of queries according to their performance is already a challenging task. For this study we followed therefore the evaluation methodology in query performance prediction and evaluate our performance prediction method as follows. Let $d_s = d_{s1}, d_{s2}, d_{s3}, d_{s23}$ be the list of the 23 detectors of the system s for the respective concepts with $P(d_{s1}) \geq ... \geq P(d_{s23})$ where $P()$ is the P@96 or the average precision measure based on the TRECVid ground truth. We now measured the correlation of the ranking of detectors according to our predicted performance $\hat{P}()$ with the golden standard d_s. We use the Kendall tau to measure rank correlation.

4.2 Results

In the following, we describe the results that we obtained from the above-described experiments.

Annotation Time: Figure 2 (a) shows the distribution of the annotation time spend for a set of 24 images.[3] The average annotation time was 18.17 s (corresponding to a mean of 0.75 s per image). For the individual annotations, we measured a mean of 1.34 s per image. We found that the difference between

---

[3] We measured the pure annotation time, excluding the time to load the images.

these two means was statistical significant according to a two-sided student t-test ($p < 0.05$).

Annotation Quality: Figure 2 (b) shows the difference between the estimated number of images in a set of 24 by the annotators and the ones at NIST. We see that the distribution is centered approximately at 0 with a right long tail. Investigations revealed that one source of this difference originated from the fact that NIST annotators were able to annotated event in a video sequence, while our annotators only saw still images and therefore could not judge concepts such as "sitting down". There was no significant difference between the set and the individual method.

Annotator's Feedback: Each annotator was asked to give feedback on the their experience after executing his task. We summarize the results. Both types of annotators found their tasks easy (2 points on a 5-point Likert scale ranging from very easy to very difficult). The set annotators on average found the cognitive load lower than the single annotator. Two annotators of the set annotation task mentioned that they found their task "pleasant", while the only annotator of individual images found the task "stressful". Clearly, due to the small numbers of annotators these results can only serve as indications.

Table 1. Rank correlation between the sorted lists of detectors for the 23 concept by our predicted performance and by the performance measured using ground truth from NIST for the five detector systems investigated.

Measure	ITI-CERTH_1	CMU1_1	ecl_liris_I_1	Marburg3_2	TokyoTech Canon_1_1	ecl_liris_I_1 individual
P@96	0.803	0.604	0.650	0.589	0.853	0.600
AP	0.624	0.520	0.512	0.400	0.583	

Performance Prediction: Table 1 shows the rank correlation for the prediction of the two performance measures P@96 and average precision per detector system. We see that the predictor performs better for the set measure P@96 compared to the average precision (AP) that is estimated through the probabilistic model of labels. The right-most column shows the performance of the single annotator which annotated individual images (we did not measure the predicted AP because this would have required too many annotations).

4.3   Discussion

We now discuss the results of the conducted experiments. The three experiments concerning the annotations show that set annotations are an attractive research direction for image annotations because there is indication that they are faster and impose a lower cognitive burden on the annotator.

Compared to numbers reported in query performance prediction [6] the rank correlation of detectors within their detector systems is high. This might not be surprising because our method employs humans. The fact that the correlation is not higher, although the comparison is against systems judged by other human annotator from NIST, can be explained by the varying agreement of the annotators, see Fig. 2 (b). The correlation for individual annotations is similar.

There is a clear decline in precision performance between the P@96 measure, which is a set-base measure, and the average precision for which we employed probability-based prediction. In principle, this can be expected because we predict the P@96 measure directly from an annotator's judgment while the probabilistic method incorporates uncertainty. The largeness of the difference, however, could also stem from the fact that the submitted runs only supplied the inverted rank as a score instead of actual confidence scores. We were not able to verify this because none of the detector runs provided scores and building detectors for this purpose was out of scope of this paper.

For performance prediction, the main advantage of the set-based annotation compared to the annotation of individual images is a saving of resources. In our experiments, set-based annotations save on average 0.59 s compared to individual annotations. If we assume an annotation depth of 192 images ($= 8 \cdot 24$) this amounts to a saving of 113.28 s per detector.

5 Conclusions

In this paper, we proposed a method for detector performance prediction, which is a promising research direction that has yet to receive a significant amount of attention. Our method uses set annotations to predict the performance of concept detectors according to some evaluation measure. First we identified several dimensions that play a role in set annotations, some of which we were not able to test in our experiments. Furthermore, because some performance measures are based on labels of individual images, we developed a probabilistic model for labels and propose to use the expected measure value, based on this model, as a suitable predictor.

The experiments we conducted to evaluate our prediction approach had a relatively small scale and therefore can only serve as indicators for the accuracy of set-base performance prediction is in reality. In this relatively small sample, we found that the mean annotation time per image was significantly lower for set annotations compared to annotations based on individual images, while having a similar annotation quality. The annotators experienced a higher cognitive load for the annotation of individual images. The performance prediction for precision at 96 images, which relied only on the set annotations, was high compared to the numbers reported in related field of query performance prediction. The prediction of the average precision measure, which used our probabilistic model for labels, did not perform strongly. We attributed this to the fact that we only had access to image ranks instead of confidence scores.

This paper presented, to our knowledge, the first attempt to combine detector performance and set annotations of concepts, as a general technique. While the initial results are promising, we propose for future work to augment the human-based predictor with automatic methods derived from those in query performance prediction. Furthermore, while this paper focused on the prediction, future work should investigate how to use predictions to improve search performance.

Acknowledgments. This work was co-funded by the EU FP7 Project AXES ICT-269980 and CUbRIK ICT-287704.

A   Appendix - Optimization of Maximum Likelihood

In this section, we present a procedure for the maximization problem of finding the maximum likelihood weights for the logistic regression model defined in Sect. 3.2. The maximization problem was formulated as follows:

$$\mathbf{w}_{ml} = \underset{\mathbf{w}}{\operatorname{argmax}}\ p(H = h|\mathbf{w})$$

where h is the estimate given by the annotator. We assume that $x(\mathbf{w})$ is Gaussian distributed around a mean μ_h with variance v_h, where both depend on the annotator's estimation h and possibly the annotator himself. In this paper, we choose a simple method to come from h to μ_h by choosing $\mu_h = h$ for $1 \leq h < N$ and $\mu_0 - 1$ and $\mu_N = N - 1$ modeling the case where the annotator oversees at least one example when annotating extreme values. We ignore v_h because it does not play a role in the optimization. Therefore, for the likelihood a weight vector $\mathbf{w}$ given an annotation h, we have:

$$p(H|\mathbf{w}) = \mathcal{N}(x(\mathbf{w}), \mu_h, v_h)$$

where $\mathcal{N}$ is the Gaussian density function. Taking the log of $\mathcal{N}$ yields:

$$log(\mathcal{N}(x(\mathbf{w}), \mu_h, v_h)) = log\left(\frac{1}{\sqrt{2\pi\, v_h}}\right) + \frac{-(x(\mathbf{w}) - \mu_h)^2}{2\, v_h}.$$

By expanding $(\cdot)^2$, leaving out constant terms and factors, and multiplying by -1 to convert the maximization to a minimization problem, we get:

$$x(\mathbf{w})^2 - 2\mu_h x(\mathbf{w}) + \mu_h^2$$

By expanding the definition of the expected number of positive examples in (2), leaving out the constant μ_h^2 and combining factors we get:

$$\left(\sum_i^N \sigma_i(\mathbf{w})\right)^2 - 2\mu_h \sum_i^N \sigma_i(\mathbf{w})$$

And by expanding the square of the expectation $x(\mathbf{w})^2$:

$$y(\mathbf{w}) = \underbrace{\left(\sum_i^N \sum_j^N \sigma_i(\mathbf{w})\sigma_j(\mathbf{w})\right)}_{u_{i,j}(\mathbf{w})} - 2\mu_h \sum_i^N \sigma_i(\mathbf{w}) \tag{5}$$

To optimize this function we use the gradient decent method with the update rule:

$$\mathbf{w}^{t+1} = \mathbf{w}^t - \lambda \nabla y(\mathbf{w}^t) \tag{6}$$

where t refers to the tth iteration, λ is the "update speed" of the method (in this paper we chose $\lambda = 0.03$) and $\nabla y(\mathbf{w}^t)$ is gradient of the method with respect to $\mathbf{w}$. The gradient ∇y is the vector of partial derivatives:

$$\nabla y = \left[\frac{\partial y}{\partial w_1}, \frac{\partial y}{\partial w_2} \right] \tag{7}$$

To calculate the two partial derivations of ∇y, we start by calculating the gradient $\nabla \sigma$ which used in the second expression of (5). As an intermediate step, we give the derivation of a general sigmoid function $\sigma(s)$:

$$\sigma'(s) = \sigma(s)(1 - \sigma(s)) \tag{8}$$

Given this relationship we get the partial derivatives for $\nabla \sigma$:

$$\frac{\partial \sigma_i}{\partial w_1} = \sigma_i(\mathbf{w})(1 - \sigma_i(\mathbf{w})) \qquad \frac{\partial \sigma_i}{\partial w_2} = \sigma_i(\mathbf{w})(1 - \sigma_i(\mathbf{w}))s_i \tag{9}$$

Furthermore, for the derivation of the products of two sigmoid functions $u_{ij}(\mathbf{w}) = \sigma_i(\mathbf{w})\sigma_j(\mathbf{w})$ in (5), we use the product rule and the results from (9). For w_1 we have:

$$\frac{\partial u_{ij}}{\partial w_1} = \sigma_i(\mathbf{w})(1 - \sigma_i(\mathbf{w}))\sigma_j(\mathbf{w})$$
$$+ \sigma_i(\mathbf{w})\sigma_j(\mathbf{w})(1 - \sigma_j(\mathbf{w}))$$

and for w_2:

$$\frac{\partial u_{ij}}{\partial w_2} = [\sigma_i(\mathbf{w})(1 - \sigma_i(\mathbf{w}))s_i]\,\sigma_j(\mathbf{w})$$
$$+ \sigma_i(\mathbf{w})[\sigma_j(\mathbf{w})(1 - \sigma_j(\mathbf{w}))s_j]$$

Therefore, the partial derivatives for the gradient ∇y in (7) are:

$$\frac{\partial y}{\partial w_1} = \left(\sum_i^N \sum_j^N \frac{\partial u_{ij}}{\partial w_1} \right) - 2\mu_h \sum_i^N \frac{\partial \sigma_i}{\partial w_1}$$

and

$$\frac{\partial y}{\partial w_2} = \left(\sum_i^N \sum_j^N \frac{\partial u_{ij}}{\partial w_2} \right) - 2\mu_h \sum_i^N \frac{\partial \sigma_i}{\partial w_2}$$

Note that although quadratic in the number of images, the gradient can be calculated efficiently by memorizing the values for $\sigma_i(\mathbf{w}^t)$ for $1 \leq i \leq N$.

References

1. von Ahn, L., Dabbish, L.: Designing games with a purpose. Commun. ACM **51**(8), 58–67 (2008), http://doi.acm.org/10.1145/1378704.1378719
2. Aly, R., Hiemstra, D., de Jong, F., Apers, P.: Simulating the future of concept-based video retrieval under improved detector performance. Multimed. Tools Appl. **60**(1), 203–231 (2012)
3. Aly, R., Hiemstra, D., de Vries, A.P.: Reusing annotation labor for concept selection. In: CIVR '09: Proceedings of the International Conference on Content-Based Image and Video Retrieval. ACM, New York (2009)
4. Goldstein, E.: The perception of multiple images. Educ. Technol. Res. Dev. **23**, 34–68 (1975)
5. Hajimirza, S., Proulx, M., Izquierdo, E.: Reading users' minds from their eyes: a method for implicit image annotation. IEEE Trans. Multimed. **14**(3), 805–815 (2012)
6. Hauff, C., Azzopardi, L., Hiemstra, D., de Jong, F.: Query performance prediction: evaluation contrasted with effectiveness. In: Gurrin, C., He, Y., Kazai, G., Kruschwitz, U., Little, S., Roelleke, T., Rüger, S., van Rijsbergen, K. (eds.) ECIR 2010. LNCS, vol. 5993, pp. 204–216. Springer, Heidelberg (2010)
7. Hauff, C., Hiemstra, D., de Jong, F.: A survey of pre-retrieval query performance predictors. In: Proceedings of the 17th ACM Conference on Information and Knowledge Management. CIKM '08, pp. 1419–1420. ACM, New York (2008). http://doi.acm.org/10.1145/1458082.1458311
8. Jonassen, D.: Implications of multi-image for concept acquisition. Educ. Technol. Res. Dev. **27**(4), 291–302 (1979)
9. Over, P., Awad, G., Fiscus, J., Antonishek, B., Michel, M., Smeaton, A., Kraaij, W., Quénot, G.: TRECVID 2011 – An overview of the goals, tasks, data, evaluation mechanisms, and metrics. In: TREC 2011 Video Retrieval Evaluation Online Proceedings (TRECVid 2010). National Institute of Standards and Technology, Gaithersburg (2011)
10. Smeaton, A.F., Over, P., Kraaij, W.: Evaluation campaigns and TRECVid. In: MIR '06: Proceedings of the 8th ACM International Workshop on Multimedia Information Retrieval, pp. 321–330. ACM, New York (2006)
11. Snoek, C.G.M., Worring, M.: Are concept detector lexicons effective for video search? In: 2007 IEEE International Conference on Multimedia and Expo, pp. 1966–1969 (2007)
12. Snoek, C.G.M., Worring, M.: Concept-based video retrieval. Found. Trends Inf. Retr. **4**(2), 215–322 (2009)
13. Yang, J., Hauptmann, A.G.: (un)Reliability of video concept detection. In: CIVR '08: Proceedings of the 2008 International Conference on Content-based Image and Video Retrieval, pp. 85–94. ACM, New York (2008)
14. Yilmaz, E., Kanoulas, E., Aslam, J.: A simple and efficient sampling method for estimating AP and NDCG. In: SIGIR'08: Proceedings of the 31st Annual International ACM SIGIR Conference on Research And Development in Information Retrieval, pp. 603–610. ACM, New York (2008)
15. Yilmaz, E., Aslam, J.A.: Inferred AP: estimating average precision with incomplete judgments. In: Fifteenth ACM International Conference on Information and Knowledge Management (CIKM), pp. 102–111. ACM, New York, November 2006

Personas – The Missing Link Between User Simulations and User-Centered Design?
Linking the Persona-Based Design of Adaptive Multimedia Retrieval Systems with User Simulations

David Zellhöfer[✉]

Department of Computer Science Database and Information Systems Group,
Brandenburg Technical University Cottbus, Cottbus, Germany
david.zellhoefer@tu-cottbus.de

Abstract. In order to establish a reproducible evaluation setup in interactive information retrieval (IIR), user simulations have been suggested. Unlike the inclusion of "real" users into the evaluation loop, user simulations scale well, are not affected by learning or tiring effects of the probands, and can be conducted at low cost.

Unfortunately, the evaluation utilizing user simulations often takes place after the IIR system has been fully implemented. As such, it cannot give valuable feedback during the design phase of the system. In this paper, we propose a methodology for linking the persona-based approach from user interaction design with the field of IIR evaluation to address this problem.

To illustrate its utility, a user-centered multimedia retrieval scenario – the ImageCLEF 2012 pilot task on personal photo retrieval – is used as an example of the usage of the proposed evaluation methodology. To conclude with, we discuss the current limitations of the approach and address open issues such as the incorporation of multiple search strategies into user simulations.

1 Introduction

The evaluation of information retrieval (IR) systems, content-based image retrieval (CBIR), and multimedia information retrieval (MIR) systems is predominated by Cranfield-based approaches [1]. Such system-centric evaluation initiatives, e.g. ImageCLEF or TRECVID in the field of MIR, provide sample information needs (IN) expressed by images, text, or the like and expect MIR systems to produce rankings of the most relevant documents with respect to the IN given a specified document corpus. Commonly, relevance feedback (RF) is used to improve the results.

This rather abstract technique provides a controlled test environment in which IR systems can be compared by the means of defined retrieval metrics such as mean average precision (MAP) or the like. Hence, the Cranfield paradigm is practical and has served the research field very well over the past decades.

© Springer International Publishing Switzerland 2014
A. Nürnberger et al. (Eds.): AMR 2012, LNCS 8382, pp. 276–290, 2014.
DOI: 10.1007/978-3-319-12093-5_17

Nevertheless, this system-based approach towards evaluation has become subject of criticism over the past years, e.g. by Voorhees [2] noting that "Cranfield test collections represent too little of the user to support adaptive information retrieval research" and "completely ignore[s] the user interface". Both theoretical and practical work focussing on adaptive and interactive IR systems, e.g. by Belkin, Borlund, Hearst, Kelly, Kuhlthau, or Ingwersen, eventually led to a new field of research: interactive information retrieval (IIR). Good overviews over the field are available, e.g. [3,4]. A survey of IIR evaluation strategies is also available [5]. To (over-)simplify the field, IIR considers the search a process in which the user tries to satisfy a dynamic IN using multiple search strategies as observed e.g. by Reiterer et al. [6].

Unfortunately, focusing on users during evaluation has numerous disadvantages, e.g. high costs and time consumption [2], learning[1] and tiring effects during numerous RF iterations, scalability issues, or reproducibility problems [7]. To overcome this issue, user simulations have been suggested. Following a SIGIR workshop on the simulation of interaction in 2010 [8], papers at ECIR 2011 [9], SIGIR 2012 [10], and a broad usage at IIiX 2012 [11]; the usage of user simulations is gaining more and more acceptance in the community – although the impact of this methodology is still surprisingly low on the field of MIR.

Alas, evaluations relying on user simulations are mostly carried out with existing or fully implemented systems. That is, although the importance of users is acknowledged in IIR, the evaluation of a system and the incorporation of (simulated) users still remains at the end of the software development cycle missing important user and contextual information that can affect the development as we will point out later in this paper.

The paper is structured as follows. In the next section, we will recapitulate the user-centered design approach towards software development and how it can be used in the development of interactive and adaptive MIR systems. Relying on personas [12], we will show how to use them for system and user simulation development. Section 3 describes the usage of the proposed methodology and its effects on retrieval metrics based on the runs submitted to the ImageCLEF 2012 pilot task on personal photo retrieval. The paper then concludes pointing out future directions of research.

2   Linking Personas to User Simulations

Approaches from interaction design or software engineering that can be subsumed under the term of user-centered design (UCD) commonly refer more to a similar methodology than to a formal technique. For the scope of this paper, user-centered design [13] consists of three main principles:

1. an early focus during development on users and their tasks,

---

[1] When dealing with a system, users will learn about its capabilities, the contents of the document collection etc. and are therefore likely to adjust their behavior accordingly distorting the final evaluation results.

2. the early usage of measurements of user reactions and performance, and
3. an iterative design.

As such, the objectives of user-centered design are close to the user-centered field of IIR. This conceptual closeness has also been discovered by others, e.g. White [14] or Karlgren et al. [15].

The left side of Fig. 1 illustrates a sample approach of user-centered design using personas (see Sect. 2.1). The general starting point of UCD is the definition of *scenarios* [16]. Scenarios consist of informal work task and context descriptions in a narrative form. They are in a form that can be understood by the stakeholders (i.e. the actual group of prospected users) of the MIR system to be developed. If needed, scenarios can be augmented with storyboards that illustrate the scenario and make it easier to understand [17]. Based on theses scenarios, *use cases* can be developed that mainly focus on the user interaction with the systems [18]. Use cases, often described in UML, are directly linked to an actor (i.e., a user) in a way that they can describe completely different interaction patterns, e.g., search strategies such as browsing or directed search in the case of this paper. In order to model the user interaction in a more detailed way, i.e., to describe the interaction at a level of direct input or the like, *interaction cases* can be designed [19]. These interaction cases usually have such a fine level of granularity that they can be implemented directly.

To recapitulate, the UCD approach provides means to incorporate users into the development process of an interactive MIR system increasing the level of formalization from a scenario to an interaction case. Thus, requirements for an adaptive MIR system can be obtained and refined step-wisely from a user's perspective.

The importance of an inclusion of contextual information and actual user behavior is acknowledged, e.g., by [14] or [7]. In his work, he focuses on simulating user interactions (mainly document selection and browsing) in order to find out if a retrieval algorithm in combination with RF is increasing the performance of the system. In order to improve the realism of the simulations, additional factors such as search experience or topic familiarity affect the interactions. It is important to note that these factors are chosen arbitrarily and have not been derived from actual user studies or the like. Hence, White's approach [14] is only operating at the interaction case/user simulation loop at the bottom of Fig. 1.

In contrast, Karlgren et al. [15] discuss the issue at a higher level of abstraction (see Fig. 1). Their approach relies on use cases to overcome limitations that are hidden in Cranfield-based evaluations. We fully agree with their observation that system-centric approaches implicitly model only one use case, i.e., a user submitting a distinct query at a time without examining other ways of information access such as browsing. Arguing that this issue is intensified in MIR because directed search is only one of the possible search strategies used for information access, they [15] propose to exploit use cases modeling different user preferences and goals in order to benchmark IR systems. How these uses cases can be constructed is left open: "The choice of which hypotheses to work on is naturally a question which this model leaves up to the system engineering team[...]" [15].

This is limiting the approach's utility as the hypothetical use cases can also become extremely non-realistic if the system designer cannot communicate with the users or have a poor understanding of their needs.

2.1   Personas

Personas have been proposed by Cooper [20] to address this problem. Personas are a means to help both user interaction designers and software developers to understand the needs and goals of potential users or user groups of the system in development. They can be considered still a state-of-the-art technique commonly used in interaction design and usability engineering. Although often misunderstood, personas "are not real people, but they are based on the behaviors and motivations of real people we have observed and represent them throughout the design process. They are *composite archetypes* based on behavioral data gathered from many actual users encountered in ethnographic interviews" [12]. Hence, personas can be used to model user groups including their behavioral patterns as they could have been observed in a real-world context. This data can be enriched by surveys, research literature, or user interviews to form a more complete picture. Regarding the UCD approach, personas are located above the aforementioned steps (see Fig. 1) as they define potential usage scenarios, contexts, and specific user interface needs. For instance, a professional news researcher will certainly use different search strategies or query types in a news archive than a teenager looking for recreational video clips on the internet.

2.2   Establishing the Link Between Personas and User Simulations

In prior work, we sketched the relation of personas to user simulations. Based on the demographic data of the assessors that generated the ground truth for a personal photograph collection – the Pythia collection [21] – we suggested to model personas. To be more precise, such personas are often called ad-hoc personas. Ad-hoc personas have been shown to help during the UCD process [12] but should be extended with additional qualitative data (such as interviews, think-aloud protocols, or user observations) from the studied usage context whenever possible [21]. However, it is known that initiatives involving larger groups of users are very cost and time intensive [2].

Usually, the demographic information of assessors is neglected because Cranfield-based benchmarks rely on averaged relevance judgements. In contrast, we decided to maintain the demographic information in order to develop both personas and user simulations based on this data.

We see a strong connection between personas as known from interaction design with user simulations from the field of IIR. User simulations are an accepted means to evaluate interactive IR systems. To achieve this, they have to rely on some sort of ground truth providing relevance judgments of documents in order to use them during (simulated) RF iterations in order to evaluate the performance of an adaptive IR system. Additionally, user simulations can be

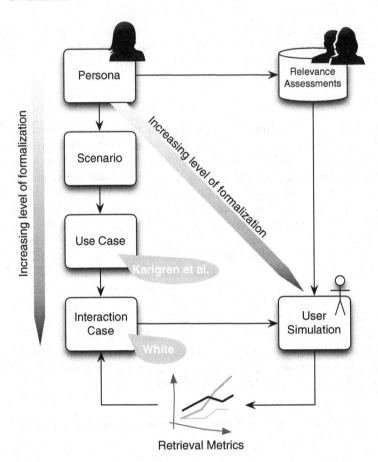

Fig. 1. Persona-based user-centered system design linked to user simulations; research foci of White [14] and Karlgren et al. [15] are indicated

extended in a way that they include interaction patterns as well, e.g., as suggested by White [14]. To obtain these interaction patterns, the designers of the user simulation have to develop a clear picture of their potential users, their goals, and usage contexts. As we have argued before, UCD is a valuable means to help system designers to cope with these challenges. By designing personas based on their users, the designers can learn about them and develop requirements step-wisely and *in cooperation*. Eventually, these requirements will lead to the formulation of interaction cases that can serve as a starting point for the development of user simulations. In order to provide a ground truth to the user simulation, persona-based relevance assessments have to be provided (see Fig. 1).

For the scope of this paper, persona-based relevance assessments are relevance assessments of real users sharing the same characteristics, e.g., the same demographics, with the persona which is associated with the current use and interaction case. In order to maintain a certain level of subjectivity regarding the relevance assessments with respect to an IN, we suggest to rely on graded relevance judgments, e.g., on a scale of 0 (irrelevant) to 3 (fully relevant) as proposed earlier [22]. In our opinion, the usage of a binary relevance scale (irrelevant/relevant) would remove to much valuable information such as the gradual perception of relevance from the ground truth eventually lowering the explanatory power and relevance granularity of the user simulations.

Finally, Fig. 1 illustrates how personas can be linked to user simulations in two ways in order to develop interactive IR systems in a user-centered manner. First, personas serve as the starting point for the UCD process that eventually results in user simulations of various interaction cases. Second, assessors sharing the same characteristics with a persona, e.g., the same behavioral data, ethnographic group, or demographics, provide (gradual) relevance assessments with respect to the tasks that have been developed in cooperation with users using scenarios and use cases. In consequence, design decisions affecting the retrieval performance and usability can be validated against personas and user simulations throughout the development of the IR system. This discriminates the approach from typical user simulation-based evaluations that are usually conducted *after* a system has been implemented.

3   Application of the Methodology

In order to illustrate the usage of the proposed methodology, first results from an experimental examination of the data from the ImageCLEF 2012 pilot task on personal photo retrieval [23] (subtask 1) will be presented in this section.

3.1   Description of the Task

The objective of the task is to find similar images to a specified visual concept or topic. Out of the 32 topics provided by the Pythia dataset [21], the 24 topics with the most relevant images in the corpus were chosen. To solve the task, 5 QBE documents are provided. All QBE documents are fully relevant. In addition to the metadata accompanying the images (e.g. GPS data), simulated browsing data (see below) representing images that might have been inspected during the search is offered. The usage of this browsing data is voluntary as the utilization of image features or metadata. For a full description of the task and the used datasets, see [23].

3.2   Acquisition of the Ground Truth

In order to obtain the ground truth, 42 assessors were asked to participate. Their core characteristics can be subsumed as follows. The majority of the assessors (28 out of 42) are male and born between 1979 and 1991 (median: 1987). Most of the assessors are students with a background in economics (26), the second largest group (13) has a background in computer science or information technology.

Regarding their level of expertise in the field of MIR or IR, 9 assessors took classes in MIR while 11 heard IR. When asked directly about their knowledge of the field the median lies at "little knowledge" with an average of 1.40, i.e., a trend towards considering themselves as an "informed outsiders" [23].

The assessors could judge the relevance of an image with respect to a topic on a graded scale ranging from 0 (irrelevant) to 3 (fully relevant). All assessors had to judge all documents regarding a topic. The topics were associated with the assessors by random. In average 2.69 topics were evaluated per assessor (standard deviation: 1.60). The individual assessments were saved separately in order to maintain them for later usage as described above. A more detailed description of the demographics, other characteristics, and the survey each assessor had to fill out is available in [21].

Calculation of the Ground Truth for Each Topic. Based on the individual assessments, an averaged ground truth (reflecting an "average" persona) has been calculated. First, the frequency of each graded relevance judgement (out of an interval from 0 (irrelevant) to 3 (fully relevant)) was counted per image and topic. Based on these relevance judgment frequencies, an estimation value was calculated and rounded. The rounded estimation value of the relevance of an image regarding a topic was then used as the averaged graded relevance assessment for this image. In consequence, each image could be associated with a graded relevance judgment for each topic. In principle, the resulting "average" persona-based ground truth resembles a traditional Cranfield-based one that has been extended with graded relevance assessments [23].

Generating Browsing Information. As we could not obtain real browsing information, it had to be generated artificially. Using the graded relevance assessments, multiple images were chosen as browsing images. The provided browsed images have a relevance grade ranging from 1 to 2, i.e., they are judged neither irrelevant nor fully relevant for a given topic. In other words, the browsing data consists of interesting images which were not fully relevant for the modeled user which caused him or her to proceed with the search.

3.3 Results with Respect to Different Personas

Based on the survey data of the assessors, 6 distinct personas were created in addition to the aforementioned average persona: (1) an IR/MIR expert persona, (2) a non-expert, (3) a male persona, (4) a female persona, (5) a persona with an IT background, and (6) a persona with a non-IT background. Please note that these personas are not meant to be complete. They have been chosen for illustration purposes only because a discussion of all possible personas which can be derived from the user interview data would exceed the page limitations. For instance, additional personas could be created on basis of the daily internet usage or photograph taking behavior. Separate ground truths were then created for each persona based on the graded relevance judgments of the associated assessors. Because of the randomized association of assessors and topics, it could not be guaranteed that all document-topic combinations were assessed by members of the distinct persona groups. Thus, missing assessments were taken from averaged ground truth (see above).

Table 1. Comparison of the average retrieval performance over all personas

Group	Run ID	P@20 Mean	P@20 Std. Dev.	NDCG@20 Mean	NDCG@20 Std. Dev.
UniCagliari	Run_1_1	0.5491	0.0137	0.4838	0.0090
UniCagliari	Run_1_2	0.6810	0.0096	0.5472	0.0040
UniCagliari	Run_3_2	0.3887	0.0082	0.3466	0.0051
KIDS	IOOA4	0.5253	0.0099	0.4548	0.0039
KIDS	IOMA0	0.5961	0.0113	0.4876	0.0037
KIDS	OBMA0	0.5485	0.0206	0.4049	0.0046
KIDS	OBOA0	0.6042	0.0224	0.4820	0.0045
KIDS	IBMA0	0.6670	0.0176	0.5435	0.0081
REGIMvid	run1	0.6783	0.0357	0.4536	0.0083
REGIMvid	run2	0.6783	0.0357	0.4537	0.0084
REGIMvid	run3	0.6783	0.0357	0.4527	0.0086
REGIMvid	run4	0.6824	0.0356	0.4540	0.0082
REGIMvid	run5	0.6783	0.0357	0.4526	0.0086

Table 1 lists the average retrieval performance (using the arithmetic mean) over all personas and standard deviation of the runs submitted to the pilot task. Because of the focus of this paper on the proposed methodology, we will omit a full discussion of the used retrieval algorithms by the workgroups participating in the ImageCLEF pilot task. Instead, we will roughly sketch their core idea and refer to the original publications. The group KIDS [24] used a multimodal clustering method to solve the task, the University of Cagliari tested their interactive CBIR system with relevance feedback support [25], while REGIMvid's approach [26] is based on fuzzy logic [27] in order to retrieve a highly diverse set of documents.

As measurements, precision @ 20 and NDCG @ 20 are used[2] because the first 20 images of the results can be easily displayed to a user without the need of scrolling or the like. In order to utilize the provided graded relevance judgements, NDCG is used. The NDCG metric [22] has been shown to be "a useful user-centered measure of system effectiveness" [28] in comparison to other common retrieval metrics. The core idea behind NDCG is to apply "a discount factor

---

[2] For the results presented in this paper, the precision @ 20 and NDCG implementations of `trec_eval` version 9.0 with standard discount settings are used.

to the relevance scores in order to devaluate late-retrieved documents" [22]. In other words, the metric rewards highly relevant documents at the first positions in the result ranking and punishes systems retrieving less relevant documents at the first places.

Figures 2 and 4 illustrate the impact of different personas on the retrieval quality in comparison to the average persona measured in precision @ 20 and NDCG @ 20. The y-axis of the figures indicate the values of the retrieval metrics while the x-axis lists the results of the different personas.

Generally speaking, the impact of the different personas on the NDCG metric is rather low, while it affects precision @ 20 much stronger. In particular, the runs Run_1_1 and Run_3_2 submitted by the University of Cagliari (Fig. 2) show a clearly varying retrieval performance if one compares the different personas with the average persona and ground truth. As expected, many runs perform differently per persona indicating that different retrieval engines or algorithms should be used for different user groups represented by the respective persona.

Interestingly, some runs (and therefore different retrieval techniques) give stable results no matter which persona is used, e.g. Run_1_2 (Fig. 2) or IOMA0 (Fig. 4), while others are only stable for certain persona. For an example, see the female persona in REGIMvid's run (Fig. 4, shaded graphs). This phenomenon indicates clearly that different retrieval techniques may show different utility for various user groups. How to compensate these effects remains future work for the participants of the task.

Table 1 shows that the standard deviation is relatively small amongst the examined runs. We ascribe this to the way missing assessments were added from the averaged ground truth. Hence, personas with a lot of missing assessments tend to converge on the average persona. If this effect caused the relatively stable NDCG metric remains an issue worth further research.

3.4 A Sample User Simulation

As the participants of the pilot task were only required to release their result rankings, their systems cannot be used for RF driven by user simulations. Instead, we present the results of a sample implementation based on the ground truth of the average persona to illustrate the utility of the suggested approach. In contrast to the pilot task, the user simulation uses one QBE document per run and a total of 32 topics. In total, 13,602 queries are used. The mean of QBE documents per topic is 425.06 with a standard deviation of 493.82. The choice of QBE documents is based on the averaged ground truth in a way that all relevant documents serve as QBE documents. The large number of queries is motivated by [2] arguing for a large number of experiments in order to obtain resilient results about the performance of adaptive IR systems. All runs are executed with a self-implemented MIR system [29,30] based on a probabilistic query language: the commuting quantum query language (CQQL) [31]. In order to retrieve similar documents, the systems can used multiple modalities, e.g. low-level features such as color histograms, GPS data, or textual data. Additionally, it can use RF to improve or personalize the results and supports multiple search strategies such

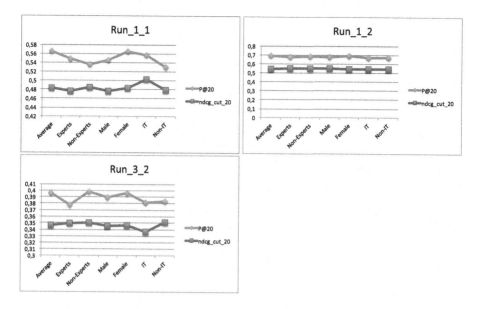

Fig. 2. Retrieval metrics per user group of the University of Cagliari's runs

as directed search and various browsing techniques. For the scope of this paper, only RF in a directed search scenario is used.

Based on the aforementioned scenario, the use case is made of a typical QBE scenario. The user inputs a sample image and the system is asked to find the most similar images from the Pythia collection [21]. The derived interaction case is reflecting a relatively lazy user. After an initial query has been submitted, the user is only willing to inspect the first 20 results. If the user detects an irrelevant image directly preceding a relevant one, the user clicks on it to move the irrelevant document behind the relevant one in the result ranking. This action is then interpreted as RF by the system expressing the new ordering

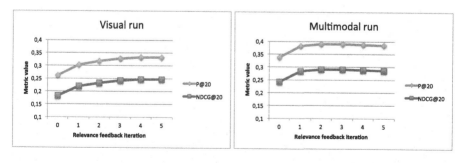

Fig. 3. Comparison of 5 relevance feedback iterations using visual and multimodal data

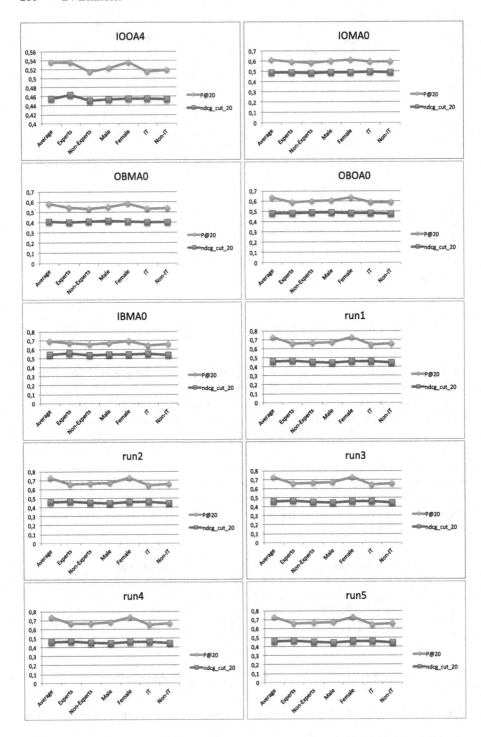

Fig. 4. Retrieval metrics per user group of KIDS lab's and REGIMvid's (shaded) runs

should be fulfilled after the next RF iteration[3]. If the user has scanned the first 20 documents, the system gets instructed to retrieve more documents fulfilling the user's preferences. After the 5th RF iteration, the user will abort the search. The user simulation is implemented accordingly to interact with the MIR system.

Please note that the user simulation is not always providing correct RF, i.e., it will not always follow the interaction case mentioned before. The aspect of erroneous RF is also discussed by [9] who use a probabilistic approach to decide whether a user simulation is giving correct or incorrect feedback. In contrast, the described user simulation is relying on a ground truth based on assessments without any corrections of wrong user input. Thus, the probability of erroneous RF is equally the same as during the relevance assessments.

Figure 3 compares the development of precision @ 20 and NDCG @ 20 over 5 RF iterations for a query using visual low-level features alone and a query exploiting multiple modalities. Again, we will neglect the discussion of the used retrieval techniques. Instead, we will focus on a possible interpretation of the graphs. Obviously, the multimodal run performs better for the average persona. Interestingly, both RF experiments show a significant improvement between the initial query (RF iteration 0) and the first RF iteration. Subsequently, the development is quickly stagnating. Concluding from this finding, it becomes obvious that the used RF mechanism might not be a good choice for the modeled interaction case, in particular if the retrieval system is using multiple modalities. As this phenomenon has been revealed during the development of the system it might be addressed without causing high costs due to a complete change of the retrieval subsystem at a late stage of the development. In consequence, the system's design should find an alternative algorithm that might fit better to the interaction case of the persona or interview the users again to find out if they are really expecting to use many RF iterations.

4   Conclusion

In this paper, we suggested a holistic approach towards the user-centered development and evaluation of adaptive and interactive IR systems. Arguing for the utilization of user simulations in the evaluation of adaptive MIR systems because of reproducibility and cost issues, we criticize that user simulations are traditionally used *after* a new system is developed. By linking user simulations with personas and multiple ground truths derived from them, we could show how to incorporate iterative IR evaluation steps into the user-centered design of IIR systems.

Furthermore, we illustrated the usage of the proposed methodology using different submitted runs to the ImageCLEF 2012 pilot task on personal photo retrieval. The results show that different personas will experience varying retrieval performances of the same retrieval system. Additionally, we describe how a sample user simulation can be implemented following the principles of UCD and how

---

[3] A detailed description of the approach as given in [32] falls out of the focus of this paper.

it can be linked to a persona-specific ground truth in order to conduct retrieval quality measurements during the design phase of an exemplary MIR system.

Although the methodology's utility could be generally in principle, further questions remain unsolved.

First, it is necessary to determine the influence of the augmentation strategy used to replace missing assessments. As said before, this strategy might have had a strong impact onto the retrieval metrics shown in Table 1.

Second, further search strategies have to be modeled as interaction cases and implemented as user simulations. In addition, interaction cases with changing search strategies, e.g. a transition between exploratory and directed search, have to be tested as there are strong indicators that they are often occurring during interactive search sessions [6]. Unfortunately, the simulation of browsing poses a challenge to us because search session data that could be exploited to generate browsing data (e.g. suggested by [10]) is not available yet for the described test collection. Alternatively, the needed data could be obtained by carrying out user observations. Hence, near future studies can only be based on artificially generated data. If this has an impact on the expressiveness of the user simulation remains subject to future research.

Third, the presented work assumes that all personas work will use the same queries but vary in their relevance assessments. Obviously, the queries will also vary in the real world [7]. To address this fact would require a much deeper study of the user information needs than the discussed study can provide. Hence, to design a full user simulation that includes queries, the user's current situation and dynamic information need should be modeled as well. Eventually, this requires a understanding of the user's search task and cognitive processes which is not trivial [33] but – without doubt – worth further studies. Nevertheless, we believe that the proposed methodology takes a step in the right direction.

To conclude with, the discussed approach linking personas with user simulations cannot easily be generalized. While this is true for most user simulations and due to their strong connection to the simulated work tasks and use cases, we can provide a combination of a document collection with different persona-specific ground truths relying on graded relevance assessments that might help other researchers to evaluate their systems in a user-centered manner leaving the grounds of the system-centric Cranfield paradigm.

Acknowledgments. This research was supported by a grant of the Federal Ministry of Education and Research (Grant Number 03FO3072).

References

1. Cleverdon, W.C.: Aslib Cranfield Research Project: Report on the Testing and Analysis of an Investigation into the Comparative Efficiency of Indexing Systems. Cranfield, USA (1962)
2. Voorhees, M.E.: On test collections for adaptive information retrieval. Inf. Process. Manage. **44**(6), 1879–1885 (2008)

3. Ingwersen, P., Järvelin, K.: The Turn: Integration of Information Seeking and Retrieval in Context. Springer, Dordrecht (2005)
4. Hearst, A.M.: Search User Interfaces. Cambridge Univ. Press, Cambridge (2009)
5. Kelly, D.: Methods for evaluating interactive information retrieval systems with users. Found. Trends Inf. Retr. **3**, 1–224 (2009)
6. Reiterer, H., Mußler, G., Mann, M.T., Handschuh, S.: INSYDER - an information assistant for business intelligence. In: Proceedings of the 23rd Annual International ACM SIGIR Conference on Research and Development in Information Retrieval, SIGIR '00, pp. 112–119. ACM (2000)
7. Borlund, P.: The IIR evaluation model: a framework for evaluation of interactive information retrieval systems. Inf. Res. 8(3) (2003)
8. Azzopardi, L., Järvelin, K., Kamps, J., Smucker, D.M. (eds.): Report on the SIGIR 2010 Workshop on the Simulation of Interaction, vol. 44. ACM, New York (2011)
9. Baskaya, F., Keskustalo, H., Järvelin, K.: Simulating simple and fallible relevance feedback. In: Clough, P., Foley, C., Gurrin, C., Jones, G.J.F., Kraaij, W., Lee, H., Mudoch, V. (eds.) ECIR 2011. LNCS, vol. 6611, pp. 593–604. Springer, Heidelberg (2011)
10. Baskaya, F., Keskustalo, H., Järvelin, K.: Time drives interaction: simulating sessions in diverse searching environments. In: Proceedings of the 35th International ACM SIGIR Conference on Research and Development in Information Retrieval, SIGIR '12, pp. 105–114. ACM (2012)
11. Kamps, J., Wessel Kraaij, Fuhr, N. (eds.): Proceedings of the 4th Information Interaction in Context Symposium, IIIX '12. ACM, New York (2012)
12. Cooper, A., Reimann, R., Cronin, D.: About face 3: The essentials of interaction design, Completely rev. and updated. edn. Wiley, Indianapolis, Ind (2007)
13. Gould, D.J., Lewis, C.: Designing for Usability: Key Principles and What Designers Think: Human-Computer Interaction. Morgan Kaufmann Publishers Inc., San Francisco (1987)
14. White, W.R.: Contextual simulations for information retrieval evaluation. In: SIGIR 2004: Information Retrieval in Context Workshop (2004)
15. Karlgren, J., Järvelin, A., Eriksson, G., Hansen, P.: Use cases as a component of information access evaluation. In: Proceedings of the 2011 Workshop on Data InfrastructurEs for Supporting Information Retrieval Evaluation, DESIRE '11, pp. 19–24. ACM (2011)
16. Carroll, M.J.: Introduction to this special issue on "Scenario-Based System Development". Interact. Comput. **13**(1), 41–42 (2000)
17. Landay, A.J., Myers, A.B.: Sketching storyboards to illustrate interface behaviors. In: Conference Companion on Human Factors in Computing Systems: Common Ground, CHI '96, pp. 193–194. ACM (1996)
18. Jacobson, L., Christerson, M., Jonsson, P., Övergaard, G.: Object-Oriented Software Engineering: A Use Case Driven Approach. Addison-Wesley, Reading (1992)
19. Schlegel, T., Raschke, M.: Interaction-cases: model-based description of complex interactions in use cases. In: Proceedings of the IADIS International Conferences Interfaces and Human Computer Interaction 2010 and Game and Entertainment Technologies 2010, pp. 195–202 (2010)
20. Cooper, A.: The Inmates Are Running the Asylum. Macmillan Publishing Co. Inc., Indianapolis (1999)
21. Zellhöfer, D.: An extensible personal photograph collection for graded relevance assessments and user simulation. In: Ip, H.S.H., Rui, Y. (eds.) ICMR '12: Proceedings of the 2nd ACM International Conference on Multimedia Retrieval, ICMR '12. ACM (2012)

22. Järvelin, K., Kekäläinen, J.: Cumulated gain-based evaluation of IR techniques. ACM Trans. Inf. Syst. **20**(4), 422–446 (2002)
23. Zellhöfer, D.: Overview of the personal photo retrieval pilot task at imageCLEF 2012. In: Forner, P., Karlgren, J., Womser-Hacker, C. (eds.) CLEF 2012 Evaluation Labs and Workshop (2012)
24. Ku, C.W., Chien, B.C., Chen, G.B., Gaou, L.J., Huang, R.S., Wang, S.E.: KIDS lab at ImageCLEF 2012 personal photo retrieval. In: Forner, P., Karlgren, J., Womser-Hacker, C. (eds.) CLEF 2012 Evaluation Labs and Workshop (2012)
25. Tronci, R., Piras, L., Murgia, G., Giacinto, G.: Image hunter at ImageCLEF 2012 personal photo retrieval task. In: Forner, P., Karlgren, J., Womser-Hacker, C. (eds.) CLEF 2012 Evaluation Labs and Workshop (2012)
26. Feki, G., Ksibi, A., Ammar, B.A., Amar, B.C.: REGIMvid at ImageCLEF2012: improving diversity in personal photo ranking using fuzzy logic. In: Forner, P., Karlgren, J., Womser-Hacker, C. (eds.) CLEF 2012 Evaluation Labs and Workshop (2012)
27. Zadeh, A.L.: Fuzzy logic. IEEE Comput. **21**(4), 83–93 (1988)
28. Carterette, B.: System effectiveness, user models, and user utility: a conceptual framework for investigation. In: Proceedings of the 34th International ACM SIGIR Conference on Research and Development in Information, SIGIR '11, pp. 903–912. ACM (2011)
29. Zellhöfer, D., Bertram, M., Böttcher, T., Schmidt, C., Tillmann, C., Schmitt, I.: PythiaSearch - A multiple search strategy-supportive multimedia retrieval system. In: Proceedings of the 2nd ACM International Conference on Multimedia Retrieval, ICMR '12. ACM (2012) (to appear)
30. Zellhöfer, D.: A permeable expert search strategy approach to multimodal retrieval. In: Kamps, J., Wessel Kraaij, Fuhr, N. (eds.) Proceedings of the 4th Information Interaction in Context Symposium, IIIX '12, pp. 62–71. ACM (2012)
31. Schmitt, I.: QQL: A DB&IR query language. VLDB J. **17**(1), 39–56 (2008)
32. Zellhöfer, D., Schmitt, I.: A preference-based approach for interactive weight learning: learning weights within a logic-based query language. Distrib. Parallel Databases **27**(1), 31–51 (2009)
33. Cole, J.M.: Simulation of the IIR user: beyond the automagic. In: Azzopardi, L., Järvelin, K., Kamps, J., Smucker, D.M. (eds.) Report on the SIGIR 2010 Workshop on the Simulation of Interaction, vol. 44, pp. 1–2. ACM (2011)

Author Index

Printed in the United States
By Bookmasters